S0-AIV-155

Electric Circuits: Theory and Engineering Applications

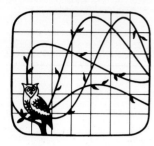

HRW
Series in
Electrical and
Computer Engineering

M. E. Van Valkenburg, Series Editor

L. S. Bobrow   ELEMENTARY LINEAR CIRCUIT ANALYSIS

C. H. Durney, L. D. Harris, C. L. Alley   ELECTRIC CIRCUITS: THEORY AND ENGINEERING APPLICATIONS

G. H. Hostetter, C. J. Savant, Jr., R. T. Stefani   DESIGN OF FEEDBACK CONTROL SYSTEMS

S. Karni and W. J. Byatt   MATHEMATICAL METHODS IN CONTINUOUS AND DISCRETE SYSTEMS

B. C. Kuo   DIGITAL CONTROL SYSTEMS

A. Papoulis   CIRCUITS AND SYSTEMS: A MODERN APPROACH

A. S. Sedra and K. C. Smith   MICROELECTRONIC CIRCUITS

M. E. Van Valkenburg   ANALOG FILTER DESIGN

Electric Circuits: Theory and Engineering Applications

CARL H. DURNEY
L. DALE HARRIS
CHARLES L. ALLEY
University of Utah

Holt, Rinehart and Winston

New York   Chicago   San Francisco   Philadelphia
Montreal   Toronto   London   Sydney   Tokyo
Mexico City   Rio de Janeiro   Madrid

Copyright © 1982 CBS College Publishing
All rights reserved.
Address correspondence to:
383 Madison Avenue, New York, NY 10017

Library of Congress Cataloging in Publication Data

Durney, Carl H., 1931–
  Electric circuits

  Includes index.
  1. Electric circuits. I. Harris, L. Dale. II. Alley
Charles L. III. Title.
TK454.D87   621.319'2   81–7086
ISBN 0–03–057951–1   AACR2

Printed in the United States of America

  2 3   144   9 8 7 6 5 4 3 2 1

CBS COLLEGE PUBLISHING
Holt, Rinehart and Winston
The Dryden Press
Saunders College Publishing

To Gordon and Maxine
To Esther
To Mildred

Preface

This textbook is designed for a typical sophomore-junior electrical engineering electric circuits course sequence. Although its content is somewhat traditional, the book differs significantly from many others in the following ways. The overall emphasis of the book is on the connection between electric circuit theory and the professional practice of electrical engineering, without slighting theoretical aspects of the subject. This includes emphasis on modeling, which many textbooks often omit, emphasis on physical interpretation, on qualitative understanding, on the use of graphical methods, and on the use of alternate procedures for obtaining solutions. Another important feature is the explanation and use of consistency checks. The student is required to examine the consistency or validity of his or her methods and solutions. Also, realistic examples are used to build student motivation. Design problems are mixed with problems of analysis.

CONNECTION WITH THE REAL WORLD

In practice, the engineering analytical procedure is to start with a real system, construct a suitable model, apply fundamental principles to the model, obtain desired results for the model, and then apply these results to the real system, allowing for the limitation of the model. The modeling step is often omitted from circuits texts, which frequently contain only analysis of models, with no connection to real systems. The student is simply given a model and asked to find specified voltages or currents related to that model. Admittedly modeling is difficult to teach, but in this book we have tried to include some modeling steps, beginning with the introduction of the concepts of modeling in Chapter 1 and continuing with the illustration of the modeling process throughout the remainder of the text. For example, in Chapter 2 the modeling process is illustrated in a discussion of automobile batteries. The importance of modeling is stressed in Chapter 3, where the difference between a capacitor (a device) and capacitance (a property) is pointed out, and capacitors and inductors are modeled by appropriate combinations of resistance, capacitance, and inductance. In addition, we have included more examples of the design of real devices than are found in most circuits textbooks. At the same time, we have tried not to sacrifice rigor or thoroughness of mathematical developments.

PHYSICAL INTERPRETATION

We are convinced that most original work, such as developing a new procedure or designing a new device, is based on physical insight or physical interpretation, as opposed to strict manipulation of mathematical relations. Experienced engineers think in terms of impedances, time constants, and other physical interpretations, and when confronted with a problem situation, they will begin to think of approximate, simpler cases, applying the fundamental principles qualitatively to get an understanding of the system. However, a typical undergraduate student will often blindly begin to write loop or nodal equations when confronted with the same problem situation. We have tried to encourage the students to learn physical interpretation and qualitative behavior, such as in the characteristics of *RC* and *RL* circuits in Chapter 3.

Along with this emphasis on physical interpretation and qualitative understanding, we have stressed graphical methods and estimates of results. Although students often seem reluctant to use graphical techniques, experienced problem solvers use them frequently.

CONSISTENCY CHECKS

Throughout the text we have tried to help students learn how to make consistency checks. By consistency checks we mean checking results for consistency with known qualitative results, with physical understanding, with results from alternate solutions, and with special cases (such as when a resistance approaches zero or infinity). Although many instructors teach students how to make consistency checks in one way or another, we feel that it is a significant advantage to have the concept formalized and included in the textbook. We hope this will help students not to write down an answer like 1.2×10^{16} A without thinking about it twice.

ALTERNATE SOLUTION PROCEDURES

Many problems in this textbook require students to solve a particular problem by a second or third approach. This helps the students to understand better the professional practice of engineering in which alternate solutions abound. We believe it is important that: (1) students compare the merits of two or more approaches to the solution of a problem, and (2) they also learn to validate their first solution through an alternate approach.

FROM THE SPECIFIC TO THE GENERAL

The natural order of learning is from the specific to the general. A student cannot comprehend a generalization without first understanding at least one specific case of that generalization. Consequently we have tried to introduce fundamental principles in terms of a specific case, develop a generalization, and then get the student to practice applying those generalizations in other specific cases.

WRITTEN TO THE STUDENT FOR THE STUDENT

We have tried to meet the student's needs and match the level of comprehension at every point in the text. We have avoided abstract rigor where it seemingly would confuse the student. On the other hand, we have attempted to be precise. We have tried to develop a style that is informal and conversational, hoping thus to achieve better communication with the student. In some instances, the explanations may seem overly detailed and tedious to an experienced educator, but we believe that this detail is necessary in writing for the less experienced student.

BUILDING THE STUDENT'S MOTIVATION TO LEARN

According to learning psychologists, one is motivated to learn only that which one perceives to be useful to oneself. Through text discussion and selected problems, we have attempted to build the student's perception of his or her need to learn. Many of the text problems relate to real-world situations that we hope will seem significant to students. In order to be motivated to learn a particular concept, one must have both the necessary self-confidence and the need to know. In 169 problems in this text, the student is required to make some kind of a consistency check on the solution. This process of making consistency checks and validations on solutions to challenging problems builds the student's confidence in his or her own learning ability.

THE PROBLEMS

Some of the problems are integrated into the reading material with the intent that the student will pause from reading to solve a particular problem at the time the problem is confronted. In this way the problem and its solution become a very important part of the learning process. A few of the problems are merely exercises in the mechanics of developing some equations of the text material; most of the problems are intended to be challenging and thought-provoking. Approximately 51 problems are primarily design problems and approximately 353 are analysis problems. Most of the design situations require the student to find R, L, or C parameters to yield a specified performance. Approximately 35 problems require the student to model the particular circuit configuration.

TOPICAL CONTENT

Chapter 1 serves as a very brief review and an introduction to the remainder of the text, including the introduction of the concept of modeling. In Chapter 2 the basic techniques of circuit analysis and design are developed for resistive circuits. This gives the student opportunity to learn the basic techniques in terms of the simpler resistive circuits, and later to learn how to apply the same techniques to more complicated circuits. A simple treatment of operational amplifiers is included to allow use of more interesting applications and examples. In Chapter 3 the concepts of capacitance and inductance are developed, along with the basic techniques of analysis and design of first-order RC and RL circuits. Physical interpretation of these circuits and their characteristic behavior are stressed.

Analysis and design of circuits with sinusoidal forcing functions are introduced in Chapter 4. The basic phasor transform for obtaining steady-state solutions is derived and applied in a way that we believe gives the student a more coherent understanding of the concepts than would be obtained through conventional treatments. The concepts of average and effective values and average power are also developed.

Techniques for the solution of second-order equations in the time domain are developed in Chapter 5. Laplace transforms are not introduced until later because we feel that it is easier for the student to obtain good physical understanding of the behavior of second-order circuits if the analysis is done in the time domain. In Chapter 6 the complete response of circuits is treated, along with the application of the standard network analysis techniques to more complicated circuits.

Fourier analysis, including both Fourier series and a brief treatment of Fourier integrals, is given in Chapter 7. Laplace transforms are introduced in Chapter 8, and Chapter 9 consists of a treatment of coupled circuits. The analysis of polyphase circuits is developed in Chapter 10.

The text is intended for a beginning course in circuit theory and electrical engineering. It should be suitable for a three quarter sequence, or, with a selection of topics, for a two quarter sequence. Sections preceded by an asterisk are optional and can be omitted without loss of continuity, since subsequent material does not depend on these sections. Students using the book should have had a course in differential equations, and a course including Laplace transforms would be helpful.

ACKNOWLEDGMENT

We are grateful for the stimulation provided by colleagues and for many suggestions of students who have studied from the manuscript form of this book. We express sincere gratitude to Mrs. Marian Swenson, who efficiently directed reproduction of the manuscript, to Mrs. Ruth Eichers, who so skillfully typed the bulk of the manuscript through many revisions, and to Mrs. Doris Bartsch, who sometimes rescued us with additional expert typing, for their truly outstanding work, and especially for their thoughtful consideration throughout a long and difficult task.

Carl H. Durney
L. Dale Harris
Charles L. Alley

Contents

Chapter 1

Prologue

Introduction What is customarily called electric circuit theory lies at the very foundation of electrical engineering. The fundamental principles and concepts of circuit theory are widely used in the design and analysis of countless kinds of electric devices and systems. Thoughts and speech of electrical engineers the world over are pervaded by the concepts of circuit theory. The concept of impedance that you may recall from your physics, for example, has become second nature to electrical engineers. Furthermore, the concepts and techniques used in circuit theory are often useful in related areas. For example, the concept of impedance is carried over into acoustical wave and microwave theories, notably in describing wave impedances. Analogies of electric circuit concepts and techniques are also found in nonelectrical areas, such as in mechanical systems, where the behavior of a mass-spring system is directly analogous to the behavior of an electric circuit. Also, the methods of analysis and problem solution, so elegantly applied to circuits, has broad application in almost all engineering efforts within and outside electrical engineering. The pages that follow will guide you in acquiring essential abilities in analyzing, problem solving, and designing.

It is difficult to overstate the importance of the study of electric circuit theory to the electrical engineering student, not only because of the foundation that it provides the student for a career in electrical engineering, but also because the classical techniques used in circuit theory are used in many other areas. Circuit theory is an exciting subject, and we wish you well in your study of it.

1.1 OBJECTIVES AND METHODS

The principal objective of this book is to help you first comprehend, and then gain the ability to apply, the fundamental laws and techniques of circuit theory in solving real engineering problems and to help you develop a foundation for work in more advanced courses.

Since we want to help you acquire ability and self-confidence in solving real problems (we mean solving problems in the broad sense of the term), perhaps we should discuss the general procedure of solving problems, just to be sure that you know what we have in mind. The process of solving engineering problems can be categorized into the following general steps:

1. Represent the actual physical system involved by a *model*, which usually contains mathematics symbols and often graphics.
2. Apply fundamental laws to analyze the model or design an idealized device based on the model.
3. Decide how well the characteristics of the physical system can be described in terms of the characteristics of the model. The characteristics of the model and the actual physical system are never exactly the same because the physical system always has properties (such as noise and nonlinearities) that are not present in the model. Try to account for the differences in some way.
4. Verify the predicted behavior or characteristics of the system by comparison with measured data, by an alternative approach, or by some other consistency check.

In your courses, so far, you have probably found that the second of these steps was emphasized. But the first, third, and fourth steps are also essential in engineering practice. A person's effectiveness with these latter steps is improved largely through personal experiences. This text is overtly designed to help you develop your confidence and ability in solving engineering problems that require the application of these four steps. We have tried, wherever possible, to point out the modeling steps, to include realistic numbers in the examples, to emphasize physical interpretation and the difference between the model and the physical system, to point out the limitations of models, and to check results for consistency.

Since you learn only what you practice, we have tried to follow this general pattern: introducing principles and techniques in terms of specific examples, then developing the generalization, and then giving you practice in applying the generalization. At the outset, we would like to point out some important generalizations. Much of the engineering work that involves circuit theory requires solving for voltages and currents, or designing circuits to produce specified voltages and currents. It is remarkable that voltages and currents can always be found by applying only the following few fundamental laws:

1. Kirchhoff's voltage law.
2. Kirchhoff's current law.
3. Voltage-current relations for resistance, inductance, and capacitance.

In principle, if you know enough mathematics, you can solve for any voltages and currents by applying only these laws. However, in practice, you will need to learn a number of additional concepts and techniques related to the application of these laws. In Chapter 2 we introduce these laws and their application to resistive circuits, the simplest kind. Then we introduce other principles and techniques in the remainder of the book, and proceed to apply the laws to more complex circuitry, from time to time introducing more advanced techniques.

We realize that what we have said in this section may not be completely clear to you as a beginning student of circuit theory, but we hope that you will refer back to this section from time to time during your study of this book.

1.2   MODELING

Since the modeling step in problem solving is a crucial step in solving real problems, we introduce the ideas of modeling in this section, hoping that an initial example will help make the process clearer.

The basic idea of modeling is to represent a real physical system by a model (which is ofttimes mathematical). The model can never represent the physical system with complete accuracy, but it is hoped that it will be accurate enough to be useful. As a simple example, suppose that we want to estimate the time required to travel from here to a city 500 miles away by automobile. The simplest model that represents the physical problem is probably an object traveling with a uniform velocity equal to the average estimated velocity of the automobile. Suppose the average velocity is estimated to be 40 miles per hour. Then with this model, we would calculate

$$t_0 = \frac{d}{v_0} = \frac{500 \text{ miles}}{40 \text{ miles/hour}} = 12.5 \text{ hours}$$

Although this model is adequate for many purposes (you have probably used it many times with satisfactory results), it is obviously a crude representation of the actual physical system. It does not include any of the variations in velocity that result from stopping at traffic lights, slowing down in congested areas, etc. Rather than make the model more complicated, we usually take the results of this simple model and compensate in some way for the limitations of the model. For example, we may leave early enough to allow extra time in reaching the destination. This process is the same one that we use in electric circuit design: We choose the simplest model that is adequate, obtain the desired results for the model, and then make allowances for the limitation of the model, realizing that the actual results obtained in the physical system will be different from those obtained from the model.

As another example, suppose that we want to analyze the fall of a skydiver before the parachute opened and determine the skydiver's velocity as a function of the amount of time from the beginning of the jump. The first step is to choose a model to represent the system.

Perhaps the simplest model that would give us approximate information is a mass falling in a vacuum. This model and its relation to the physical system is illustrated in Fig. 1.1. Note that the mass is not the skydiver, only a model representing the skydiver. In a vacuum, gravity produces the only force acting on the mass, and we can apply Newton's law (force = mass × acceleration) to the model and solve for v_{m_1}, the time-varying velocity of this mass. The results are shown in

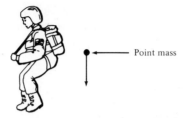

Point mass

fig. 1.1   A point mass falling in a vacuum used as a model of a skydiver.

Fig. 1.2, where v_{m_1} is the model velocity—not the velocity of the skydiver. This model has an advantage in its simplicity and could yield reasonably good accuracy for values of time that are sufficiently small, but it would yield very poor accuracy for large values of time.

Let us construct a better model that includes the approximate drag force of the atmosphere. This drag force might be modeled with sufficient accuracy for our purposes by $f_d = kV_m^2$ where f_d is the drag force, v_m is the velocity, and k is a proportionality factor that is a function of the skydiver's "shape" and "attitude" in respect to the direction of fall. We must live with some approximations in essentially all models, as in this one in particular. The drag force is not exactly proportional to the velocity squared, and furthermore the k of the skydiver can only be approximated.

After we had found the k for a sphere falling in atmosphere, then we might use a sphere of appropriate area and mass as a model of the skydiver. Then by applying Newton's laws to the model in Fig. 1.3, we get the curve shown in Fig. 1.4, which, by comparison with Fig. 1.2, shows that the drag causes the velocity to increase more slowly with time, as we would expect. Although v_{m_2} is closer to v_s, the velocity of the skydiver, than v_{m_1} is, v_{m_2} is still not the same as v_s. There are still factors that our second model does not include. The shape of the sphere is not the same as the shape of the skydiver, and any tumbling of the skydiver would not be accounted for in our model. You may be able to think of still other factors that are not accounted for by the model.

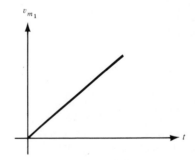

fig. 1.2   Velocity of the model (no friction).

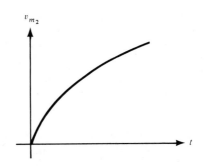

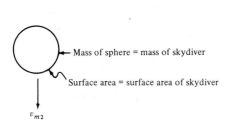

fig. 1.3 Second model of the falling skydiver in Fig. 1.1a, a sphere falling in a fluid of constant viscosity.

fig. 1.4 Velocity of the model (drag force included).

We should never use a more complicated model than is justified for the particular accuracy that is needed. Perhaps in some cases, the curve in Fig. 1.2 will be adequate, in which case we are glad not to complicate the effort with the more complicated model of Fig. 1.3. In still other situations, the model of Fig. 1.3 would not yield desired accuracy. Remember always to use the simplest model that is adequate.

It is obvious that there is a significant difference between both the models of Fig. 1.1 and Fig. 1.3 compared to the physical system (the skydiver). However, it seems that many times a person will forget the difference between a model and the physical system it represents, and thus mistakes in analysis or design will result. For example, it may serve for many purposes to model a resistor (a small carbon cylinder) as resistance that is independent of temperature and electric current. But in other situations, the model can be in serious error because the resistor may exhibit some of the properties of capacitance and/or inductance; also the resistance may be significantly dependent on current and/or temperature. If we are always given the model of an electric circuit to start with and are never asked to go through the modeling step, we will probably get in the habit of thinking of the model as a precise representation of the physical device and tend to forget the effects of temperature and other factors upon the device.

This is especially easy to do because in this case the model is a very good representation of the physical object, and the differences between this particular model and the object it represents are almost always unimportant.

PROBLEM

1.1 Write equations and/or sentences that describe two different models that represent a brass cylinder suspended from a screen-door spring and oscillating as a result of an initial downward displacement given to the cylinder. First describe a *simple* model that you believe would give a reasonably accurate history of position of the cylinder as a function of time during the first period of its motion. Then modify this first model to give improved accuracy of position at t_1, where $t_1 > 20T$ and T is the period of the position. Finding an acceptable model requires judgment. Do *not* expect to find *the* unique solution to this problem.

1.3 REVIEW OF CHARGE, CURRENT, VOLTAGE, AND POWER

Even though you have worked with charge, current, voltage, and power in your physics classes, a review of the basic concepts and relationships might be helpful.

Charge

You have learned in your physics courses that electric charge experiences a force if in an electric field, and that the movement of charge represents electric current. One might think of electric charge as "quantity" somewhat analogous to gallons of water. The elementary models used to represent charge are charged particles called electrons and protons. An electron has a charge of -1.6021×10^{-19} coulomb, where the *coulomb* is the basic unit of charge (abbreviated C). A proton has a charge of $+1.6021 \times 10^{-19}$ coulomb. An electron has a mass of 9.1091×10^{-31} kilogram; the proton is much more massive, having a mass of 1.6725×10^{-27} kilogram. Other particles called neutrons have no charge at all, but have the same mass as a proton. Of course, the particle models that we have described are only elementary models, but they are sufficient for our purposes. In physics courses you will study about many other kinds of particles; and if you study quantum mechanics, you will be working with wave functions instead of the kinds of particles described above.

Current

Electric current is defined as flow of charge; that is,

$$i = \frac{dq}{dt} \tag{1.1}$$

where q is charge in coulombs, t is time in seconds, and i is current in amperes. By integrating (1.1), we get another useful relation between current and charge

$$Q = \int_{t_1}^{t_2} i \, dt \tag{1.2}$$

where the letter[1] Q in (1.2) refers to the charge that has passed a point during the time interval $(t_2 - t_1)$. Current is simply a measure of the rate of flow of charge with time, but the definition is a little complicated because charge can be both positive and negative and because the definition of current also includes direction. We will use an example to help make the concepts of current clear.

Figure 1.5 illustrates charged particles, both positive and negative, moving past a checkpoint. We would like to find the resulting current in the direction indicated by the arrow (to the right), assuming that all of the particles passed the checkpoint in an interval of 10^{-21} second. For small changes in q and t, we can approximate

---

[1] Lowercase letters are used for quantities that are functions of time; capital letters are used for quantities that do not vary with time. This rule will be discussed in detail later.

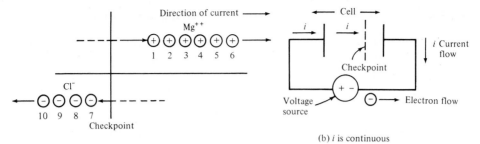

(a) Positive and negative charges in motion

(b) *i* is continuous

fig. 1.5   *Charged particles moving past a checkpoint.*

(1.1) by

$$i \approx \frac{\Delta q}{\Delta t} \tag{1.3}$$

where Δq and Δt are small changes. Next, we must decide what Δq is in terms of the particles shown in Fig. 1.5.

From (1.2) we can see that Δq is the net charge passing the checkpoint to the right in time interval $\Delta t = (t_2 - t_1)$. Negative charges moving to the left contribute to Δq as do positive charges moving to the right.

The electrolytic separation of magnesium for chlorine in a melted solution of $MgCl_2$ might illustrate the situation of Fig. 1.5. Each positive charge in Fig. 1.5 is a magnesium ion having a charge of $+2 \times 1.6 \times 10^{-19}$ C, while each negative charge is a chloride ion having a charge of -1.6×10^{-19} C. You should not be concerned with the fact that, in this particular illustration, more positive ions pass the checkpoint than do negative ions. At a different checkpoint in the electrolyte, more negative ions than positive ions would pass.

Now let us assume that the charges of Fig. 1.5 pass the checkpoint in $\Delta t = 10^{-21}$ second. The net charge that passes the checkpoint is

$$\Delta q = \left(\overset{Mg^{++}}{2 \times (+6)} + \overset{Cl^-}{(-1) \times (-4)} \right) \times 1.6 \times 10^{-19} = 2.56 \times 10^{-18}$$

Then the current is

$$i = \frac{\Delta q}{\Delta t} = \frac{2.56 \times 10^{-18}}{10^{-21}} = 2{,}560 \text{ A}$$

where A is the abbreviation for amperes.

In certain *dynamic* situations, charge can accumulate in a capacitor (electrical reservoir), but in many other *steady* situations (such as the one under consideration here), the current is continuous; see Fig. 1.5b. The current external to the cell flows in solid conductors consisting of wires and a voltage source or generator. Only electrons flow in these solid conductors. At any checkpoint within the electrolyte of the cell or in the external conductor path, the current will have the same value of

2,560 A. Notice that electrons flow in the external conductor from the left to the right plates (or sides) of the electrolyte.

PROBLEM

1.2 In the situation of Fig. 1.5, where the current is 2,560 A, calculate the number of electrons that pass any checkpoint on the external conductor in 1 microsecond (10^{-6} second). Remember that the charge of one electron is 1.6×10^{-19} C. The answer, of course, must be consistent with $i = \Delta q / \Delta t$.

Throughout this textbook, you will be required to make consistency checks on your problem solutions, as was suggested in Prob. 1.2. The responsible, independent, practicing engineer continually finds techniques to build confidence in his or her problem solutions. Consider Prob. 1.3.

PROBLEM

1.3 You may have found the solution to Prob. 1.2 through $i = \Delta q / \Delta t$. If that is the case, then the solution to the following must be consistent with the solution to Prob. 1.2. Magnesium ions are formed in the solution and migrate to the right plate. Chloride ions are formed in the solution and migrate to the left plate. For each magnesium ion, two chloride ions are formed. So in this steady situation and during the time interval of 10^{-21} second, how many magnesium ions reach the right plate and how many chloride ions reach the left plate? These two numbers should be consistent with the current that exists everywhere and also with your solution to Prob. 1.2.

It is important to note that we can also say that the current flowing in the cell to the left in Fig. 1.5 is $-2,560$ A. In gases and in solutions, current is composed of the flow of both positive and negative charges; in conductors, however, current is due almost solely to the movement of negative charges, electrons. Nevertheless, in practice we seldom need to distinguish between the flow of positive and negative charges in circuit analysis because we usually just want to find the net current. Consequently, it is usually easier to think of current in terms of the flow of positive charges, regardless of whether or not negative charges are contributing to the current.

You will see when you study solid-state electronics that in some "semiconductor" materials, a current may have one component due to electron flow and another component due to "hole flow." The latter might be modeled as a positive ion flow even though a "hole" has some but not all of the properties of an ion. What is important here is that there are two distinct charge flows that contribute to the total current.

Ampere hour, like coulomb, is a unit of electric charge. The capacity of an electric cell or battery to deliver charge is usually specified in ampere hours (A · h). Charge has dimensions of current × time. A coulomb, of course, is an ampere second.

PROBLEM

1.4 The General Electric rechargeable nickel cadmium cell, size C, Model GC2, is rated at 1.2 A · h. This is the equivalent of how many kilocoulombs (kC)?

Answer : 4.320 kC

Voltage

In our discussion of charge and current, we did not say anything about the forces required to move the charges and thus cause current flow. These forces, which are described in terms of electric fields (as you will find out when you study electromagnetic field theory), are not of direct interest to us in circuit theory. Instead of working with electric fields, we will work with *potential difference*, a related quantity. Since the unit of potential difference is the *volt*, the term potential difference is often shortened to *voltage.*

By definition, the potential difference between two points is the work or energy required to move a charge of one coulomb from one of the two points to the other. If the potential difference between two points is zero, then no current can flow through any resistance between those points. Potential difference is analogous to the potential energy of a mass in a gravitational field. A stationary ball on a level floor will not be induced by gravity to roll because the ball has no potential energy with respect to the ground, but a ball placed on the side of a hill will roll down the hill because of the potential energy it possesses. Also, the potential energy of the ball at one point on the side of the hill is equal to the work required to carry the ball up the hill.

At this point we want to introduce the sign convention that we will use for potential difference throughout the text. The "plus" sign is at a point higher in potential than the "minus" sign. Where current encounters resistance, flow must be from a point of higher potential to a point of lower potential, much like water flowing down a hill; see Fig. 1.6a. On the other hand, a voltage source causes

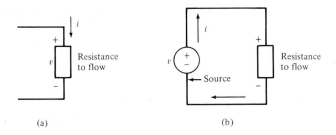

(a)   (b)

fig. 1.6   *Sign conventions for potential difference v.*

current to flow from a point of lower potential of the source to a point of higher potential if the situation is like that of Fig. 1.6b. This is much like a pump forcing water uphill. Of course, the situations of Fig. 1.6 are highly restricted but help us to begin to develop our concepts of sign conventions. More general situations will be treated as we proceed through the text. For example, if two or more voltage sources are present, then the current may be forced to flow into (rather than out of) the positive terminal of a particular source.

Power and Energy

Power is defined as the rate at which energy is transported or transformed. A relationship among voltage, current, and power can be obtained by starting with the definition for potential difference. Since potential difference is energy per unit charge, the amount of energy dw required to move a differential amount of charge dq through a potential difference of v is

$$dw = v \, dq \tag{1.4}$$

where w is energy in watt seconds (or joules), v is in volts, and q is in coulombs. Hence dw/dt is

$$\frac{dw}{dt} = v \, \frac{dq}{dt}$$

We define power as

$$p = \frac{dw}{dt} \tag{1.5}$$

where p is power in watts and t is time in seconds. Using the definition of current in (1.1), we get

$$p = vi \tag{1.6}$$

Since both v and i can be either positive or negative, p can also be either positive or negative. The meaning of the sign of p is shown in Fig. 1.7. Positive

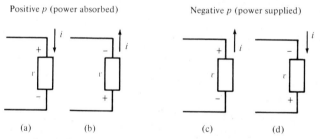

Positive p (power absorbed)   Negative p (power supplied)

(a)   (b)   (c)   (d)

fig. 1.7   *Meaning of the algebraic sign of power.*

power means power absorbed, such as in heating a resistor. Negative power, on the other hand, means power is being supplied by some kind of energy source. Positive power into the "box" would remain if both the sign of v and the direction of i were changed as in Fig. 1.7b. Similarly, negative power into the "box" would still occur if both the sign of v and the direction of i were reversed as in Fig. 1.7d.

The concept of energy was used in (1.4) to define voltage or potential. But our interest in energy extends far beyond this rather academic situation. Another important relationship is

$$w = \int_{t_1}^{t} p \, dt = \int_{t_1}^{t} vi \, dt \tag{1.7}$$

where w is in watt seconds, p in watts, and t in seconds, and where w in (1.7) means the energy delivered or absorbed in the time interval from t_1 to t. Notice that t_1 is a particular value of time and t is variable. Then in this case w is a function of t, and therefore lowercase w is used.

PROBLEM

1.5 Is it the proper use of capital W in writing

$$W = \int_{t_1}^{t_2} p \, dt$$

Why?

In situations where voltage is constant,

$$w = \int_{t_1}^{t} Vi \, dt = V \int_{t_1}^{t} i \, dt = V(q - Q_1) \tag{1.8}$$

where q is the charge at t and Q_1 is the charge at t_1. The quantity $(q - Q_1)$ has the meaning of the charge that passes a particular checkpoint in the time interval from t_1 to t. Apply (1.8) to a battery assuming its terminal voltage is constant; see Fig. 1.8. Note that i as shown is positive, q is positive, and energy w leaving the battery to flow into the "box" is positive.

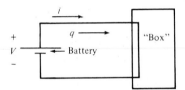

fig. 1.8 *Battery delivers energy w.*

Example 1.1

The rechargeable cell of Prob. 1.4 has a capacity, when fully charged, to deliver 1.2 ampere hours, or 4,320 coulombs. Find its full stored energy where we assume that its terminal voltage throughout discharge is 1.25 volts. In (1.8), Q_1 and t_1 are each equal to zero at the beginning of the discharge, and $q = 4,320$ coulombs. $W = 4,320 \times 1.25 = 5,400$ watt seconds. For a consistency check, follow a different route, as follows. The cell will deliver 1.2 amperes for a duration of one hour. During this hour, the current is 1.2 amperes and the voltage is 1.25 volts. Therefore, $P = VI = 1.25 \times 1.2 = 1.5$ watts. Then the energy is $W = P \times t = 1.5 \times 3,600 = 5,400$ watt seconds.

● ● ●

The material in this section is meant only to provide you with a review and the basic concepts that we expect you to have learned before starting a study of this book. Beginning with the next chapter, we will systematically develop the methods of circuit analysis in detail. Before proceeding with that, we will conclude this chapter with a discussion of units and symbols.

1.4 UNITS AND SYMBOLS

In this book we use the SI system of units, which was adopted by the Eleventh General Conference on Weights and Measures held in Paris in 1970. SI is an internationally agreed-upon abbreviation for Système International d'Unités (International System of Units). Table 1.1 lists the six fundamental units of the SI system,

TABLE 1.1 SI Basic Units

Quantity	Unit	Symbol
Length	meter	m
Mass	kilogram	kg
Time	second	s
Electric current	ampere	A
Temperature	Kelvin	K
Luminous intensity	candela	cd

each defined in terms of a specified standard. For example, the *kilogram* is the mass of a particular cylinder of platinum-iridium alloy that is preserved in a vault at Sèvres, France, by the International Bureau of Weights and Measures. The *meter* is defined in terms of an orange-red line in the spectrum of krypton 86.

Some units derived from these basic units are given in Table 1.2. In Table 1.3

TABLE 1.2 Some SI Derived Units

Quantity	Unit	Symbol	Relation to Other Units
Force	newton	N	$kg \cdot m/s^2$
Work or energy	joule (or watt second)	J	$N \cdot m$
Power	watt	W	J/s
Charge	coulomb	C	$A \cdot s$
Potential difference	volt	V	J/C
Electrical capacitance	farad	F	$A \cdot s/V$
Inductance	henry	H	$V \cdot s/A$
Frequency	hertz	Hz	s^{-1}

you will find a list of the standard prefixes for multiples and submultiples of units. For example, 10^{-3} volt would be written as 1 mV and spoken of as one millivolt. Likewise, 3×10^3 hertz would be written 3 kHz and spoken of as three kilohertz. You should note the pronunciation of the prefixes and form the habit early of pronouncing them correctly, thus avoiding learning some incorrect pronunciations that you might hear. Note particularly the pronunciation of giga, which is often mispronounced.

In our communication with you and in your communication with others, it is

TABLE 1.3   *SI Standard Prefixes*

Factor	Prefix	Pronunciation	Symbol
10^{12}	tera	tĕr′ ȧ	T
10^{9}	giga	jĭ′ gȧ	G
10^{6}	mega	mĕg′ ȧ	M
10^{3}	kilo	kĭl′ ō	k
10^{2}	hecto	hĕk′ tō	h
10	deka	dĕk′ ȧ	da
10^{-1}	deci	dĕs′ ĭ	d
10^{-2}	centi	sĕn′ tĭ	c
10^{-3}	milli	mĭl′ ĭ	m
10^{-6}	micro	mī′ krō	μ
10^{-9}	nano	năn′ ō	n
10^{-12}	pico	pē′ cō	p
10^{-15}	femto	fĕm′ tō	f
10^{-18}	atto	ăt′ tō	a

very important to have clearly defined and well-understood symbols. As you study this book, you will find specific symbols defined as needed, then used precisely thereafter. Once these symbols have been defined in the book, it is important that all of us "stick together" in the manner that we interpret and use them.

Even though the symbols are defined as needed, we would like to present one general rule now: A lowercase letter will be used to represent a quantity that varies with time. For example,

$$p = 10t^2 \tag{1.9}$$

where p represents power (in watts) and t is time (in seconds). On the other hand, capital letters are used for quantities that do not vary with time, as in

$$p = P_a \sin \omega t + P_b \tag{1.10}$$

where P_a and P_b do not vary with time. The relationship among p, P_a, and P_b is shown in Fig. 1.9. Note that Fig. 1.9 also displays an appropriate use of symbols.

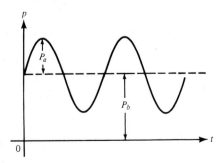

fig. 1.9   *p as a function of time.*

In another equation

$$p_c = p_d \sin at + p_e \tag{1.11}$$

the message in the lowercase letters is that p_c, p_d, and p_e vary with time in some way.

Example 1.2
Is it appropriate to write $P = vi$?

Solution
No, because P refers to power, which does not vary with time, and the use of the lowercase symbols i and v imply variation with time.

• • •

Example 1.3
A circuit of some composition is in the "box" shown in Fig. 1.10. Current

fig. 1.10   Circuit for Example 1.3.

$i = 10e^{-100t}$ amperes flows in the indicated direction and voltage $V = 20$ volts appears at the terminals as indicated.

   a.   Are symbols used correctly in writing $p = Vi$?

Solution
The symbols are used correctly. The product of voltage V and current i is power p. Because i is a variable with respect to time, the product is a variable, thus p rather than P.

   b.   What is the equation of the power absorbed in the box?

Solution

$$p = Vi = 200e^{-100t} \text{ watts}$$

• • •

Example 1.4
A moderately large steam electric plant generates 1.05 GW. How many million residential refrigerator motors would this power supply if each used 200 watts simultaneously?

Solution

$$1.05 \text{ GW} = 1.05 \times 10^9 \text{ W}$$

$$P = 1.05 \times 10^9 \text{ W} = n \times 200 \text{ W}$$

$$n = \frac{1.05 \times 10^9}{200} = 5.25 \times 10^6 = 5.25 \text{ million refrigerators}$$

● ● ●

1.5   DIMENSIONALLY CONSISTENT EQUATIONS

Although in this book we do not intend to make a big fuss over what is known as dimensional analysis, it is very important to make quick mental checks continually on derived equations to assure that terms that are *added* or *equated* are always of the same dimension. We must never attempt to add or equate meters to seconds and to other quantities that are not dimensionally equal. By giving some thought to the dimensions of the various terms in the equation, we can easily avoid some serious "bloopers."

It may or may not have occurred to you that the argument x of e^x, sin x, cos x, sinh x, and other such functions must be dimensionless when such arguments relate to the physical world. As an example, the definition of an angle in radians is the ratio of a subtended arc to the appropriate radius. So the argument of sin x, cos x, and similar trigonometric functions has no dimensions. It is perhaps sufficient to state, without formal proof, that both the function itself and its argument must be dimensionless for such functions as e^x, sin x, cos x, sinh x, and others.

Example 1.5
Is it appropriate to write $p = P_1 \times t$? No, it is *not* if both p and P_1 each represent power and therefore $(P_1 \times t)$ is not dimensionally equal to power. It *is* appropriate to write $p = A \times t$ if A has dimensions of power/time.

● ● ●

Example 1.6
Is it dimensionally correct to write $i_1 = i_2 \sin(I_3/I_4) + I_5/3$ where i and I are used to represent current? The argument (I_3/I_4) is dimensionless and the sine of this argument is also dimensionless. Therefore each of the three terms i_1, $i_2 \sin (I_3/I_4)$, and $I_5/3$ has dimensions of current if the 3 is a dimensionless number. It follows that this equation is dimensionally correct.

● ● ●

Example 1.7
Is it appropriate to write $i = A \sin at$? Only if A has dimensions of current and a has dimensions of $(1/t)$.

● ● ●

Example 1.8

In the following equation, the p's and P's are in watts, and are functions of time and are independent of time, respectively. In each of these, state whether the equation is dimensionally correct or not. Where incorrect, give the reasons. If correct, state the special conditions. $p^n = P_1 p_1 + B P_2$.

The term $P_1 p_1$ has dimensions of power squared, so the equation would be dimensionally correct only if $n = 2$ and B has the dimensions of power. The term $P_1 p_1$ is a variable with respect to time; therefore, the p on the left-hand side should be lowercase.

• • •

PROBLEMS

1.6 In each of the following situations, the p's and P's are powers, variable and constant, respectively. State whether the p or P on the left of each of the following equations is used correctly. State whether each equation is dimensionally correct or not. Give reasons and special conditions.

   a. $P_3 = P_1 e^{-at} + P_2$

   b. $p = \dfrac{P_1 P_2}{P_2} + p_3$

   c. $p = P_1 e^{-p_1 t} + 10 p_2 + 10 P_2$

   d. $p = P_1^{0.1t} + P_2^{p_2}$

   e. $p^2 = P_1 p_1 + p_1^2$

   f. $p = P_1^2 / P_2 + P_3$

1.7 At a particular checkpoint in the semiconductor material of a transistor, 93 percent of the current is due to electrons and the remaining 7 percent due to "holes" (positive carriers). The charge of the "hole" is equal to the magnitude of the electronic charge. Calculate the number of electrons and the number of "holes" that pass this checkpoint in one millisecond if the current is equal to one milliampere.

1.8 It has been proposed that it might be feasible to build a 0.2-GW wind-driven electric generator. Here, the wind turbine would include 80-meter airfoils mounted vertically on horizontally moving cars that travel on a continuous oval track of 15 km in length. The estimated cost of construction of this plant is $1,000 per kilowatt of capacity (about the same as a coal-fired steam electric generating plant).

   a. Estimate the cost of constructing this plant.

   b. Even at the best wind power sites, there will be periods when the electric power generation will be too low to supply the need, so the installation of a storage battery as an auxiliary power source is considered. The battery would be charged during high winds and discharged when the wind power is inadequate. Assume that the battery is to have an energy capacity sufficient to deliver 0.10 GW over a period of 20 hours. Assume that the cost of this battery per kWh is the same as an automobile battery rated at 50 A · h, 12 V, and costing $50. Determine the cost of this battery.

   c. If this auxiliary battery is assembled to deliver 30 kV, what is its coulomb capacity?

Answers
(a) Plant cost = $200 million
(b) Battery cost = $166.7 million
(c) 240 MC

1.9 In a metallic conductor such as copper, the electrons move or "drift" at a surprisingly low velocity. This is *not* a contradiction with the fact that the velocity of electric wave propagation is extremely high by comparison; these are two very different phenomena. In copper, there are about 8.4×10^{22} electrons/cm^3 free to "drift" or move. One might select a copper conductor for a particular application so that the current density in the cross section of the conductor is 400 A/cm^2.

a. Under these conditions, calculate the "drift" velocity of the electrons.

b. If the current in a conductor on a solid-state chip is 10 μA at this current density, calculate the two dimensions of the cross section of this conductor if its cross section is rectangular and one dimension is five times the other.

Ohm's Law and Kirchhoff's Laws

Introduction From your study of electricity in physics and elsewhere, you have acquired some familiarity with electric networks that include three components: resistances, capacitances, and inductances. This text in its entirety will help you to become highly skilled in analyzing such networks, but Chapter 2 is constrained to networks of resistances only. Other chapters beginning with Chapter 3 will lead you through the concepts of networks that employ all three of these circuit elements. By studying Chapter 2, you will learn how to apply Ohm's law and Kirchhoff's two laws to networks where the properties of capacitance and inductance are not present. This constraint will help you grasp some basic ideas in their simplest form.

Ohm's law applies to resistance only; Kirchhoff's two laws apply to networks, in general, where inductance and capacitance are also involved. You will see the restricted nature of Ohm's law, and sense the general nature of Kirchhoff's two laws.

A Thermistor Thermometer

In this chapter, as in some others, we use an illustrative problem as a means of introducing the principles and techniques to be explained in the chapter. This will help you to sense the importance of the material to be studied, to gain added perspective and insight as you learn, and to make the connection between principles and real applications. After introducing the problem, we digress to develop the basic principles and techniques in a systematic way, returning later in the chapter to the solution of the problem as an example of how these principles and techniques are applied.

In numerous applications ranging from patients in hospitals to pot roasts in microwave ovens, it is important to make temperature measurements, often in the form of continuous recordings. For example, in one new area of research, that of treating cancer by microwave heating, the temperature of the cancer tissue must be maintained at particular temperatures such as 43°C (normal body temperature is 37°C) for periods of approximately an hour. Without some method of measuring the temperature of the tissue, it is not possible to keep the tissue at the desired temperature. A commonly used technique is to insert a tiny thermistor (a component that changes resistance significantly with temperature) into the tissue by means of a hypodermic needle. The temperature is then monitored by measuring the current or voltage in a network similar to a Wheatstone bridge where this current or voltage is sensitive to the change of the thermistor resistance, which in turn is sensitive to its temperature. Such a system has the advantage that the temperature information is contained in an electric signal that can be connected to a recorder when a continuous recording of the temperature is desired, and also can be transmitted to other locations in applications requiring the monitoring of temperature at a remote location, such as when using an external meter to display the temperature inside a freezer. Although there are many other ways to measure temperature, thermistor methods are among the most common. Since the thermistor thermometer is a good illustration of circuit design and analysis, we shall use it as a basis for developing some of the basic concepts and techniques of resistive network design.

In this example, we will design a thermistor thermometer that you can build from a 1.5-V penlight battery, a $0-50$-μA ammeter, a common thermistor, and three resistors. A thermistor is very inexpensive, but the cost of a typical $0-50$-μA ammeter is not necessarily insignificant. However, you could design the circuit to work with an ammeter or voltmeter that you already have, although you might need more voltage than one penlight battery can furnish, depending on the sensitivity of the meter that you use. Also, the improved circuit described near the end of the chapter uses a less expensive meter. You are encouraged to buy or "scrounge" the parts and construct your own thermometer. If you do not wish to spend the money to do so, you might be able to construct a "breadboard" model of the thermometer using parts in your school laboratory.

The commonly used circuit shown in Fig. 2.1 is known as a Wheatstone bridge

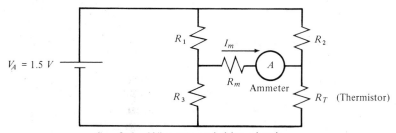

fig. 2.1   Wheatstone bridge circuit.

circuit, after the man who invented it. When $R_1/R_3 = R_2/R_T$, the current I_m through the ammeter is zero, and hence the bridge is said to be *balanced*. Small changes in the resistance of any of the resistors cause unbalance and a nonzero current through the meter. Thus if R_T is a thermistor, with a resistance that is a strong function of temperature, I_m will be an indication of temperature, and the face of the meter can be calibrated in degrees. In other applications, the ammeter might be replaced or supplemented by a recorder. Before we can describe the design of the thermometer circuit, we must digress for the major part of the chapter into the development of the basic principles and concepts that we will use. Then we will return to complete the design.

2.1   DEFINING i AND v

In analyzing a circuit like the Wheatstone bridge, we must deal with currents in various branches, in resistors, and in other circuit elements. We will use the symbol i (lowercase i implies that the current is a function of time) to represent a current in a particular part of the circuit. The position of the symbol i on a diagram and its associated arrow give the precise meaning of the symbol. For example, by the diagram in Fig. 2.2, i_3 is defined as a current that flows from a to b in that branch of

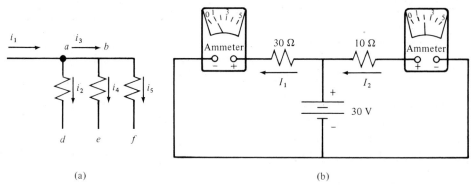

(a)   (b)

fig. 2.2   Currents in a network.

the circuit. It is extremely important that you understand that the arrows do *not* necessarily show the direction of the actual flow of the respective currents. The arrows only serve to help define what is meant by the symbol i. For example, in the dc circuit of Fig. 2.2b, $I_1 = 1$ A as indicated by the direction of the arrow and the reading of the ammeter (the ammeter reads upscale when current flows into the positive $(+)$ terminal and out of the negative $(-)$ terminal). On the other hand, $I_2 = -3$ A because the arrow defining I_2 is opposite in direction to the actual current flow, as indicated by the ammeter. Capital letters are used to designate quantities that do not vary with time.

Similarly, a positive value of i_3 in Fig. 2.2a represents a current in this branch flowing from a to b; a negative value of i_3 represents a current flowing from b to a. In many situations with which we will be concerned, the currents as well as voltages will be functions of time in such a manner that the current flow at a particular time might be from a to b (i_3 positive), and at a later instant the current might have reversed so that i_3 is negative.

Just as the arrow is used in defining a particular i that represents a current, polarity marks (+ and −) are needed to help define a voltage difference. Figure 2.3a shows how voltage differences might be defined for the Wheatstone bridge circuit of Fig. 2.1. The diagram defines V_1 as the *potential difference* of b in respect to c, or voltage of b above c, because the + is adjacent to b and the − is adjacent to c. The + and − signs define V_3 as the voltage of 0 above c, and V_T as the voltage of d above 0. These polarity marks are required to give precise meaning to the symbol V and do *not* necessarily give the sign (+ or −) of a particular voltage difference at a particular time. For example, in the dc circuit of Fig. 2.3b, $V_1 = 10$ V,

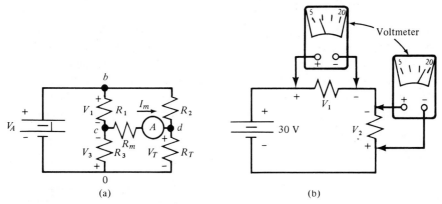

fig. 2.3 Voltages in a network.

as indicated by the + and − marks defining V_1 and the reading of the voltmeter (the voltmeter reads upscale when the + terminal on the voltmeter is connected to a point of higher potential than the − terminal). But $V_2 = -20$ V, as indicated by the + and − signs defining V_2 and by the reading of the voltmeter.

We can immediately establish the polarity of a particular voltage at any time as soon as we know the sign of a value of v at that instant. For example, if V_A, Fig. 2.3a, were replaced by a sinusoidal generator so that V_1 became v_1, then if at a particular time v_1 has a negative value, c is positive in respect to b at that instant, or c is at a higher potential than b.

Example 2.1

A generator delivers the sinusoidal voltage $v_1 = 100 \sin(\omega t + 35°)$ volts to a network of resistances R_1, R_2, and R_3, and capacitance C, as shown in Fig. 2.4.

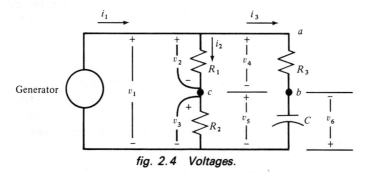

fig. 2.4 Voltages.

Ultimately we will be able to find the equations of the voltages v_2, v_3, v_4, and v_5 given v_1, R_1, R_2, R_3, and C. Our problem at the moment, however, is to find the numerical values of v_2, v_3, v_4, v_5 at various values of time when given the equations for these voltages. Finding these numerical values will help us think clearly about the method of defining a voltage.

The following equations are given:

$$v_1 = 100 \sin(\omega t + 35°) \text{ volts}$$

$$v_2 = 20 \sin(\omega t + 35°) \text{ volts}$$

$$v_3 = 80 \sin(\omega t + 35°) \text{ volts}$$

$$v_4 = 60 \sin(\omega t + 88.13°) \text{ volts}$$

$$v_5 = 80 \sin(\omega t - 1.87°) \text{ volts}$$

a.   Fill in the following table showing values of v_1, v_2, v_3, v_4, v_5, and v_6 at the indicated values of ωt.

Solution
At $\omega t = 0$,

$$v_1 = 100 \sin 35° = 57.36 \text{ volts}$$

$$v_2 = 20 \sin 35° = 11.47 \text{ volts}$$

$$v_3 = 80 \sin 35° = 45.89 \text{ volts}$$

$$v_4 = 60 \sin 88.13° = 59.97 \text{ volts}$$

$$v_5 = 80 \sin (-1.87°) = -2.61 \text{ volts}$$

$$v_6 = -v_5 = 2.61 \text{ volts}$$

At $\omega t = 130°$,

$$v_1 = 100 \sin 165° = 25.88 \text{ volts}$$

$$v_4 = 60 \sin 218.13 = -37.05 \text{ volts}$$

ωt	v_1	v_2	v_3	v_4	v_5	v_6
0	57.36	11.47	45.89	59.97	−2.61	2.61
130	25.88			−37.05		
200	−81.92					

In Prob. 2.1 you will be required to complete this table.

b.   Find some consistency check that can be applied to some of the values calculated for this table.

Solution

Intuitively we can see that $v_1 = v_2 + v_3$. Voltages v_2 and v_3 are, in a sense, merely components of v_1. This fact is also an application of Kirchhoff's voltage law, which we will study in detail very soon.

We see that at $\omega t = 0$, $v_2 + v_3 = 11.47 + 45.89 = 57.36$, which in turn is equal to v_1. So our values of v_1, v_2, and v_3 are consistent. In a similar vein, $v_4 + v_5 = 59.97 + (-2.61) = 57.36 = v_1$.

• • •

PROBLEM

2.1 a. Finish filling the table of Example 2.1.
     b. Wherever practical, run consistency checks on your solution to (a).

2.2 OHM'S LAW

As the flow of a fluid (liquid or gas) in a pipe encounters resistance to that flow, electrical flow or current encounters resistance. This concept of electrical resistance is embedded in Ohm's law,[1] expressed mathematically as

$$i = \frac{v}{R} \qquad \text{or} \qquad v = iR \tag{2.1}$$

where i is current in amperes and v the voltage in volts across a resistance of R ohms. It is important to realize that, depending on the direction assigned to i compared to the polarity marks assigned to v, it might be that $i = -v/R$ or $v = -iR$.

For example, in Fig. 2.5,

$$v_1 = i_1 R_1 \qquad v_1 = -i_2 R_1 \qquad v_2 = -i_1 R_2 \qquad v_2 = i_2 R_2 \tag{2.2}$$

fig. 2.5 Sign of i and v, where $i_1 = -i_2$.

It is equally important to remember that in Fig. 2.5 at the instant when v_1 is positive, i_1 is positive, and when i_2 is negative, v_2 is also negative. It might be helpful for you to note that in a resistor, the true or actual current (not assigned current direction) must flow "downhill," or the true current must flow *into* the + terminal of the resistor and leave the − terminal (see Section 1.4).

---

[1] Ohm's law, for the purposes of this book, is axiomatic. In your laboratory you will experience many, many instances supporting its validity.

Example 2.2

In the circuit of Fig. 2.6, $i = 10e^{-2t}$ A. We wish to find equations for v_1 and v_2 and plot i, v_1, and v_2. You might wonder why v_2 is assigned the polarity marks, com-

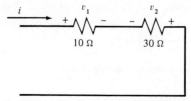

fig. 2.6   Circuit for Example 2.2 and Prob. 2.2.

pared to v_1, shown here. Ultimately you will see the need for the freedom to define arbitrarily a voltage or a current, with its associated polarity marks, for a particular situation.

Solution

$$v_1 = +10i = 100e^{-2t} \text{ V}$$

and

$$v_2 = -30i = -300e^{-2t} \text{ V}$$

See Fig. 2.7 for the plots of i, v_1, and v_2.

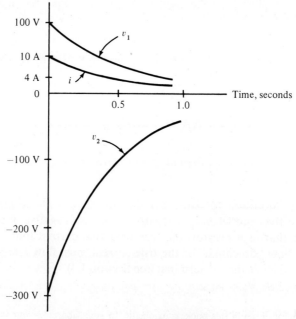

fig. 2.7 i, v_1, v_2 of Fig. 2.6.

PROBLEM

2.2   In the circuit of Fig. 2.6, $i = -10e^{-200t} \sin 1,000t$.

   a.   Plot i, v_1, and v_2. Calculate just a few points.

   b.   Where $t_1 < t_2 < (\pi/500)$ and $t_1 > 0$, find the values of t_1 and t_2 such that the following statements are true:

$$i \text{ is negative} \qquad \text{for } 0 < t < t_1$$
$$i \text{ is positive} \qquad \text{for } t_1 < t < t_2$$

       In other words, find a region of time in which i is negative and another region in which i is positive.

   c.   In the two regions of time found in (b), indicate the sign ($+$ or $-$) of v_1 and of v_2.

   d.   For a consistency check, note that i and v_1 must always have the same sign, whereas v_2 is of the opposite sign with respect to i or v_1.

2.3   LINEAR AND NONLINEAR RESISTANCE

In perhaps the most common situations (like the Wheatstone bridge circuit of Fig. 2.1), resistances are either truly—or are assumed to be—independent of current i or voltage v. For these situations, we say resistance is linear. However, there are many other situations, especially in electronics, where resistance is clearly a function of i or v.

The volt-ampere characteristics of two different linear resistances are displayed in Fig. 2.8. From Ohm's law, we see that the resistance of each is equal to the slope of its volt-ampere function. We see that $R_1 = 5 \ \Omega$, and $R_2 = 10 \ \Omega$. These two resistances are linear because R (or v/i) is independent of current (or voltage). The name "linear" comes from the straight-line characteristics of the v-i plots, as in Fig. 2.8.

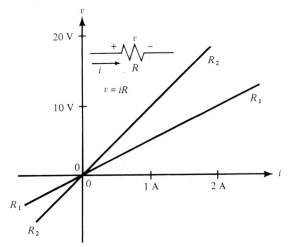

fig. 2.8   Volt-ampere characteristic of two linear resistances.

In contrast to a linear resistance, the volt-ampere characteristic of a Zener diode, an electronic device, is displayed in Fig. 2.9. For this device, the ratio v/i is

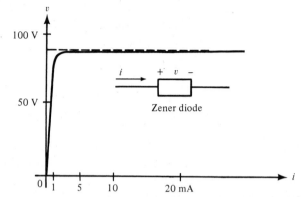

fig. 2.9   Volt-ampere characteristic of a nonlinear resistance.

definitely a function of current i or voltage v, and the diode is said to be *nonlinear* because its volt-ampere characteristic is not a straight line. One way to handle a nonlinear resistance is to approximate its volt-ampere characteristic by two or more straight lines.

PROBLEM

2.3 a. On the volt-ampere characteristic of Fig. 2.9, draw two straight, intersecting lines that together approximate this characteristic. Assume that these two straight lines intersect at 1.0 mA.

b. What is your best estimate of the linear-resistance model of the diode for $i < 1.0$ mA?

c. For any applied voltage where $i < 1$ mA, your model should yield nearly the same current as shown in Fig. 2.9. Check at several voltages.

One might be tempted to assume that a reasonable model of the diode for $i > 1.0$ mA is another linear resistance equal to the slope of the straight-line approximation in this region, but this turns out to be grossly inaccurate, as we will see when we solve Prob. 2.3.

The study of circuits that employ nonlinear elements is much more complicated than the study of linear circuits. This text is almost exclusively concerned with linear circuits, but occasionally we will consider some nonlinear situations to remind ourselves that nonlinear circuits exist and, in many cases, must be given some treatment beyond the bounds of linear analysis. Furthermore, as is suggested by Prob. 2.3, frequently a nonlinear resistor *can* be modeled with sufficient accuracy by linear resistance. We will see more of this.

2.4   INDEPENDENT VOLTAGE AND INDEPENDENT CURRENT SOURCES

Batteries, generators, and other devices supply voltages and currents to networks. In modeling (see Section 1.2) such devices, we make use of independent voltage sources, as represented graphically in Fig. 2.10. The independent voltage source is a *concept*

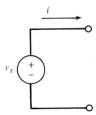

fig. 2.10   The graphical representation of a voltage source.

and not a real-world device. It delivers a voltage that is completely independent of the network to which it is connected. An independent voltage source might be a reasonably good, but not a precise, model of your automobile battery. If it is satisfactory to assume that the voltage delivered by the battery is independent of the battery current, then we could think of the independent voltage source as a satisfactory model of the battery. We will soon see how to model more precisely the battery and other devices that deliver current and voltage.

Some of our electronic devices tend to deliver a current somewhat independent of the network to which it is connected. There is a motivation here to model this device approximately by a current source. The independent current source, also a concept and not a device, is one that delivers a current completely independent of the network to which it is connected. Such a source (Fig. 2.11) may seem strange to you, but nevertheless the independent current source has important conceptual value, as does the independent voltage source.

fig. 2.11   The graphical representation of a current source.

We will soon see how we can use the concept of these two independent sources to help us model or represent practical, real-world batteries, generators, electronic amplifiers, and other devices that deliver current and voltage.

Example 2.3

The volt-ampere characteristic of a small automobile battery might be as shown in Fig. 2.12. Note the logarithmic scale on the abscissa. The ordinate and abscissa of this plot are the terminal voltage and terminal current, respectively, of the battery. For the terminal current to be 1.0 A, the load resistance would be 12 Ω. When the terminal current is 10 A, the terminal voltage is 11.7 V and the load resistance is $11.7/10 = 1.17$ Ω.

In what region of current is the voltage source of Fig. 2.10 a reasonably good model of this battery? A voltage source is a precise model of the battery as long as

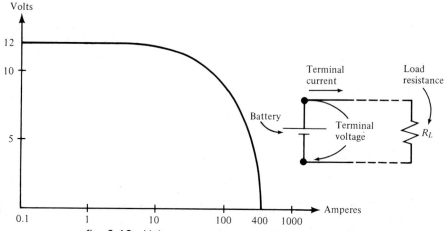

fig. 2.12   Volt- ampere characteristic of a car battery.

the terminal current does not affect the terminal voltage. Because the terminal voltage of this battery is nearly independent of current in the region of $0 < i < 10$, the voltage source, for nearly all purposes, would be a rather good model of the battery in this region. For $i > 100$ A, the voltage source is a very poor model of the battery because, in this region, terminal voltage varies "widely" with current. We will soon learn how to use a resistance in combination with a voltage source to rather accurately model a battery such as this over the complete range of its current or voltage.

• • •

2.5   KIRCHHOFF'S VOLTAGE LAW

Kirchhoff's voltage law is an extremely powerful principle that has general application to all electric networks. You have probably performed experiments in your physics courses that support its validity. What is even more important, intuitively we can accept it very readily as in Example 2.1b. We will treat some related ideas first and then state the law later.

Consider a network consisting of an independent voltage source that forces the current i through each of three resistances as shown in Fig. 2.13a. Then by applying Ohm's law, each of the separate voltages can be written in terms of the current as

$$v_1 = iR_1 \qquad v_2 = iR_2 \qquad v_3 = iR_3 \qquad (2.3)$$

From the fundamental definition of voltage or potential, dW_1, the amount of energy absorbed when a differential amount of charge dq is moved from a to b (Fig. 2.13a), is $dW_1 = v_1 \, dq$. A similar relation holds for each of the other potential differences in the circuit. In the case of the resistors, dW is the electric energy consumed or converted to heat in forcing charge dq through the voltage v. In the circuit of Fig. 2.13a, $(v_s \, dq)$ is the energy supplied by the source, and this energy

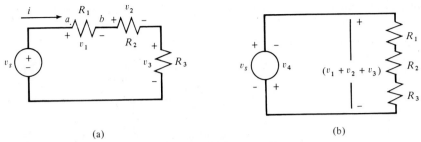

(a)   (b)

fig. 2.13   Voltages add.

is the total energy consumed by the three resistors. Then, of course, the energy supplied by the source must be equal to the total energy consumed, or $v_s \, dq = v_1 \, dq + v_2 \, dq + v_3 \, dq$. This leads us to a particular application of Kirchhoff's voltage law,

$$v_s = v_1 + v_2 + v_3 \tag{2.4}$$

or

$$-v_s + v_1 + v_2 + v_3 = 0 \tag{2.5}$$

Figure 2.14a will help us to visualize the general statement of Kirchhoff's voltage law. We need not ask what is represented by each of the rectangles: a resistance, capacitance, inductance, or some other passive element; a voltage source; or even an *open circuit*.

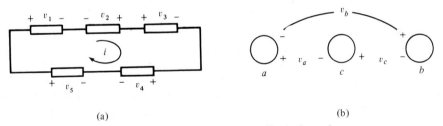

(a)   (b)

fig. 2.14   Illustrating Kirchhoff's voltage law.

Only resistance can *absorb* average energy, but inductance and capacitance can momentarily store energy from, then later *deliver* energy back to, the circuit or system. For dq flowing in the direction of the current as indicated in Fig. 2.14a, $v_1 \, dq$ is an energy *absorbed* from the system by the v_1 element, whereas $v_2 \, dq$ is an energy *delivered* to the system (see Fig. 1.7). It follows that

$$v_1 \, dq - v_2 \, dq + v_3 \, dq + v_4 \, dq - v_5 \, dq = 0$$

or

$$v_1 - v_2 + v_3 + v_4 - v_5 = 0 \tag{2.6}$$

Equation (2.6) illustrates an application of Kirchhoff's voltage law to a particular situation. In general:

> Kirchhoff's voltage law states that when the correct sign is applied to each voltage, the algebraic sum of all the separate voltages around any closed path must be equal to zero.

Kirchhoff's voltage law applies to any closed path, independently of whether a current follows that path or not. To illustrate, consider the three cylindrical transmission line conductors of Fig. 2.14b. Here we are looking at the end view of three cylindrical conductors a, b, and c. The potential of conductor a in respect to c is given by v_a, and similarly for the other two voltages v_b and v_c. Kirchhoff's voltage law says that the sum of these three voltages around the closed path $cabc$ is equal to zero, as follows:

$$v_a + v_b + v_c = 0 \qquad (2.7)$$

even though $cabc$ is not a conducting path.

The validity of Kirchhoff's voltage law should seem intuitively correct because we experience many other similar situations. Consider the sum of elevation differences (analogous to voltage differences) in a closed path. The elevation difference from a point on your ceiling to a point on your roof plus the elevation difference between the roof and the floor plus the elevation difference between the floor and ceiling must be equal to zero. Of course each elevation difference must be assigned the appropriate sign: ceiling to roof—positive, roof to floor—negative. Many other similar situations are equally intuitive. For example, the sum of temperature differences around a closed path must be equal to zero.

Example 2.4
Given that (see Fig. 2.15)

$$v_s = 25 \sin \omega t \text{ V}$$

$$v_1 = 10 \sin \omega t \text{ V}$$

$$R_1 = 20 \, \Omega \qquad \text{and} \qquad R_2 = 5 \, \Omega$$

let us find equations for v_2 and v_3 and then make some checks for consistency of this solution.

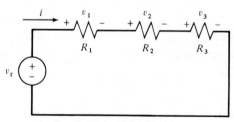

fig. 2.15   Series circuit.

Solution
From Ohm's law,

$$i = \frac{v_1}{R_1} = \frac{10 \sin \omega t}{20} = 0.5 \sin \omega t$$

and

$$v_2 = iR_2 = 5 \times 0.5 \sin \omega t = 2.5 \sin \omega t$$

From Kirchhoff's voltage law,

$$v_s = v_1 + v_2 + v_3$$

or

$$v_3 = v_s - v_1 - v_2 = (25 - 10 - 2.5)\sin \omega t$$

so

$$v_3 = 12.5 \sin \omega t$$

Now let us check our solution to see whether it is consistent with other facts we know must be true. First, we know that Ohm's law requires that

$$R_3 = \frac{v_3}{i} = \frac{12.5 \sin \omega t}{0.5 \sin \omega t} = 25$$

If our solution is correct, it must be consistent with Kirchhoff's voltage law: In other words, our solution must satisfy

$$v_2 = v_1 + v_2 + v_3 = i(R_1 + R_2 + R_3)$$

Substituting our solution into this expression gives

$$25 \sin \omega t = (0.5 \sin \omega t) \times (20 + 5 + 25) = 25 \sin \omega t$$

and we see that our solution is consistent with Kirchhoff's voltage law, as it must be if it is correct.

• • •

Checking the solution in Example 2.4 for consistency with Kirchhoff's current law is one example of a general procedure that is important for you to learn, that of checking your results for consistency in every way that you can. The ultimate check for validity of a result is a physical measurement, but by checking your result for consistency with known relations, you can often find mistakes. Even though a result is consistent with some known facts, it could still be incorrect, but the more ways that you can check it for consistency, the more confident you can be in its correctness. To help you learn this procedure, we will illustrate consistency checks throughout the text, and you will be required to make consistency checks on your work.

PROBLEM
2.4 a. In the circuit of Fig. 2.16, find I_{10}.
  b. Check the consistency of your solution to (a).

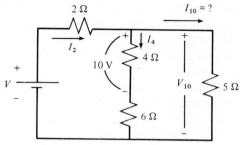

fig. 2.16   Circuit of Prob. 2.4.

Example 2.5

When an automobile battery delivers zero current, its terminal voltage is 12.8 V; at a current of 100 A, its terminal voltage is 11.6 V. We want to calculate battery current when a 0.075-Ω resistor is connected across the battery terminals.

Solution

One approach to the problem is first to find a circuit model of the battery. To do this, we write an equation of V_b, terminal battery voltage, as a function of its current. With the limited data available, it seems sensible to assume that $V_b = f(I)$ is a straight line through the two data points. The V_0 intercept of the straight line is 12.8 and the slope is $(11.6 - 12.8)/100 = -0.012$, as shown in Fig. 2.17. Then it follows that

$$V_b = 12.8 - 0.012I \tag{2.8}$$

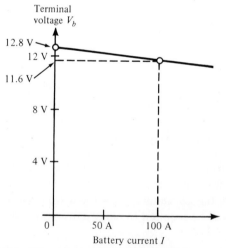

fig. 2.17   *The volt-ampere characteristics of a battery.*

Notice that in (2.8), each term has the dimensions of voltage; therefore, the number 0.012 must have the dimensions of ohms. This slope as calculated above is the ratio of volts/amperes, which in turn is ohms. So we have good reason to believe that there exists a circuit for which (2.8) holds. Figure 2.18 shows that circuit, for when one applies Kirchhoff's voltage law to this circuit, (2.8) follows. One may think of either (2.8) or the equivalent circuit of Fig. 2.18 as a model of the battery. Notice that the model includes an independent voltage source ($V_s = 12.8$ V) and a so-called *internal* resistance of 0.012 Ω. Before proceeding to the final solution, we should certainly make a consistency check on the model. The model yields $V_b = V_s = 12.8$ V

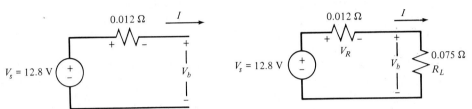

fig. 2.18   Model or equivalent circuit of battery.   fig. 2.19   The complete battery circuit.

at $I = 0$ and $V_b = 12.8 - 100 \times 0.012 = 11.6$ V at $I = 100$ A. Since these values agree with the known data, we see that it is consistent to that extent, giving us confidence in its correctness. Next we combine the model of the battery with the 0.075-Ω resistance to calculate the desired current, as shown in Fig. 2.19. Applying Kirchhoff's law gives

$$V_s = V_R + V_b$$

From Ohm's law,

$$V_R = I \times 0.012 \text{ V}$$

$$V_b = I \times 0.075 \text{ V}$$

Substituting these latter two expressions into the former and solving for I gives us

$$V_s = I(0.012 + 0.075) = 0.087I \text{ V}$$

$$I = \frac{V_s}{0.087} = \frac{12.8}{0.087} = 147.1 \text{ A}$$

As a consistency check, we note that our solution does satisfy Kirchhoff's voltage law:

$$12.8 = 147.1 \times 0.012 + 147.1 \times 0.075$$

$$12.8 = 12.8$$

● ● ●

PROBLEMS

2.5 a. For the region where $i > 1.0$ mA, develop a circuit model for the Zener diode of Fig. 2.9. *Suggestions:* Assume your model will have sufficient accuracy if in this region $(i > 1.0$ mA) the volt-ampere characteristic is a straight line whose intercept and slope are not zero. Estimate the best numerical values that you can from Fig. 2.9 and proceed. Write the equation of this straight line. This equation is a mathematical model and it can lead you to the desired circuit model. Note the dimensions of each term in the equation, and then show numerical values on the circuit that are equivalent to the equation. On the circuit, show clearly the diode current and diode voltage.

   b. Check calculated current-voltage values from this circuit model against values read from Fig. 2.9.

2.6 The battery of Example 2.5 is used to "crank" a car where the starter current on a cold morning is 200 A when the starter is running. The "starter motor" has an internal resistance of 0.01 Ω, but it also develops an internal voltage in the direction that opposes the current and battery voltage. Develop a circuit model of this situation. Show numerical values of resistances, two "independent voltages" (battery terminal voltage and starter terminal voltage) and current. Show current and voltage polarities.

2.7 For the "battery starter" situation of Prob. 2.6, the starter current is reduced when the engine is "hot" because the starter develops a larger voltage when it runs at a higher speed. For "hot" starting, the starter current is 130 A. Assuming that the two resistances are unchanged, find the circuit model for the "hot" start.

Graphical Approach

Example 2.5 was based on the assumption that the V_b-I characteristic of the battery could be sufficiently well approximated by a straight line. Now to expand our technique, we assume that the battery volt/ampere characteristic is definitely not a straight line, as indicated in Fig. 2.20. This battery is loaded into a known resistance of R_L. See the inset diagram in Fig. 2.20a. Knowing the plot of the battery characteristic and the value of R_L, let us find the current I. Now we must remember that the battery volt-ampere characteristic is a function plotted in this figure, but we do not know the analytical equation for this function.

$$V_b = f(I) \quad \text{(for battery)} \tag{2.9}$$

The battery terminal or external voltage V_b must also be given by Ohm's law, as in

$$V_b = IR_L \tag{2.10}$$

Here we have two simultaneous equations where V_b and I are unknown. In principle, we must solve these two equations simultaneously, but we do not know the analytical function of (2.9). However, we can easily make a graphical solution of these two equations by plotting the two functions on one set of coordinates as they are in Fig. 2.20a. The intersection of the two functions, of course, gives the desired solution. For R_{L_1}, the simultaneous solution is shown at a. For a smaller load resistance R_{L_2}, the simultaneous solution is shown at b.

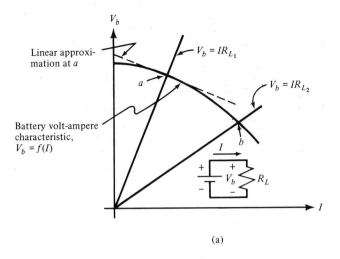

Linear approxi-
mation at a

$V_b = IR_{L_1}$

$V_b = IR_{L_2}$

Battery volt-ampere
characteristic,
$V_b = f(I)$

(a)

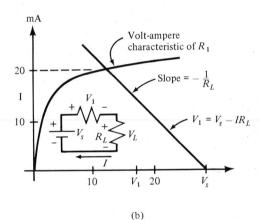

Volt-ampere
characteristic of R_1

Slope $= -\dfrac{1}{R_L}$

$V_1 = V_s - IR_L$

(b)

fig. 2.20 *Graphical solutions in nonlinear situations. (a) Nonlinear battery delivers current to a load R_L. (b) Nonlinear resistance in series with a linear resistance.*

PROBLEM

2.8 a. The volt/ampere characteristic of a particular battery is given by the values in Table 2.1. Determine by a graphical technique the battery voltage V_b and I when a resistance of 0.045 Ω is connected to the battery terminals.

  b. For this particular load resistance, one can, without error, assume that the battery characteristic is a straight line tangent to the actual battery characteristic at the intersection point a of Fig. 2.20. Then one could proceed to find the equivalent circuit that would be valid for the particular R_L. Proceed in this fashion to find the appropriate equivalent circuit. Solve this equivalent circuit for the desired current to compare with the value found in (a).

It is important to note that this equivalent circuit is precise only for the particular R_L =0.045 Ω. A different resistor calls for a different V_b and a different internal battery resistance.

TABLE 2.1 Volt-Ampere
Characteristic of a
Particular Battery

I, Amperes	V_b, Volts
0	12.8
50	12.3
100	11.5
150	10.4
200	9.2
250	7.8
300	6.1
350	4.3
400	2.4

In Prob. 2.8 and the related discussion, we treated a situation where the volt-ampere characteristic of a source was nonlinear. See the battery volt-ampere characteristic of Fig. 2.20a. Let us consider another nonlinear situation common in electronics. Transistors and other solid-state devices have volt-ampere characteristics that are very much nonlinear. The Zener diode of Fig. 2.9 is one example. Consider Fig. 2.20b as another example where R_1 is nonlinear. Notice in this plot that I is the vertical coordinate, not V as in Fig. 2.20a. The choice of which variable is used on the vertical coordinate is somewhat arbitrary, but this choice is common in electronics. Here we assume that the nonlinear device R_1 is in series with a linear resistance R_L, called the load resistance. An independent voltage source V_s is applied to the combination. See insert diagram of Fig. 2.20b. We desire to know the current and the voltages across R_1 and R_L, respectively. Applying Kirchhoff's voltage law, we see that $V_1 = V_s - IR_L$, where V_1 is the voltage across R_1 and V_s is the source voltage. This function is plotted as the straight line shown, the straight line being known as the *load line*. The volt-ampere characteristic of the nonlinear device R_1 is also a function of the same variables, V_1 and I. The intersection of these two functions gives the solution to the values of V_1 and I; then of course $V_L = V_s - V_1$.

PROBLEMS

2.9 For the situation of Fig. 2.20b, $V_s = 30$ V. Find the necessary value of R_L such that:
  a. $V_1 = 15$ volts
  b. $V_1 = 20$ volts
  c. $V_1 = 5$ volts
2.10 For the three transmission line conductors of Fig. 2.14b, two of the three line voltages are known as follows:

$$v_a = 520 \sin 377t \text{ kV} \tag{2.11}$$

$$v_b = 520 \sin(377t - 120°) \text{ kV} \tag{2.12}$$

  a. Find the equation of v_c.
  b. On one set of coordinates, plot v_a, v_b, and v_c over one period of time. To do this,

you can quickly plot v_a and v_b, noting the maximum value of 520 kV and the values of time at which the respective voltage curve passes through zero. Now to save time, you might rapidly scale values from v_a and v_b at a particular time to find the corresponding value of v_c. It may not be obvious to you, but the plot of v_c is still another single sine wave having a maximum value of 520 kV.

NOTE

You may recall from your physics, mathematics, or some other experience that the sum of any two sinusoids of the same angular velocity adds together as another sinusoid of the same angular velocity. Your plot of these three waves should support this assertion. If you wish, you may analytically add these two sinusoids to find the exact equation of the sum by applying the appropriate trigonometric identities. You may or may not recall that sinusoids can be added very simply by converting each to its equivalent phasor. You will become highly skilled in this latter technique later in this course.

2.6 KIRCHHOFF'S CURRENT LAW

To understand Kirchhoff's current law you need to understand the concept of a node. A node is merely a junction point of two or more circuit branches. Typically we are not interested in a node unless three or more branches are tied together at a junction such as a, b, c, and d of Fig. 2.21a. Two branches tied together, such as e, also form a node, but such a node is not highly significant. So throughout this text, unless otherwise specified, a node refers to a junction of three or more branches.

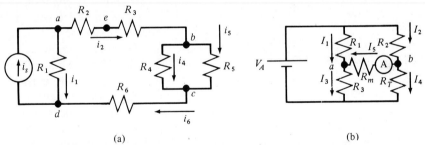

fig. 2.21   Circuits used to illustrate Kirchhoff's current law.

Electric current is incompressible flow. It follows very simply that the sum of all of the currents flowing into a node must be equal to the sum of the currents leaving a node. Since charge does not accumulate at a node, the sum of all of the currents leaving a node or all entering a node must equal zero. The application of this principle to the network of Fig. 2.21a leads to the following relationships:

$$i_s = i_1 + i_2 \qquad \text{node } a \qquad (2.13)$$

$$i_2 = i_4 + i_5 \qquad \text{node } b \qquad (2.14)$$

$$i_4 + i_5 = i_6 \qquad \text{node } c \qquad (2.15)$$

$$i_6 + i_1 = i_s \qquad \text{node } d \qquad (2.16)$$

Notice from (2.14) and (2.15) that $i_2 = i_6$. We could have concluded easily that the current in R_3 is identical to the current in R_6 intuitively when we visualize electric currents as incompressible flows.

The incompressible-flow idea can also be expressed differently: The sum of all currents flowing into a node must be zero. This could be a disturbing statement if one assumed that all current flows were positive, or alternately, all current flows were negative. But when any particular current can be positive or can be negative, then the concept that the sum of all currents flowing into a node must add to zero is palatable. In the case of node a of Fig. 2.21a, the currents entering the node are i_s, $-i_1$, and $-i_2$; that is,

$$i_s + (-i_1) + (-i_2) = 0 \qquad (2.17)$$

Of course, (2.13) and (2.17) are identical equations.

> Kirchhoff's current law states that the sum of all of the currents flowing into a node (or the sum of all of the currents leaving a node) must add to zero.

The concept of a node being a *point* places unnecessary restrictions on the significance of Kirchhoff's current law. This law could be stated in a broader sense—namely, that the sum of all of the currents entering any finite closed volume must add to zero.[2] The sum of all the currents entering a box, a building, or any other enclosure must add to zero. An enclosure, real or imaginary, might include a complex network with many conductors or leads passing from the network through the enclosure to the outside world. When we sum all of the currents leaving (or entering), then this sum must add to zero.

Another example of applying Kirchhoff's current law is the Wheatstone bridge circuit, shown again in Fig. 2.21b. At node a, I_1 and I_5 are entering the node and I_3 is leaving, so

$$I_1 + I_5 = I_3$$

Similarly, at node b,

$$I_2 = I_4 + I_5$$

Example 2.6

a. For the situation of Prob. 2.4 (see Fig. 2.16) find I_2 by applying Kirchhoff's current law to an appropriate node.

Solution
From Prob. 2.4 we found that $I_4 = 2.5$ A and $I_{10} = 5$ A. Therefore, $I_2 = I_4 + I_{10} = 2.5 + 5 = 7.5$ A.

b. Find V.

---

[2] This is not exactly true for a finite volume such as a building because the building might accumulate a charge or give up a charge. This capacitance effect would be negligible in many, but not all, situations.

Solution

Apply Kirchhoff's voltage law to the left-hand loop. $V = 2I_2 + 10I_4 = 15 + 25 = 40$ volts.

   c.   To check the consistency of these values of I_2, I_{10}, and V, apply Kirchhoff's voltage law to the outside loop. $V = 2I_2 + 5I_{10}$; that is, $(2 \times 7.5) + (5 \times 5) = 40 = V$. Check.

NOTE

In finding I_{10} for Prob. 2.4, you probably applied Kirchhoff's voltage law to the right-hand loop. Then later in finding V, Kirchhoff's voltage was applied to the left-hand loop. We also applied Kirchhoff's current law to the node to find I_2. Then to check the consistency of the solution, it was wise to apply Kirchhoff's voltage law to the outside loop, a loop not previously used in the solution.

● ● ●

PROBLEMS

2.11   In the circuit of Fig. 2.22, the following two currents are known:

$$i_1 = 2 + 5t - 3e^{-4t} \text{ A} \tag{2.18}$$

$$i_2 = 8 + t \text{ A} \tag{2.19}$$

Find the equation for i_3.

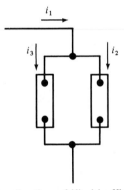

fig. 2.22   *Application of Kirchhoff's current law.*

2.12   It is not difficult to design circuits that will separate sinusoidal currents from steady currents, as illustrated in the enclosure of Fig. 2.23. The enclosure is such that i_3 has the

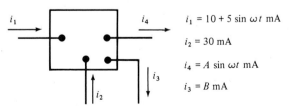

$i_1 = 10 + 5 \sin \omega t \text{ mA}$

$i_2 = 30 \text{ mA}$

$i_4 = A \sin \omega t \text{ mA}$

$i_3 = B \text{ mA}$

fig. 2.23   *Illustrating Kirchhoff's current law.*

steady component B with no sinusoidal component, and i_4 has no steady component. Find the numerical values of A and B.

2.7 RESISTANCES IN SERIES

Two resistances or other elements are said to be in series when they each carry identically the same current. In the circuit of Fig. 2.21a, R_2, R_3, and R_6 are in series because each conducts the current i_2. Of course, $i_2 = i_6$. In this circuit, conceivably i_1 might have the same numerical magnitude as i_2. The two currents could not be identical in the sense that the same electrons flow through R_1 that also flow through R_2. Therefore, these two resistances would not be in series even though the two currents are equal in magnitude. In the circuit in Fig. 2.21b, no one resistance is in series with any other resistance, because no two resistances carry the same identical currents.

We will now demonstrate how a network can be simplified where two or more resistances are in series. Again consider the network of Fig. 2.21a. Apply Kirchhoff's voltage law to the closed path through R_1, R_2, R_3, R_4, and R_6. It follows immediately that

$$i_2 R_2 + i_2 R_3 + i_4 R_4 + i_2 R_6 - i_1 R_1 = 0 \qquad (2.20)$$

By collecting terms, we get

$$i_2(R_2 + R_3 + R_6) + i_4 R_4 - i_1 R_1 = 0 \qquad (2.21)$$

It follows that

$$i_2 R_{236} + i_4 R_4 - i_1 R_1 = 0 \qquad (2.22)$$

where

$$R_{236} = R_2 + R_3 + R_6 \qquad (2.23)$$

Equation (2.22) tells us that we can replace the three resistances R_2, R_3, R_6 by the new equivalent resistance $R_{236} = R_2 + R_3 + R_6$. So the network of Fig. 2.21a can be reduced or simplified to that of Fig. 2.24 in the sense that (2.22) is true for the circuit of Fig. 2.24. The two networks are often said to be equivalent. But this is only partially true. Notice that some of the detail of the network of Fig. 2.21 is lost and is

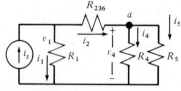

fig. 2.24   Circuit simplification.

not present in Fig. 2.24. For example, the voltages across R_2, across R_3, and across R_6 do not explicitly appear in the network of Fig. 2.24. However, the two networks are equivalent in respect to i_1, i_2, i_4, i_5, v_1, and v_4.

PROBLEM

2.13 For the network of Fig. 2.25, the following numerical values are known:

$$R_1 = R_5 = R_9 = 2 \ \Omega$$

$$R_2 = R_3 = R_8 = 3 \ \Omega$$

$$R_4 = R_6 = R_7 = 4 \ \Omega$$

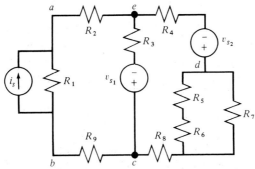

fig. 2.25 Simplify by adding resistances in series.

a. Combine resistances that are in series leading to the simplest possible equivalent of this network of Fig. 2.25. Show the numerical value of each resistance. Even though you may recall how to combine resistances in parallel, you are not allowed to do so for this problem.

b. Draw the diagram of this simpler network.

c. On your new simpler network, identify the nodes a, b, c, d, and e corresponding to those same nodes in Fig. 2.25. Are the two networks equivalent in respect to any branch current? Explain. Are the two networks equivalent in respect to the following: v_{ab}, v_{ac}, v_{ec}? Explain.

2.8 RESISTANCES IN PARALLEL

Two resistances are in parallel when identically the same voltage is impressed across each. In Fig. 2.24, R_4 is in parallel with R_5. In Fig. 2.25, no two resistances are in parallel with each other. However, the series combination of R_5 and R_6 combined into a single resistance is in parallel with R_7. In Fig. 2.21b, no two resistances are in parallel with each other. Like series resistances, parallel resistances can also be combined into a single equivalent resistance. To show how this works in respect to the circuit of Fig. 2.24, we apply Kirchhoff's current law to node a.

$$i_2 = i_4 + i_5 = \frac{v_a}{R_4} + \frac{v_a}{R_5} = v_a\left(\frac{1}{R_4} + \frac{1}{R_5}\right) = \frac{v_a}{R_t} \tag{2.24}$$

where the equivalent resistance R_t is defined by

$$\frac{1}{R_t} = \frac{1}{R_4} + \frac{1}{R_5} \tag{2.25}$$

PROBLEMS

2.14 Prove that two resistances R_1 and R_2 in parallel have an equivalent resistance R_t given by

$$R_t = \frac{R_1 R_2}{R_1 + R_2} \qquad (2.26)$$

2.15 Find an equivalent of the network of Fig. 2.25 that has the minimum number of resistance elements. Do this by successively combining resistances in series and resistances in series and resistances in parallel for maximum simplification. Show the numerical value of each of the resistances in the resultant network.

2.16 Determine whether your "equivalent circuit" of Prob. 2.15 is equivalent to that of Fig. 2.25 with respect to:

   a. v_{ab}.
   b. v_{ac}.

The equivalent resistance of n resistances in parallel is easily found by applying Kirchhoff's current law to the network of Fig. 2.26.

$$i_t = i_1 + i_2 + i_3 + \cdots + i_n \qquad (2.27)$$

then

$$i_t = \frac{v}{R_t} = \frac{v}{R_1} + \frac{v}{R_2} + \frac{v}{R_3} + \cdots + \frac{v}{R_n} \qquad (2.28)$$

or

$$\frac{1}{R_t} = \frac{1}{R_1} + \frac{1}{R_2} + \frac{1}{R_3} + \cdots + \frac{1}{R_n} \qquad (2.29)$$

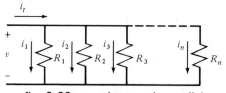

fig. 2.26   n resistances in parallel.

The total resistance R_t is easily calculated on pocket calculators that will compute the reciprocal of a number in one manual step. For two parallel resistances, using your calculator, (2.25) is a better form of calculation than (2.26) because, in the latter case, both R_1 and R_2 must be stored or entered twice.

Example 2.7
Three resistances—11, 9, and 7 Ω, respectively—are in parallel. Use the pocket calculator to calculate the separate reciprocals and their sum, and then finally the total resistance as suggested by (2.29).

Solution
Take the reciprocal of 11 and add to the reciprocal of 9 to the reciprocal of 7. Then take the reciprocal of the sum of the separate reciprocals. This leads to a parallel resistance of 2.90 Ω.

● ● ●

Conductances Compared to Resistances

The conductance of a resistor element in units of mhos is defined as the reciprocal of its resistance in ohms; that is,

$$G = \frac{1}{R} \qquad (2.30)$$

where G is the conductance in mhos and R is the resistance in ohms. (The *mho*, also referred to as the *siemens*, is the reciprocal of the ohm.) Then the equivalent of (2.29), stated in terms of mhos is

$$G_t = G_1 + G_2 + G_3 + \cdots + G_n \qquad (2.31)$$

where G_t is the total equivalent conductance of n conductances in parallel and where $G_1, G_2, G_3, \ldots, G_n$ are the separate conductances.

PROBLEMS
2.17   Prove (2.31).
2.18   Prove that the total equivalent conductance G_t of n conductances in series is given by

$$\frac{1}{G_t} = \frac{1}{G_1} + \frac{1}{G_2} + \frac{1}{G_3} + \cdots + \frac{1}{G_n} \qquad (2.32)$$

where $G_1, G_2, G_3, \ldots, G_n$ are the separate conductances.

The following shorthand notation will be used as a convenience. The symbol $(R_1 \| R_2)$ will mean the equivalent resistance of R_1 and R_2 in parallel; $(R_1 \| R_2 \| R_3 \| R_4)$ is the equivalent of R_1, R_2, R_3, and R_4 in parallel.

Estimating Total Resistance or Conductance

It is important to develop reasonable skills in estimating the total resistance or conductance of either a series or a parallel combination of resistances. In many situations, an estimate is all that is needed; furthermore, when the precise calculation is needed, a simple mental approximation can uncover a gross error in the attempt at a precise calculation. Even in an attempt to calculate the total resistance of a series combination precisely, we might let a gross error go undetected because we failed to make a simple approximate addition. The total resistance of parallel combination of resistance requires more mental effort to estimate. The following relations for two resistances R_1 and R_2 in parallel are very useful:

1.   Where $R_1 = R_2$, $(R_1 \| R_2) = R_1/2$.
2.   Where $R_1 \rightarrow 0$, $(R_1 \| R_2) \rightarrow 0$.

3.   Where $R_1 \rightarrow \infty, (R_1 \| R_2) \rightarrow R_2$.
4.   Where $R_1 < R_2, (R_1 \| R_2) < R_1$.

PROBLEMS
2.19   Prove each of the above four relations for parallel resistances.
2.20   Figure 2.27 graphically displays these four relations. Calculate at least three points on this curve.

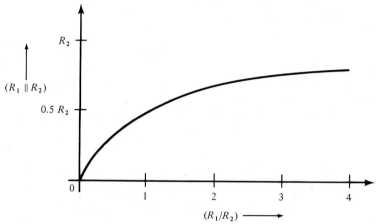

fig. 2.27   Resistance of R_1 and R_2 in parallel.

You are encouraged to develop your judgment and skill in estimating solutions to many of your problems, even though a precise calculation is needed. As you form the habit of verifying your own analyses by these approximations and other techniques, you will build your self-confidence in your accuracy.

Example 2.8
From Fig. 2.28,
    a.   First estimate the numerical values of I_1, I_2, and I_3.
    b.   Then calculate these three currents.

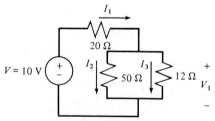

fig. 2.28   Circuit for Example 2.8.

Solution

a. $(50\|12)$ is about 10. R_t is about $10 + 20$ or 30Ω. I_1 is approximately 10/30 or 0.33 A. I_3 is about four times I_2, and $I_3 + I_2$ must be I_1 or approximately 0.33 A. I_2 is estimated to be about 0.065 and I_3 to be about 0.27.

b. $(50\|12) = 9.677 \Omega$
$R_t = 20 + 50\|12 = 29.68 \Omega$
$I_1 = V/R_t = 10/29.68 = 0.3370$ A
$V_1 = I_1(50\|12) = 0.3370 \times 9.677 = 3.261$ V
$I_2 = V_1/50 = 3.261/50 = 0.0652$ A
$I_3 = V_1/12 = 3.261/12 = 0.2718$ A

• • •

PROBLEM

2.21 In the circuit of Fig. 2.29, $v = 100t$ V.
   a. Estimate v_6.
   b. Then calculate v_6 precisely.

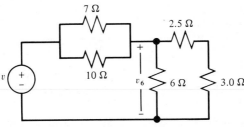

fig. 2.29   Circuit for Prob. 2.21.

Estimating Currents or Voltages as a Function of One of the Parameters

Ofttimes we need to understand how a particular current or voltage varies as a function of the resistance or conductance of a particular element. Example 2.9 illustrates how one might quickly estimate such a function without detailed calculation.

Example 2.9

Given that $i_s = 10e^{-8t}$ in Fig. 2.30, and that R_3 varies from 0 to ∞. We know that $i_2 = Ae^{-8t}$, where A is unknown at this point. Without detailed calculations, we wish to prepare an approximate plot of $A = f(R_3)$.

Solution

The value of R_1 does not in any way influence the solution because the current from a current source is independent of the network to which it is connected.

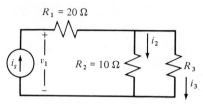

fig. 2.30   Circuit for Example 2.9.

It is easy to see that as R_3 increases, i_2 must increase because i_3 decreases. Look at the two limiting situations: $R_3 \rightarrow 0$ and $R_3 \rightarrow \infty$. As $R_3 \rightarrow 0$, $i_3 \rightarrow i_s$, and therefore $i_2 \rightarrow 0$. As $R_3 \rightarrow \infty$, $i_3 \rightarrow 0$ and $i_2 \rightarrow i_s$. Another point on the function is easily reached. When $R_3 = R_2 = 10\ \Omega$, $A = 10/2 = 5$. With these data, the approximate plot of the function is easily sketched, as shown in Fig. 2.31.

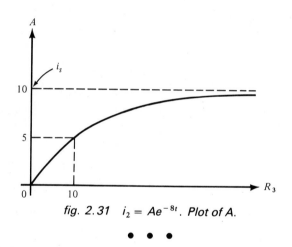

fig. 2.31 $i_2 = Ae^{-8t}$. Plot of A.

● ● ●

PROBLEMS

2.22 For the situation of Example 2.9, develop the equation for $A = f(R_3)$ and calculate a few points on the plot of this function to compare with the estimated plot shown in the solution of this example.

2.23 In the situation of Example 2.9, $v_1 = Be^{-8t}$. Notice that the value of R_1 *does* influence the value of v_1.

a. Without detailed calculations, prepare an approximate plot of $B = f(R_3)$. Look to the limiting values as $R_3 \rightarrow 0$ and as $R_3 \rightarrow \infty$.

b. Develop the equation for $B = f(R_3)$ and compare its values with your estimate of (a).

2.24 In the network of Fig. 2.32, $V_s = 10$ V.

a. With simple calculations at the most, prepare an approximate plot of V_2 as a function of R_1 as it varies from 0 to 200 ohms. (*Suggestion:*   Apply Kirchhoff's voltage law to the closed loop that involves V_1, V_2, and V_3.)

b. Develop an equation for $V_2 = f(R_1)$.

c. Compare your precise and approximate solutions.

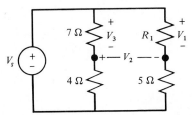

fig. 2.32   Circuit for Prob. 2.21.

2.9   VOLTAGE DIVISION

The situation continually reoccurs in the series circuit like that of Fig. 2.33 where one needs to calculate one of the voltages in terms of either of the other two. It is

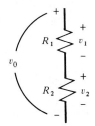

fig. 2.33   Voltage division.

easily seen that

$$\frac{v_1}{v_2} = \frac{iR_1}{iR_2} = \frac{R_1}{R_2} \tag{2.33}$$

and that

$$\frac{v_1}{v_0} = \frac{iR_1}{i(R_1 + R_2)} = \frac{R_1}{R_1 + R_2} \quad \text{and} \quad \frac{v_2}{v_0} = \frac{R_2}{R_1 + R_2} \tag{2.34}$$

Equations (2.33) and (2.34) are sometimes called the voltage divider equations.

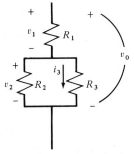

fig. 2.34   Voltage division.

Example 2.10
In the circuit of Fig. 2.34, find v_0 as a function of v_1, R_1, R_2, and R_3. Apply the voltage divider equation.

Solution

$$\frac{v_0}{v_1} = \frac{R_1 + (R_2\|R_3)}{R_1} \quad \text{or} \quad v_o = \frac{v_1(R_1 + (R_2\|R_3))}{R_1}$$

• • •

PROBLEMS

2.25 In the circuit of Fig. 2.34 and without preliminary writing, write the explicit equation for i_3 as a function of v_0, R_1, R_2, and R_3. Of course, $i_3 = v_2/R_3$, so you might apply the voltage divider idea to first find v_2 and then i_3. Use, if you wish, the symbolism of $(R_2\|R_3)$.

2.26 Consider the network of Fig. 2.34. Develop an equation for $i_3 = f(v_0, R_1, R_3, R_2)$, where R_2 is varied and where v_0, R_1, and R_3 are constant. The term $R_2\|R_3$ may not appear in this equation. Reduce the expression to the simplest form. Build your confidence in your equation by first using it to predict i_3 for $R_2 = 0$, and then i_3 for $R_2 = \infty$. Then ignore your equation while you predict i_3 for the two simplified circuits, first where $R_2 = 0$, and then where $R_2 = \infty$. This procedure should give a strong indication of the consistency of your developed equation.

2.10 CURRENT DIVISION

How the current divides into two parallel paths is a recurring situation and deserves some attention. For the typical case shown in Fig. 2.35, we can write the following relations:

$$v = i_1 R_1 = i_2 R_2 \qquad \text{or} \qquad \frac{i_1}{i_2} = \frac{R_2}{R_1} \tag{2.35}$$

$$v_0 = \frac{i_0 R_1 R_2}{R_1 + R_2} = i_1 R_1 \qquad \text{or} \qquad \frac{i_1}{i_0} = \frac{R_2}{R_1 + R_2} \tag{2.36}$$

Notice from (2.36) that the current in one branch is an increasing function of resistance in the other branch. This should not be surprising.

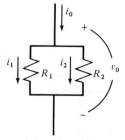

fig. 2.35 Current division.

PROBLEM

2.27   From Fig. 2.36,

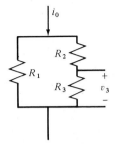

fig. 2.36   Circuit for Prob. 2.27.

a.   Without any preliminary writing, write an equation that gives v_3 in terms of i_0, R_1, R_2, and R_3. Base your approach on the current divider equation.

b.   Check your solution to (a) by an alternate approach where you base your solution on the voltage divider equation.

2.11   SOLUTION BY CIRCUIT REDUCTION

We have combined resistances in series and in parallel to simplify a circuit analysis. We have also seen how the divider ideas can expedite our solutions. These ideas will now be applied and expanded.

Example 2.11
In the network of Fig. 2.37, we would like to:

a.   Estimate V_2 where $I_s = 10$ A.

b.   Then calculate V_2 precisely.

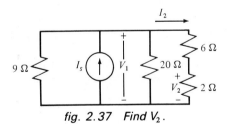

fig. 2.37   Find V_2.

Solution

a.   First we reduce the network by combining the 9- and 20-Ω resistances, as shown in Fig. 2.38. Then we apply the current divider equation to calculate I_2. The term $(9\|20)$ is estimated to be 6.

$$I_2 = \frac{10(9\|20)}{(2+6)+(9\|20)} \approx \frac{60}{14} \approx 4.1 \text{ A} \qquad (2.37)$$

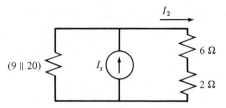

fig. 2.38   Circuit reduced from Fig. 2.37.

Then it follows that

$$V_2 \approx 2 \times 4.1 \approx 8.2 \text{ V} \tag{2.38}$$

b.   For the precise solution, $(9\|20) = 6.207 \ \Omega$, then

$$V_2 = \left(\frac{10 \times 6.207}{8 + 6.207}\right)2 = 8.738 \text{ V}$$

• • •

PROBLEM

2.28   In the network of Fig. 2.39. $V_s = 100$ V.
  a.   By applying circuit reduction and divider ideas only, find all of the branch currents.
  b.   Having calculated branch currents, apply Kirchhoff's voltage law to at least two closed paths, and apply Kirchhoff's current law to two nodes to be assured that you have the correct solution.

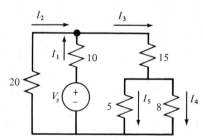

fig. 2.39   Find branch currents.

2.12   THÉVENIN'S THEOREM

In Section 2.4 we developed a model for a battery by placing a resistance in series with a voltage source. We found that this model accurately represents the device, providing that the volt-ampere characteristics of the devices are linear. Thévenin's theorem generalizes this modeling procedure for any linear network containing voltage and current sources. As we will see later in the text, Thévenin's theorem applies to a linear network involving capacitance and inductance elements, as well as resistance elements, but at this point we apply the theorem to resistive networks only. Imagine a general network combination of any number of voltage sources, any number of current sources, and any number of resistance elements having two terminals of special interest such as the *a-b* terminals of Fig. 2.40a. Thévenin's theorem states that there are values of v_{oc} and R_s, respectively, such that the network of Fig. 2.40b is the equivalent or model of the general network of Fig. 2.40a. The performances of the general network of Fig. 2.40a and that of the Thévenin model of Fig. 2.40b will be identical at the *a-b* terminals of each.

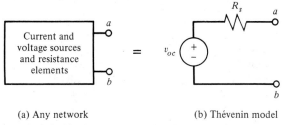

(a) Any network   (b) Thévenin model

fig. 2.40   Illustration of Thévenin's theorem.

Example 2.12
Consider the network of Fig. 2.41a.

a.   Where we wish to study terminal voltage v_{ab} as a function of both v and R where all other circuit parameters V_a, V_b, I_a, R_1, R_2, R_3, R_4, and R_5 are constant. The use of Thévenin's theorem greatly simplifies this problem as follows. The theorem tells us that we can model that constant part of the network as in Fig. 2.40b, so that a simpler model of the whole network is as shown in Fig. 2.41b.

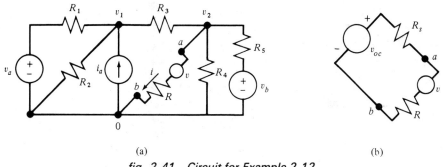

(a)   (b)

fig. 2.41   Circuit for Example 2.12.

b.   As soon as we have found the respective values of v_{oc} and R_s, it will then be very simple to study v_{ab} as a function of both R and v.

• • •

The model of Fig. 2.40b may be obtained by the following two steps:

1.   Either measure or calculate the voltage at the open terminals *a-b* of the constant network of Fig. 2.40a. This voltage is known as the open-circuit voltage, v_{oc} of the Thévenin model of Fig. 2.40b.

2.   Kill the sources in the constant network of Fig. 2.40a and either measure or calculate the resistance at the terminals *a-b*. To *kill* a source, we make the voltage of a voltage source equal to zero and the current of a current source equal to zero.

The proof of the general equivalence of the Thévenin model found by these two steps is not given here. However, in solving the associated problems, you will validate the procedure for a number of particular situations. The sources may be time-varying, such as sinusoidal, or dc.

Example 2.13

Let us calculate the Thévenin model of the circuit to the left of *a-b* terminals in Fig. 2.42. Let us assume that we are interested in the current in R_L for various values of

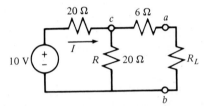

fig. 2.42   Circuit for Example 2.13.

R_L. With R_L removed, the voltage v_{oc} is 5 V. In order to determine the resistance looking into the terminals *a-b*, we must kill the 10 volt source or reduce the 10 volts to zero volts. This places the two 20-Ω resistances in parallel. Then the resistance looking into the terminals *a-b* is

$$R_s = R_{ab} = 6 + (20\|20) = 6 + 10 = 16 \ \Omega \qquad (2.39)$$

The Thévenin model for Fig. 2.42 is shown in Fig. 2.43. Using the Thévenin model

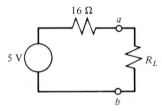

fig. 2.43   Thévenin model of Fig. 2.42.

and the voltage divider principle, the voltage across R_L is

$$V_{ab} = \frac{5R_L}{R_L + 16}$$

Check this Thévenin solution by the application of circuit reductions to the original network of Fig. 2.42.

$$V_{ab} = \frac{R_L}{(R_L + 6)} \times V_{cb} = \frac{R_L}{(R_L + 6)} \times \frac{20\|(R_L + 6)}{20 + 20\|(R_L + 6)} \times 10$$

$$V_{ab} = \frac{R_L}{(R_L + 6)} \times \frac{\dfrac{20(R_L + 6)}{20 + (R_L + 6)} \times 10}{20 + \dfrac{20(R_L + 6)}{20 + (R_L + 6)}} = \frac{10R_L}{20 + (R_L + 6) + (R_L + 6)}$$

$$V_{ab} = \frac{10R_L}{2R_L + 32} = \frac{5R_L}{R_L + 16} \qquad \text{Check}$$

$\bullet \quad \bullet \quad \bullet$

PROBLEMS

2.29 Solve for the voltage across the R_L resistor in Fig. 2.42 using a different reduction approach than that of Example 2.13. For example, first find I, then the current I_{ab}. Manipulate the algebra of this solution to make it agree with the Thévenin solution for V_{ab}.

2.30 Short-circuit the terminals a-b in both Figs. 2.42 and 2.43 and determine the short-circuit currents in both circuits. Of course these two should be equal.

2.31 Select the 20-Ω resistance R in Fig. 2.42 as R_L instead of the right-hand resistance and draw the Thévenin model for the circuit. Determine the current through R from both the circuit of Fig. 2.42 and your Thévenin model as a consistency check.

Example 2.14

This example will demonstrate how Thévenin's theorem can give insight into some situations. In the network of Fig. 2.44 we desire $V_R = f(R)$.

a. Without detailed calculations, prepare an approximate plot of V_R as a function of R.

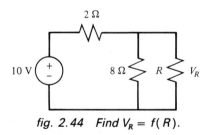

fig. 2.44 Find $V_R = f(R)$.

Solution

We can see easily by the application of the voltage divider concept that

$$V_R(0) = 0 \qquad \text{and} \qquad V_R(\infty) = \frac{8}{10} \times 10 = 8 \text{ V}$$

Then when R is about 2.5, $(2.5\|8)$ is approximately 2 and thus V_R is approximately $10/2 = 5$ V; see Fig. 2.45.

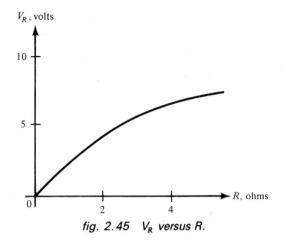

fig. 2.45 V_R versus R.

b.   Without using Thévenin's theorem, develop the equation of $V_R = f(R)$ and calculate $V_R = f(2.5)$.

Solution

$$V_R = \frac{(R\|8)10}{2 + (R\|8)}$$

$$V_R(2.5) = \frac{(2.5\|8)10}{2 + (2.5\|8)} = \frac{1.905 \times 10}{2 + 1.905} = 4.878 \text{ V}$$

c.   Find the Thévenin equivalent of the circuit (see Fig. 2.46).

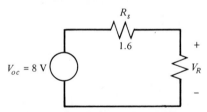

fig. 2.46   Thévenin equivalent.

Solution

$$V_{oc} = \frac{8}{10} \times 10 = 8 \text{ V} \qquad \text{and} \qquad R_s = (2\|8) = 1.6 \ \Omega$$

From this Thévenin model we see that $V_R = 8R/(1.6 + R)$.

d.   Calculate $V_R(1)$ and $V_R(5)$ by the expression developed in (b) and also in (c).

e.   Manipulate the equation for V_R in (b) so that it has identically the same form as the corresponding equation for V_R in (c).

• • •

2.13 *NORTON'S THEOREM*

In Section 2.4 we introduced the concept of a current source that delivers a current that is independent of the associated circuitry. When a resistance, or conductance, is placed in parallel with this current source, the combination may accurately represent a real device that delivers current and voltage. Although this current source model is most often used to represent EMF (that is, voltage) sources with high internal resistance, Norton proposed that any linear, active, two-terminal network may be represented by a current source in parallel with a single resistance, as illustrated in Fig. 2.47. The element values for the *Norton* equivalent circuit of Fig. 2.47b may be obtained by either measurement or calculation, as follows:

1. Connect a short circuit across the terminals *a-b* in Fig. 2.47a and calculate the current that will flow through the short. This current is I_{sc}. The measurement of I_{sc} may be made in the laboratory by connecting an ammeter having negligible resistance directly between points *a* and *b*. This measurement must be made with great caution because the ammeter will "burn out" if it is not capable of safely carrying the current I_{sc}.

2. Kill the active sources and measure the resistance looking into the terminals *a-b* with R_L removed. The measured resistance is R_s. This resistance may be calculated in a complex circuit having known element values by killing the current or voltage sources.

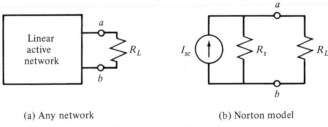

| (a) Any network | (b) Norton model |

fig. 2.47 Illustration of Norton's theorem.

Example 2.15

Let us obtain the Norton equivalent circuit, or model, for the network given in Fig. 2.48. We have selected the 20-Ω resistor as the desired load R_L. The short-circuit

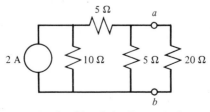

fig. 2.48 Circuit for Example 2.15.

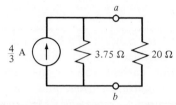

fig. 2.49   Norton model for the circuit of Fig. 2.42.

current I_s is obtained by shorting terminals *a-b*. Using the current-division rule, the current that flows through this short circuit is

$$I_{sc} = \frac{2(10)}{10 + 5} = \frac{4}{3} \text{ A} \qquad (2.40)$$

The resistance looking into the terminals *a-b* with the 20-Ω resistor removed and the 2-A current source killed (replaced by an open circuit) is $(5\|15) = 3.75 \ \Omega$. This Norton's model is shown in Fig. 2.49.

• • •

PROBLEMS

2.32 Determine the current through the 20-Ω resistor in both Figs. 2.48 and 2.49 and compare them as a validity check for the Norton's circuit.

2.33 a. Choose the 5-Ω resistance between *a* and *b* in Fig. 2.48 as the load resistance R_L and determine the element values for a Norton equivalent circuit.

   b. Calculate the current in the horizontal 5-Ω resistance from both the Norton reduction and also by a Thévenin reduction. These two, of course, must be consistent.

2.34 a. Draw a Norton equivalent circuit and determine the element values for the circuit of Fig. 2.44, using the 8-Ω resistor as R_L with $R = 10 \ \Omega$.

   b. Calculate the voltage across the 8-Ω resistance by both the Norton reduction and a Thévenin reduction.

We may transform either a current or voltage source model to a Thévenin (voltage source) or a Norton (current source) model. The equivalence of these models may be demonstrated by connecting a general load resistance R_L across the source terminals and writing the expressions for the voltage across R_L for each circuit. For example, consider Fig. 2.50a. For the voltage model, using voltage division,

$$v_L = \frac{v_s R_L}{R_L + R_s} \qquad (2.41)$$

For the Norton equivalent,

$$v_L = \frac{v_s}{R_s} \left(\frac{R_L R_s}{R_L + R_s} \right) = \frac{v_s R_L}{R_L + R_s} \qquad (2.42)$$

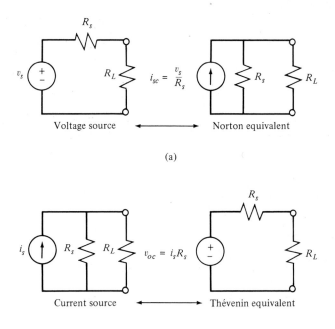

Voltage source ⟷ Norton equivalent

(a)

Current source ⟷ Thévenin equivalent

(b)

fig. 2.50 (a) Conversion of a voltage model to a current model. (b) Conversion of a current model to a voltage model.

PROBLEM

2.35 Show that the voltages across a general load resistor R_L are equal for the two sources of Fig. 2.50b.

The preceding example and problem demonstrate that a voltage model may be transformed into a current model or vice versa without altering the voltages or currents in the *external*, or load, parts of the circuit. However, if we were to calculate the power dissipation inside the generator, we may find a vast difference in the two models. As an example, let us consider the model of the automobile battery given in Fig. 2.18 and repeated in Fig. 2.51 for convenience.

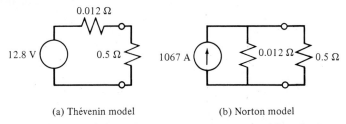

(a) Thévenin model   (b) Norton model

fig. 2.51 Models of the automobile battery of Fig. 2.18.

Example 2.16

The Thévenin model of a typical automobile battery connected to a 0.5-Ω resistance representing the lights and heater is given in Fig. 2.51a. The current in this circuit is

$$I = \frac{12.8 \text{ V}}{0.512 \text{ Ω}} = 25 \text{ A}$$

The power dissipated inside the battery with this load is

$$P = I^2 R_s = (25)^2 (0.012) = 7.5 \text{ W}$$

On the other hand, the current through R_s in the Norton model of Fig. 2.51b, using current division, is

$$I = \frac{1,067(0.5)}{0.5 + 0.012} = 1,042 \text{ A}$$

The hypothetical power dissipated in this source is

$$P = (1,042)^2 (0.012) = 13,029 \text{ W}$$

Since the Thévenin model quite accurately represents the physical makeup of the battery, the 7.5-W dissipation determined for this model in Example 2.16 represents the actual dissipation quite accurately. However, the 13,029-W hypothetical dissipation in the Norton's model is grossly inaccurate and would cause the battery to self-destruct immediately. Therefore, when device dissipation or efficiency is under consideration, we must use a model that accurately represents the physical makeup of the device in order to obtain meaningful values. This unreal loss in the Norton model of the battery reminds us of the fact that the equivalence of either the Norton or Thévenin model is guaranteed to be valid only at the R_L terminals. You should briefly review the fact that these equivalences were required only at the R_L terminals (see Section 2.12).

Thévenin and Norton models are not restricted to networks having a single source. In fact, the network may include both current and voltage sources as shown in Fig. 2.52. The only requirements are that the elements are linear and there are only two terminals.

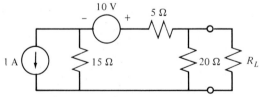

fig. 2.52   *Network containing both a current source and a voltage source.*

• • •

PROBLEMS

2.36   Draw a Thévenin model for the circuit of Fig. 2.52, including the element values. *Hint :* Transform the Norton source to a Thévenin source.

2.37   Draw a Norton model for the circuit of Fig. 2.52, including element values.

The preceding examples and exercise should convince us that Thévenin's and Norton's theorems provide simple solutions to complex networks as well as tidy ways to represent them.

2.14   THE SUPERPOSITION THEOREM

Another theorem that may simplify the solution of a complex linear network having two or more active (power) sources is the *superposition theorem*. This theorem states that where all resistances are linear, the *response to several sources is equal to the sum of the responses to each source, with all other sources killed.* Commonly in electronics and in other situations, resistances are nonlinear; that is, resistance is a function of current or voltage. Superposition is not vigorously valid in these cases, but still *might* be used to obtain a reasonably accurate solution. *Caution:* Apply superposition with care.

Example 2.17
Let us use the superposition theorem to determine the current I through the 5-Ω resistance in Fig. 2.53. Let us first turn the current source off and determine the

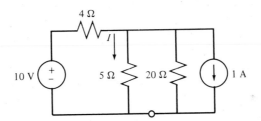

fig. 2.53   Circuit for Example 2.17.

voltage across the 5-Ω resistance. Since $5\ \Omega \parallel 20\ \Omega = 4\ \Omega$, the voltage across the 5-Ω resistance is

$$V = \frac{10(4)}{8} = 5 \text{ V}$$

and the resulting current component is

$$I_1 = \frac{5 \text{ V}}{5\ \Omega} = 1 \text{ A}$$

We now kill the 10-V source and determine the component of current caused by the current source. Since $(4\parallel20) = 3\frac{1}{3}\ \Omega$, we may use the current division rule to determine the current component I_2 in the 5-Ω resistance due to the current source; that is,

$$I_2 = -\frac{1(3\frac{1}{3})}{5 + 3\frac{1}{3}} = -0.40 \text{ A}$$

Then the total current through the 5-Ω resistor is

$$I = I_1 + I_2 = 1 \text{ A} - 0.40 \text{ A} = 0.60 \text{ A}$$

• • •

PROBLEMS

2.38   Transform the Thévenin equivalent on the left in Fig. 2.53 to a Norton equivalent and then solve for the current through the 5-Ω resistor as a validity check on the superposition theorem.

2.39   In the circuit of Fig. 2.52, let $R_L = 20 \text{ }\Omega$ and use the superposition theorem to solve for the current through R_L. Use the Thévenin model obtained in Prob. 2.36 to validate your results.

2.15   SOLUTION BY BRANCH CURRENTS

Solution by circuit reductions is a powerful approach that can be applied to many, but not all, situations. For example, a network might be rather complex without any two elements being in series or in parallel, such as the one in Fig. 2.54. One cannot

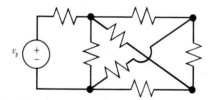

fig. 2.54   Circuit with no parallel elements and no series elements.

find, in this network, two resistances in series with each other or two resistances in parallel with each other. Consequently, the simplification techniques previously considered are not useful here.

Consider Fig. 2.55, another network in which there are no resistances in parallel or in series. This is a classic circuit that you probably have solved before in your physics courses or elsewhere. This circuit could be solved by one of the previous techniques, but the method of loop currents is especially applicable to this situation and will be developed here.

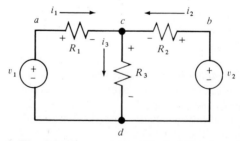

fig. 2.55   Branch-current method of circuit analysis.

Initially, we will analyze the circuit in Fig. 2.55 using branch currents. To apply the *branch current* method to the problem of finding three branch currents i_1, i_2, and i_3 in the network of Fig. 2.55:

1. Assign currents to the three branches. The direction assigned to each current is arbitrary. Show currents and assigned directions on the diagram.
2. Indicate polarities of the voltage across each resistance. These polarity marks must be consistent with assigned current directions. In a resistance, current flows internally from $+$ to $-$.
3. Apply Kirchhoff's voltage law to a closed path such as *acd*. Proceed around the closed loop in an arbitrary direction—clockwise in this example. Record source voltages such as v_1 on the left of the equal sign giving its value a $+$ sign if proceeding from $-$ to $+$. Record voltages across resistances on the right of the equal sign and assign a $+$ to a resistance voltage when proceeding from $+$ to $-$. It follows for closed path *acd*,

$$v_1 = R_1 i_1 + 0 i_2 + R_3 i_3 \qquad (2.43)$$

In a similar manner, we apply Kirchhoff's voltage law to path *bcd*:

$$v_2 = 0 i_1 + R_2 i_2 + R_3 i_3 \qquad (2.44)$$

Then we apply Kirchhoff's current law to node *c*:

$$0 = i_1 + i_2 - i_3 \qquad (2.45)$$

Now these three simultaneous equations can be solved for the three branch currents i_1, i_2, and i_3.

Note the three steps taken in order to write the necessary branch current equations. You must be especially careful in establishing the sign associated with each term in these equations. Carefully executing these three steps will assist you in minimizing the error.

Three equations are needed for this situation, and each must be independent of the other two. Methods of assuring that the necessary number of equations are independent will be treated later. It should be apparent, however, that the application of Kirchhoff's current law to node *d* leads to identically the same equation as (2.45). It may not be quite so obvious, but it is true, that the application of Kirchhoff's law to the outside loop (through v_1 and v_2) is not independent of (2.43) and (2.44).

PROBLEMS

2.40 In the network of Fig. 2.55, $v_1 = 2e^{-4t}$ V and $v_2 = 10 \sin 1{,}000t$ V, $R_1 = 10 \ \Omega$, $R_2 = 20 \ \Omega$, and $R_3 = 30 \ \Omega$. Solve for the three currents i_1, i_2, and i_3. Check your solution to make sure that the three currents add properly at node *c*. Also make sure that all of the voltages sum to zero in one of the closed paths—perhaps the outside path.

2.41 In the network of Fig. 2.56, find v_1, where $v_s = 10$ V and $i_s = 1$ A.
   a. Use the *branch-current* approach.
   b. Using whatever techniques seem to be appropriate, devise some consistency checks for your solution. You may wish to convert the Thévenin circuit consisting of v_s and the 5-Ω resistance to a Norton equivalent, then proceed to v_1.

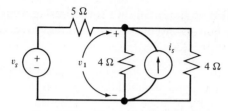

fig. 2.56   Circuit for Prob. 2.41.

2.16   LOOP-CURRENT METHOD

The branch-current method applied to the circuit of Fig. 2.55 leads to a family of three simultaneous equations. One of these equations (2.45) was an expression of Kirchhoff's current law and may be used to eliminate one of the unknown currents from the voltage equations, (2.43) and (2.44). For example, let us express i_3 explicitly, using (2.45) in terms of i_1 and i_2 :

$$i_3 = i_1 + i_2 \qquad (2.46)$$

Now let us use this expression to eliminate i_3 from (2.43) and (2.44) to obtain

$$v_1 = R_1 i_1 + R_3(i_1 + i_2) \qquad (2.47)$$

$$v_2 = R_2(i_2) + R_3(i_1 + i_2) \qquad (2.48)$$

These equations may be rearranged into a more convenient form as follows:

$$v_1 = (R_1 + R_3)i_1 + R_3 i_2 \qquad (2.49)$$

$$v_2 = R_3 i_1 + (R_2 + R_3)i_2 \qquad (2.50)$$

Now if we wished to proceed with the solution by using determinants to obtain i_1 and i_2, we would encounter only determinants of second order rather than third order. Then after i_1 and i_2 have been found, we can easily find i_3 from ($i_1 + i_2$).

We can implement the line of thought that leads to (2.49) and (2.50) in a highly organized way by assigning what are known as *loop currents*, as shown in Fig. 2.57.

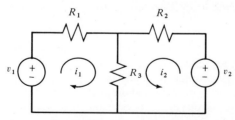

fig. 2.57   Loop currents.

Here the loop currents i_1 and i_2 are the previous branch currents i_1 and i_2, respectively, but we think of each of these currents passing through branch 3 to close on themselves. After the two loop currents are found, the current through R_3 is found from $i_1 + i_2$. The loop currents yield a neatly organized approach to the writing of the simultaneous equations. We may write Kirchhoff's voltage equation for the closed path traversed by a given loop as the sum of the voltage contributions from each loop current considered individually. This approach to obtaining loop voltages is analogous to the use of the superposition theorem for obtaining branch currents in multisource networks. For example, let us write Kirchhoff's voltage equations for the circuit of Fig. 2.57 as follows:

$$v_1 = (R_1 + R_3)i_1 + R_3 i_2 \tag{2.51}$$

$$v_2 = R_3 i_1 + (R_2 + R_3)i_2 \tag{2.52}$$

In (2.51), we may think of the term $(R_1 + R_3)i_1$ as the voltage in the i_1 loop due to i_1 alone, and the term $R_3 i_2$ as the voltage in this same loop due to i_2 alone. Observe that (2.51) and (2.52) are identical, respectively, to (2.49) and (2.50).

The writing of loop equations may be further systematized by using the following form for (2.51) and (2.52):

$$v_1 = R_{11}i_1 + R_{12}i_2 \tag{2.53}$$

$$v_2 = R_{21}i_1 + R_{22}I_2 \tag{2.54}$$

where

$R_{11} = R_1 + R_3$ is the total *self-resistance* encountered by the current i_1 in loop 1 with loop 2 open, or disconnected

$R_{12} = R_3$ is the *mutual resistance* between loop 1 and loop 2, or the resistance these two loops share in common

$R_{21} = R_3$ is the *mutual resistance* between loop 2 and loop 1; in linear circuits, R_{21} is always equal to R_{12}

$R_{22} = R_2 + R_3$ is the total *self-resistance* of loop 2 with loop 1 open or disconnected

The order of the subscripts on the R coefficients of (2.53) and (2.54) is related to the order in which the terms appear in these equations. This organization pattern simplifies our thinking and improves our accuracy in writing these simultaneous loop equations. All three terms of (2.53) are said to be in the first row, whereas those of (2.54) are in the second row. Of the terms on the right-hand side of the equal signs, the i_1 terms are in the first column, while the i_2 terms are in the second column. Now notice that the first (left) subscript on any R coefficient refers to its row position, and the second (right) refers to its column position. Furthermore, the first subscript on any R coefficient is identical to its v subscript, and the second R coefficient is identical to its i subscripts. This systematic notation is generally used to identify the elements in matrices and facilitates the writing of the loop equation by inspection. For example, let us write the loop equations for the n-loop circuit using a matrix notation. Equations (2.55), (2.56), and (2.57) apply to a network described by n loop equations; however, any R_{jk} coefficient or any loop-impressed voltage v_j may be negative depending on the network and depending on how loop current i_k is assigned.

$$v_1 = R_{11}i_1 + R_{12}i_2 + R_{13}i_3 + \cdots + R_{1n}i_n \qquad (2.55)$$

$$v_2 = R_{21}i_1 + R_{22}i_2 + R_{23}i_3 + \cdots + R_{2n}i_n \qquad (2.56)$$

$$\vdots \qquad\qquad\qquad \vdots$$

$$v_n = R_{n1}i_1 + R_{n2}i_2 + R_{n3}i_3 + \cdots + R_{nn}i_n \qquad (2.57)$$

See Fig. 2.58 as an example. The loop currents are arbitrarily assigned as shown. As discussed relative to the branch current method, when in doubt it is well to assign polarities of voltages due to each loop current in each resistance, as in Fig. 2.58.

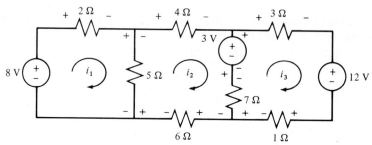

fig. 2.58   A circuit having three independent loops.

PROBLEM

2.42 Show that for the network of Fig. 2.58, the R and v coefficients of (2.55), (2.56), and (2.57) are as follows: $v_1 = 8$, $v_2 = -3$, $v_3 = 3 - 12$, $R_{11} = 2 + 5$, $R_{22} = 4 + 5 + 6 + 7$, $R_{33} = 3 + 1 + 7$, $R_{12} = R_{21} = -5$, $R_{23} = R_{32} = -7$, $R_{13} = R_{31} = 0$.

If you hesitated or had trouble in solving Prob. 2.42, perhaps you should review your techniques for determining the signs of each term in the equation that comes from applying Kirchhoff's voltage equation to a closed loop.

PROBLEM

2.43 Suppose that in Fig. 2.58 the current in the third window is assigned to flow in the counterclockwise direction rather than the clockwise direction as shown. Let this new current be i'_3.

  a. Now show that $v'_3 = 12 - 3$, $R'_{33} = 3 + 1 + 7$, and $R'_{23} = R'_{32} = 7$.

  b. From your R and v coefficients that apply to the third window, in each case (clockwise i_3 and counterclockwise i'_3), write the voltage equation showing numerical values of these coefficients. Show that these two equations are consistent with the fact that $i_3 = -i'_3$.

Assignment of Appropriate Loop Currents

There is considerable freedom in the manner in which the loop currents are assigned. One very common pattern is to fill each *window* with a loop current, as illustrated in Figs. 2.58 and 2.59. The assignment of one loop current to each *window*

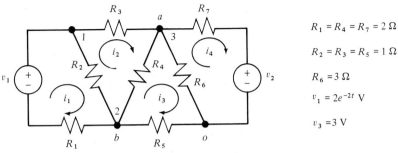

$R_1 = R_4 = R_7 = 2\ \Omega$

$R_2 = R_3 = R_5 = 1\ \Omega$

$R_6 = 3\ \Omega$

$v_1 = 2e^{-2t}$ V

$v_3 = 3$ V

fig. 2.59   Loop currents.

is a rather good procedure for many situations. Sometimes there is a good case for assigning loop currents different from one current per *window*. Problem 2.44 illustrates this point. Here we wish to reassign loop currents of Fig. 2.59 so that a particular loop current is identical to the branch current in R_6. Removing i_3 alone achieves this goal but creates a contradiction; namely, i_4 is both the branch current in R_7 and also in R_6. So i_3 might be reassigned around the loop R_4, R_7, v_2, and R_5. Of course, we might have alternately solved our problem by leaving i_3 alone and reassigning i_4. This brings us to guidelines for assigning loop currents.

To assign the minimum required loop currents:

1. There must be at least one loop current in each branch.
2. A particular loop current cannot be the only loop current in two or more branches.

Appendix B treats the assignment of loop currents in a more sophisticated manner than we have done here. These two guidelines are fail-safe, but they are sometimes difficult to apply in complex situations.

PROBLEM

2.44 We desire the voltage v_{ao} in the network of Fig. 2.59.
   a.   Reassign i_3 so that the only loop current in R_6 is i_4.
   b.   Find v_{ao} by first solving for i_4 by the method of loop currents.
   c.   Also find i_3. Then using your values of i_3 and i_4, show that the voltage in the i_4 window sums to zero.

Loop Currents Where There Are Current Sources

The examples so far have treated situations with voltage sources but no current sources. Consider the effect of adding a current source to the network of Fig. 2.57, as shown in Fig. 2.60. Here i, v_a, v_b, R_1, R_2, and R_3 are known; i_1 and i_2 are unknown. Proceed by assigning a closed path for the known current source i. Notice that the addition of the known current i did not change the number of unknown currents (i_1, i_2). We apply Kirchhoff's voltage law to the two loops described by the

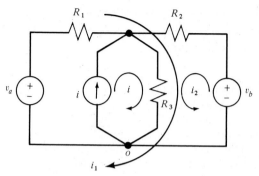

fig. 2.60   Two voltage sources, one current source.

unknown currents i_1 and i_2, respectively. Notice that for our purposes here, one does not attempt to sum the voltages in a loop that includes the current source because the voltage across the current source is a function of the whole network and is therefore unknown until the unknown currents have been found.

PROBLEMS

2.45 Show that the R's and v's in the format of (2.55), (2.56), (2.57), and the network of Fig. 2.60 are as follows: $v_1 = v_a - iR_3$, $v_2 = v_b - iR_3$, $R_{11} = R_1 + R_3$, $R_{22} = R_2 + R_3$, and $R_{12} = R_{23} = R_3$.

2.46 As a consistency check, solve Prob. 2.45 by transforming the current source and R_3 into a Thévenin's model.

Avoid Brute Force

We should be thoughtful about the manner in which we assign loop currents, or otherwise we can complicate our lives unnecessarily. Consider the circuit of Fig. 2.61. Suppose we wish to find the voltage v_{ao} by the application of loop currents.

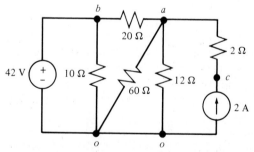

fig. 2.61   Simplify before assigning loop currents.

One might "rush in where angels fear to tread" and fill each *window* with a loop current, and then become involved in the solution of four simultaneous equations. The thoughtful person would notice that R_{bo} in no way influences the voltage V_{bo}.

Furthermore, the current entering node *a* from the right is 2 A, independently of the 2-Ω resistance through which it flows. Then, in addition, the 60-Ω and 12-Ω resistances can be combined to an equivalent of 10 Ω.

PROBLEMS

2.47 a. In the circuit of Fig. 2.61, combine the 12-Ω and 60-Ω resistances into one single equivalent resistance, then assign one unknown loop current and also assign the known 2-A current to a loop. Then write a single equation involving the unknown loop current, which leads immediately to v_{ao}.

   b. Make a consistency check on your solution by applying Kirchhoff's voltage law or Kirchhoff's current law, or a combination in some way, to demonstrate that you have the correct solution.

   c. Determine the voltage V_{co} and note that this voltage *is* dependent upon the 2-Ω resistance.

   d. Find the current in the 42-V source and note that this current *is* dependent upon the 10-Ω resistance.

2.48 For purposes of finding V_{ao} in Fig. 2.61, this circuit reduces to Fig. 2.62.

   a. Apply the principle of superposition to find V_{ao}.

   b. Convert the combination of 42 V and 20 Ω to a Norton equivalent and then solve for V_{ao}.

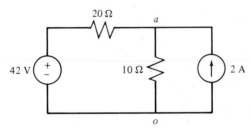

fig. 2.62 Simplified circuit for Prob. 2.48.

2.17 THE NODAL METHOD

The nodal method is another powerful technique for solving complex networks. You will ultimately see the advantages and disadvantages of various approaches: branch currents, loop currents, and nodes. For example, for the circuit in Fig. 2.57, we found that three simultaneous equations were required to find the three unknown currents by the branch current method. By the loop current method, we found that only two simultaneous equations were required. By the nodal method, only one equation need be manipulated for this particular circuit, as we show next.

Figure 2.57 is repeated here as Fig. 2.63 for convenience. Let node *d* be a *reference* node, to which other voltages will be referred. Both v_1 and v_2 are already referred to this node. Let v_c have the meaning of the voltage of *c* with respect to *d* or with respect to the reference.

We will apply Kirchhoff's current law at node *c* but write the currents in terms of the known voltages v_1 and v_2 and the unknown nodal voltage v_c. Each of these

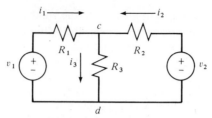

fig. 2.63   Circuit of Fig. 2.57.

three currents is given by

$$i_1 = \frac{v_1 - v_c}{R_1} \qquad i_2 = \frac{v_2 - v_c}{R_2} \qquad i_3 = \frac{v_c}{R_3} \tag{2.58}$$

Then applying Kirchhoff's current law at c gives us

$$i_1 + i_2 = i_3 \qquad \text{or} \qquad \frac{v_1 - v_c}{R_1} + \frac{v_2 - v_c}{R_2} = \frac{v_c}{R_3} \tag{2.59}$$

Collecting the terms results in

$$v_c\left(\frac{1}{R_1} + \frac{1}{R_2} + \frac{1}{R_3}\right) = \frac{v_1}{R_1} + \frac{v_2}{R_2} \tag{2.60}$$

Now we can solve this equation for v_c, the unknown nodal voltage. Having determined this voltage, we can immediately solve for the three currents. It is important to note that we did not have to solve simultaneous equations to arrive at this solution, an obvious advantage of the nodal method for this particular circuit.

To arrive at (2.60), we started with the resistance values of each element. If we had started with conductance values, the solution would have been even more elegant. Where the separate conductances are G_1, G_2, and G_3, the nodal voltage is given by

$$v_c = \frac{v_1 G_1 + v_2 G_2}{G_1 + G_2 + G_3} \tag{2.61}$$

PROBLEM

2.49   Prove (2.61) by using Norton's theorem to replace the voltage sources with current sources of Fig. 2.63.

Let us now consider the circuit of Fig. 2.64, which has two nodes, 1 and 2, in addition to the reference node 0. It has known current sources I_A and I_B and known conductances G_1, G_2, and G_3. We need to find the voltage V_1 at node 1 and the voltage V_2 at node 2, both with reference to the common node 0. Then, as before, we may find all the branch currents by simply applying Ohm's law. Summing the currents at node 1,

$$I_A = V_1 G_1 + (V_1 - V_2)G_3 \tag{2.62}$$

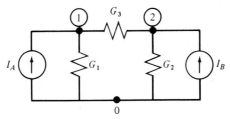

fig. 2.64   *A circuit that has two nodes in addition to the common node 0.*

Summing the currents at node 2,

$$I_B = V_2 G_2 - (V_1 - V_2)G_3 \tag{2.63}$$

These equations may be rearranged as follows:

$$I_A = (G_1 + G_3)V_1 - G_3 V_2 \tag{2.64}$$

$$I_B = -G_3 V_1 + (G_2 + G_3)V_2 \tag{2.65}$$

Notice that I_A and I_B are known currents that *enter* the nodes 1 and 2, respectively, whereas all of the other terms—$(G_1 + G_3)V_1$, and so on—are currents that leave the nodes. Observe the striking similarity between these nodal equations and the loop equations (2.49) and (2.50). As we did with the loop equations, let us use the matrix subscripts and rewrite (2.64) and (2.65) as

$$I_1 = G_{11}V_1 - G_{12} V_2 \tag{2.66}$$

$$I_2 = -G_{21}V_1 + G_{22} V_2 \tag{2.67}$$

where
$I_1 = I_A$
$I_2 = I_B$
G_{11} = *self-conductance* $(G_1 + G_3)$ from node 1 to all other nodes
G_{12} = *mutual conductance* G_3 between node 1 and node 2
G_{22} = *self-conductance* $(G_2 + G_3)$ from node 2 to all other nodes
G_{21} = *mutual conductance* between node 2 and node 1 and is the same as G_{12} in any passive network

Equations (2.66) and (2.67) are in matrix format, where known currents that enter the nodes are on the left of the equal signs; G_{11}, G_{12}, and so on are known conductances; V_1 and V_2 are unknown nodal voltages. These equations may be better understood if we view them as a superposition solution. For example, (2.66) states that the net current *out* of node 1 is the sum of all the currents that would flow out as a result of V_1 acting alone or when all of the other sources $(V_2, I_A,$ and I_B in this case) are *killed*, plus the current that would flow out of node 1 as a result of V_2 acting alone when sources V_1, I_A, and I_B are killed. In this case, the assumed (or known) node 2 voltage is positive, so this latter current component actually flows into node 1, and therefore its sign is negative. Notice that we must think of nodal voltages as source voltages even though there is no generator behind each.

The matrix equations may be written more generally by assigning positive signs to all of the coefficients and then determining the actual sign of each coefficient after a consistent set of polarities have been assigned to the node voltages. For example, if (2.66) and (2.67) had been written with positive signs preceding G_{12} and G_{21}, then $G_{12} = G_{21} = -G_3$. The self-conductance coefficients such as G_{11} and G_{22} are always positive.

The n nodal equations in matrix format may be written for a circuit with n independent nodes, plus a common node, as follows:

Currents entering node   Currents leaving node

$$I_1 = G_{11}V_1 + G_{12}V_2 + G_{13}V_3 + \cdots + G_{1n}V_n \qquad (2.68)$$
$$I_2 = G_{21}V_1 + G_{22}V_2 + G_{23}V_3 + \cdots + G_{2n}V_n \qquad (2.69)$$
$$\vdots$$
$$I_n = G_{n1}V_1 + G_{n2}V_2 + G_{n3}V_3 + \cdots + G_{nn}V_n \qquad (2.70)$$

The I's of (2.68), (2.69), and (2.70) are current elements, each of which might have either sign. These elements are usually the independent current sources of a Norton's equivalent generator with the internal conductance G_s absorbed in the self-conductance of the node. The current source I_k for any node k is zero if there is no power source connected directly to that node.

To illustrate the nodal approach further, consider the network of Fig. 2.65. We desire to find the nodal voltages V_1 and V_2. In the first step toward finding these

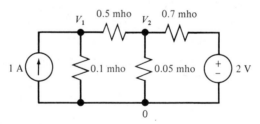

fig. 2.65   *Example of a two-node situation.*

voltages, we seek the two matrix equations having the format of (2.68), (2.69), and (2.70). Soon we will learn how to write the G coefficients and I elements on inspection much like we do the R's and V's of the loop equations in the matrix format. But to make sure we understand the basic applications of Kirchhoff's current laws at the two nodes, let us apply the more tedious, but perhaps safer, techniques as in finding (2.60). Sum the currents entering node 1:

All currents entering node V_1

$$1 + 0.5(V_2 - V_1) - 0.1V_1 = 0 \qquad (2.71)$$

or

$$1 = 0.6V_1 - 0.5V_2 \qquad (2.72)$$

For node 2,

All currents entering node V_2

$$0.7(2 - V_2) + 0.5(V_1 - V_2) - 0.05V_2 = 0 \qquad (2.73)$$

or

$$1.4 = -0.5V_1 + 1.25V_2 \qquad (2.74)$$

Thus, (2.72) and (2.74) are the two simultaneous equations for the nodal voltage V_1 and V_2.

PROBLEMS

2.50 a. Solve (2.72) and (2.74) for V_1 and V_2.

   b. From these voltages V_1 and V_2 and other known values in the circuit of Fig. 2.65, solve for all branch currents, then test to make sure that the currents sum to zero at each of the nodes.

2.51 Apply the loop current method to the circuit of Fig. 2.65 by assigning a loop current to the center and another to the right-hand windows. Solve for each of these loop currents; then find V_1 and V_2.

2.52 In Fig. 2.65, convert the Norton equivalent of the current source and the 0.1-mho conductance to an equivalent Thévenin source. This leads to a one-node problem. Solve for V_2 from one equation. Apply this known value of V_2 to the original network of Fig. 2.65 to find V_1.

Nodal Superposition

Equations (2.72) and (2.74) are in the format of the general matrix equations (2.68), (2.69), and (2.70), where $I_1 = 1$, $I_2 = 1.4$, $G_{11} = 0.6$, $G_{12} = G_{21} = -0.5$, and $G_{22} = 1.25$. By applying superposition, these values of the I's and G's could have been written by inspection without writing the preliminary equations (2.71) and (2.73). Return to the general matrix equations of (2.68) and (2.69) to see how we can do this.

In (2.68), I_1 has the meaning of the known current *entering* node 1, or 1 A in Fig. 2.65. The term $(G_{11}V_1)$ has the meaning of the current leaving node 1 due to V_1 acting alone. When V_1 is acting alone, the 1-A source is *killed* (made zero), V_2 is killed (zero in respect to reference 0), and the 2-V source is killed (made zero). So for this purpose, V_1 is impressed across the 0.1 mho and 0.5 mho in parallel—or, in other words, $G_{11} = 0.6$. The current $G_{12}V_2$ is the current leaving node 1 due to V_2 acting alone. For this purpose, V_1 is killed or at zero potential so that the current leaving node 1 due to V_2 alone is $-0.5V_2$, making $G_{12} = -0.5$. To find G_{22}, we kill the 1-A source, V_1, and the 2-V source; thus $G_{22} = (0.5 + 0.7 + 0.05) = 1.25$.

PROBLEMS

2.53 Show by applying superposition that in Fig. 2.65, $G_{21} = -0.5$ and $I_2 = 1.4$, where G_{21} and I_2 have the meanings given in (2.68), (2.69), and (2.70).

2.54 Write the two simultaneous equations for Fig. 2.65 using the G's and I's just now determined and solve these equations for V_1 and V_2. Your solution here should agree with solutions to Probs. 2.51 and 2.52.

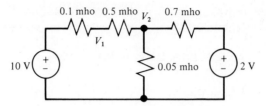

fig. 2.66   Equivalent of Fig. 2.65.

In solving Prob. 2.52, we found that an equivalent to the circuit of Fig. 2.65 is that of Fig. 2.66. To find V_2, we might treat our problem as a *one-node* situation so that

$$
\underbrace{(0.1 \text{ s } 0.5)10 + 0.7 \times 2}_{\substack{\text{Current}\\\text{entering}}} = \underbrace{[0.7 + 0.05 + (0.1 \text{ s } 0.5)]V_2}_{\substack{\text{Current}\\\text{leaving}}} \tag{2.75}
$$

where (0.1 s 0.5) means the conductance of 0.1 mho in series with 0.5 mho.

PROBLEM

2.55 a. Prove that the total conductance G_s of conductances G_1 and G_2 in series is given by

$$
G_t = \frac{1}{1/G_1 + 1/G_2} = \frac{G_1 G_2}{G_1 + G_2}
$$

b. Show that it follows that (0.1 s 0.5) = 0.08333. Then, from (2.75),

$$
V_2 = \frac{10(0.1 \text{ s } 0.5) + 0.7 \times 2}{0.7 + 0.05 + (0.1 \text{ s } 0.5)} = 2.680 \tag{2.76}
$$

With appropriate practice we could have written (2.76) without the aid of (2.75).

Notice in respect to the circuit of Fig. 2.65, we had the option of treating the network either as a one-node or a two-node situation. Using the one-node option, we applied circuit reduction to combine two conductances in series, but we avoided the solution of two simultaneous equations.

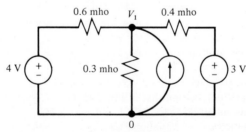

fig. 2.67   Find V_1.

Now that V_2 is known, apply the nodal approach to find V_1. You may or may not need the aid of a preliminary equation similar to (2.75) in order to write

$$V_1 = \frac{(10 \times 0.1) + (2.680 \times 0.5)}{0.1 + 0.5} = 3.900 \qquad (2.77)$$

It is important to note that we were able to write directly and explicitly the value of V_2, then the value for V_1 in the network of Fig. 2.66. Here we used the nodal-superposition approach. Problem 2.56 requires one to write directly an equation for V_1 of the network of Fig. 2.67 without previous algebraic manipulation. To do this, review (2.76), an explicit equation for V_2. Note that the denominator of the right side of this equation multiplied times V_2 yields the current leaving the node due to V_2 alone. The numerator contains the two components of current flowing into the node due to the separate voltage sources, 10 V and 2 V.

PROBLEMS

2.56 a. Without prior algebraic manipulation, write an explicit expression for V_1 of the network of Fig. 2.67.
    b. Reduce the arithmetic to a single number for the value of V_1.
    c. By the loop-current method, first find one or more loop currents, and then V_1.
    d. Find all of the branch currents and add these at node V_1. Do they sum according to Kirchhoff's laws?

2.57 For the network of Fig. 2.68, write the two simultaneous nodal equations involving V_1 and V_2. Then solve these simultaneous equations for explicit equations for V_1 and V_2. For this problem you are not allowed to use Thévenin's or Norton's theorem.

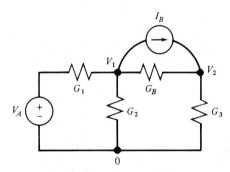

fig. 2.68   Find V_1 and V_2.

2.58 a. In the network of Fig. 2.68, use Thévenin's theorem to replace the I_B-G_B combination with the appropriate voltage-conductance combination. Treat the situation as a one-node problem to find the equation for V_1.
    b. When the equation for V_1 is known, find the equation for V_2, treating this new situation as another one-node problem.

2.59 a. Without prior fanfare, in the network of Fig. 2.69, use the nodal approach to write an explicit expression for V_1 in terms of V_A, V_B, G_1, and G_2.
    b. Without the use of the nodal approach, first find I, then V_1.

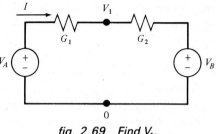

fig. 2.69   Find V_1.

2.18   DESIGN OF THE THERMISTOR THERMOMETER

With the methods that we have developed in this chapter, we are now prepared to design the thermistor thermometer described in the introduction to this chapter. You will remember that the meter current I_m (I_5 in Fig. 2.70a), and thus the meter indication, is dependent upon thermistor resistance R_T, which in turn depends upon thermistor temperature, and thus the meter can indicate temperature. We must find an equation for $I_5 = f(V_a, R_1, R_2, R_3, R_m, R_T)$. From this point, we will select specific values for fixed resistances R_1, R_2, and R_3 for acceptable thermometer performance. This equation will also give us reasons for selecting a particular meter with its R_m, and a particular voltage V_a. Finally this equation will relate meter current I_m to R_T and thus to temperature.

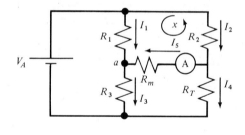

(a) Branch currents

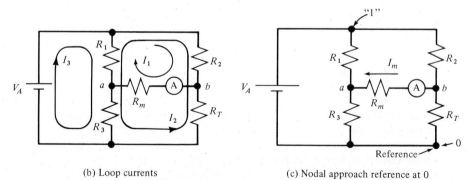

(b) Loop currents   (c) Nodal approach reference at 0

fig. 2.70   Different approaches to analyzing the bridge circuit of Fig. 2.1.

Three alternate approaches to find the equation for the meter current are suggested in Fig. 2.70. Consider first the method of "branch currents" as indicated in Fig. 2.70a. Here we need to solve for I_5, the meter current. But there are five unknown currents. Thus five simultaneous equations are required. And these must be solved for I_5. The loop current method shown in Fig. 2.70b is simpler because it requires only three simultaneous equations. The simplest approach to the bridge circuit, and the one we will use here, is the nodal approach, as illustrated in Fig. 2.70c. The reference is selected at 0. The two unknown nodal voltages are V_a and V_b, and the meter current is $(V_b - V_a)G_m$. Summing currents at node a,

$$G_1 V_a = (G_1 + G_3 + G_m)V_a - G_m V_b \tag{2.78}$$

Summing currents at node b,

$$G_2 V_a = -G_m V_a + (G_2 + G_T + G_m)V_b \tag{2.79}$$

PROBLEM
2.60 Prove both (2.78) and (2.79).

Then solving (2.78) and (2.79) for V_a, using determinants, we have

$$V_a = \frac{\begin{vmatrix} G_1 V_a & -G_m \\ G_2 V_a & (G_2 + G_T + G_m) \end{vmatrix}}{D} \tag{2.80}$$

where

$$D = \begin{vmatrix} (G_1 + G_3 + G_m) & -G_m \\ -G_m & (G_2 + G_T + G_m) \end{vmatrix}$$

$$= G_2(G_1 + G_3 + G_m) + G_T(G_1 + G_3 + G_m) + G_m(G_1 + G_3) \tag{2.81}$$

Solving for V_b,

$$V_b = \frac{\begin{vmatrix} (G_1 + G_3 + G_m) & G_1 V_a \\ -G_m & G_2 V_a \end{vmatrix}}{D} \tag{2.82}$$

Then

$$I_m = (V_b - V_a)G_m = \frac{G_m V_a (G_2 G_3 - G_1 G_T)}{D} \tag{2.83}$$

PROBLEM
2.61 Expand the appropriate determinants to prove (2.83).

Notice that D is factored here in a particular way. This gives insight into the effects of G_T and G_m upon meter currents, as we will discuss later.

Since (2.83) is a rather complex expression, we would be foolish to proceed without making some consistency checks to give us confidence in its validity. It is just too easy to make an algebraic mistake in getting a complex result like (2.83).

To make the first consistency check, we note from the circuit diagram in Fig. 2.70c that when $R_m \to \infty$, I_m must be zero because $R_m \to \infty$ corresponds to an "open circuit," through which no current can flow. If (2.83) is correct, it must be consistent with our observation that $I_m \to 0$ as $R_m \to \infty$. From (2.83), we see that as $R_m \to \infty$, $G_m \to 0$, D of (2.83) $\to$ a finite number, while the numerator of (2.83) $\to 0$, which is indeed consistent with our observation.

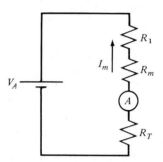

fig. 2.71   Circuit of Fig. 2.70c, reduced by letting $R_3 \to \infty$, and $R_2 \to \infty$.

Another consistency check can be made by noting that when both R_3 and $R_2 \to \infty$, the circuit reduces to the one shown in Fig. 2.71. For this simpler circuit, we can see by inspection that

$$I_m = \frac{-V_a}{R_1 + R_m + R_T} = \frac{-V_a}{1/G_1 + 1/G_m + 1/G_T}$$

Again, it must be true that (2.83) must be consistent with this result if (2.83) is indeed true. Noting that $G_3 \to 0$ and $G_2 \to 0$ corresponds to $R_3 \to \infty$, and $R_2 \to \infty$, we find from (2.83) that

$$\lim_{\substack{G_3 \to 0 \\ G_2 \to 0}} I_m = \frac{-G_m G_1 G_T V_a}{G_T G_1 + G_T G_m + G_1 G_m} = \frac{-V_a}{1/G_m + 1/G_1 + 1/G_T}$$

which is consistent.

For another check, we let $R_2 \to 0$ and $R_3 \to 0$ to get the reduced circuit shown in Fig. 2.72. This reduction may be a little harder for you to see at first, but you can tell that it is correct by noting that $R_3 \to 0$ means a is short-circuited to

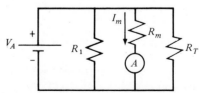

fig. 2.72   Circuit of Fig. 2.70c, reduced by letting $R_2 \to 0$ and $R_3 \to 0$.

the reference 0, and $R_2 \to 0$ means that b is short-circuited to 1. From the circuit of Fig. 2.72, we note that $I_m = V_a/R_m = G_m V_a$. Letting $G_3 \to \infty$ and $G_2 \to \infty$ in (2.83) shows again that (2.83) is consistent.

PROBLEM

2.62 Make still another consistency check on (2.83) by letting both R_1 and $R_T \to \infty$.

Now to gain further insight into our thermometer problem, refer again to (2.83). Conductance G_T is a function of the temperature to be measured. Our thermometer must have a finite range, perhaps $-10°$ to $+30°$ Celsius. Then we might select $-10°C$ to correspond to $I_5 = 0$. By our equation, $I_m = 0$ when $G_2 G_3 = G_1 G_T$. Let G_0 be the conductance of the thermistor at the low point on the meter (assumed $-10°C$ at the moment). Then $G_T = G_0 + \Delta G$.

To select G_0, we turn our attention to the particular thermistor that might be used. Let us consider using a JA41J1 thermistor. This thermistor has a resistance characteristic that is described approximately by

$$R_T = 10^4 \exp\left[3,500\left(\frac{1}{T} - \frac{1}{273}\right)\right] \qquad (2.84)$$

where T is the temperature in kelvins and R_T is the resistance in ohms. This formula would not give the precise resistance of a particular thermistor, but it will serve as an adequate approximation for our design. Using (2.84), we find that the resistance of the thermistor at $-10°C$ (263 K) is

$$R_0 = R_T(-10°C) = 10^4 \exp\left[3,500\left(\frac{1}{263} - \frac{1}{273}\right)\right] = 1.628 \times 10^4\ \Omega$$

or

$$G_0 = \frac{1}{R_0} = 6.142 \times 10^{-5}\ \text{mho}$$

At $-10°C$ we desire $I_m = 0$. This requires $G_2 G_3 = G_1 G_0$. At the moment, we select

$$G_2 = G_3 = G_1 = G_0 = 6.142 \times 10^{-5}\ \text{mho} \qquad (2.85)$$

Using (2.85) and $G_T = G_0 + \Delta G$ in (2.83) gives us

$$I_m = \frac{-G_0(\Delta G)G_m V}{G_0(2G_0 + G_m) + (G_0 + \Delta G)(2G_0 + G_m) + 2G_0 G_m} \qquad (2.86)$$

or

$$I_m = \frac{-\Delta G\, G_m V}{4G_0 + 4G_m + \Delta G(2 + G_m/G_0)} \qquad (2.87)$$

The value of G_m of a typical 50-μA ammeter might be $1/3,500 = 2.857 \times 10^{-4}$ mho. Then (2.87) becomes

$$-I_m = \frac{+2.857 \times 10^{-4}(\Delta G)V}{1.388 \times 10^{-3} + 6.652\ \Delta G} \qquad (2.88)$$

Now we use (2.84) to calculate ΔG where $G_0 = 6.142 \times 10^{-5}$ mho,

$$\Delta G = G_T - G_0 = \frac{1}{R_T} - G_0$$

This leads to the following table of values of ΔG as a function of Celsius temperature.

Let us run a consistency check on one value of meter current calculated as shown in Table 2.2. This table shows the meter current to be 25.14 μA at a temper-

TABLE 2.2   Calculated Values of ΔG and I_m

T (degrees Celsius)	ΔG (mho)	$-I_m$ (μA)
$-10°$	0	0
$-\ 5°$	1.731×10^{-5}	3.290
$0°$	3.858	6.702
$+\ 5°$	6.452	10.14
$+10°$	9.588	13.52
$+15°$	13.36	16.77
$+20°$	17.85	19.80
$+25°$	23.18	22.60
$+30°$	29.44	25.14
$+35°$	36.78	27.40
$+40°$	45.33	29.41

ature of 30°C where ΔG has been determined as 29.44×10^{-5} mho. Using the calculated meter current of 25.14 μA and applied voltage, we calculate nodal voltage V_a, and then V_b. Knowing the meter current, we can write the node voltage V_a directly (see Fig. 2.73).

$$V_a = \frac{(1 \times G_0) - (25.14 \times 10^{-6})}{2G_0} = \frac{1}{2} - \frac{25.14 \times 10^{-6}}{2 \times 6.142 \times 10^{-5}} = 0.2953 \text{ V} \quad (2.89)$$

In a similar fashion, we can write the nodal voltage V_b.

$$V_b = \frac{(1 \times G_0) + (25.14 \times 10^{-6}}{2G_0 + \Delta G}$$

$$= \frac{(6.142 \times 10^{-5}) + (25.14 \times 10^{-6})}{(2 \times 6.142 \times 10^{-5}) + (29.44 \times 10^{-5})}$$

$$= 0.2075 \text{ V} \quad (2.90)$$

$$V_a - V_b = 0.2953 - 0.2075 = 0.0878 \text{ V} \quad (2.91)$$

And this should be equal to

$$I_m \times 3{,}500 = 25.14 \times 10^{-6} \times 3{,}500 = 0.0880 \text{ V} \quad (2.92)$$

Equations (2.91) and (2.92) demonstrate good consistency.

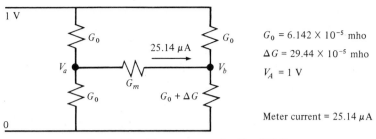

$G_0 = 6.142 \times 10^{-5}$ mho

$\Delta G = 29.44 \times 10^{-5}$ mho

$V_A = 1$ V

Meter current = 25.14 μA

fig. 2.73   Meter current for T = 30° C.

The calculated values of the thermistor characteristic and meter current tabulated in Table 2.2 are plotted on Fig. 2.74. Remember, these meter currents are for $V_a = 1$ volt. Since I_m is directly proportional to V, the values of I_m for a 1.5-V dry cell can be obtained by multiplying the values of Table 2.2 by 1.5. Once you know the R_m for your meter, you can use the design equations to get the temperature range that you desire.

In completing your thermometer, you will also want to mark the scale of the microammeter in degrees. It is desirable for the scale to be reasonably linear; that is, equal separation between marks of equal temperature. However, I_m is certainly not a linear function of temperature, as indicated by (2.83) and (2.84). It turns out, though, that nearly linear relations can be obtained by choosing proper values of R_1, R_2, and R_3. The graph of I_m versus T in Fig. 2.74 for our case shows that we have a reasonably good result, although it is not linear.

It is interesting to inquire why the curve in Fig. 2.74 is almost linear when ΔG versus T is quite nonlinear. Notice from Fig. 2.74 that the slope of ΔG versus T is an increasing function. On the other hand, (2.88) shows that the effect of ΔG in the numerator is partially canceled by the effect of ΔG in the denominator. This effect requires the slope of meter current versus ΔG to have a decreasing slope as ΔG increases. The increasing slope of the thermistor nearly cancels the decreasing slope of (2.88), resulting in the almost linear relations of I_m to T.

PROBLEMS

2.63 Equation (2.83) gives the meter current in the general situation of Fig. 2.70. Here it was assumed that $R_1 \neq R_2 \neq R_3$. Then after (2.83) was developed, we set $R_1 = R_2 = R_3 = R_0$ or $G_1 = G_2 = G_3 = G_0$ and $G_T = G_0 + \Delta G$. This leads us to (2.87). Prove that (2.87) is correct by beginning with the inset model of Fig. 2.74. Write nodal equations to find V_a and V_b and then I_5.

2.64 Make further adaptations in the design of our thermometer by assuming that we desire V_a to be 3.2 volts and meter current of 50 μA to correspond to a temperature of 30°C. Conductance G_0 is not to be changed. However, by adding resistance in series with the meter, G_m is decreased.
  a. Find the necessary value of G_m.
  b. Plot the new meter current as a function of temperature. Compare with the corresponding plot of Fig. 2.74.

Although most circuit design problems involve capacitance and inductance as well as resistance, the design of the thermistor thermometer is a good example of the

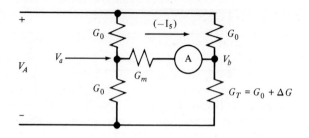

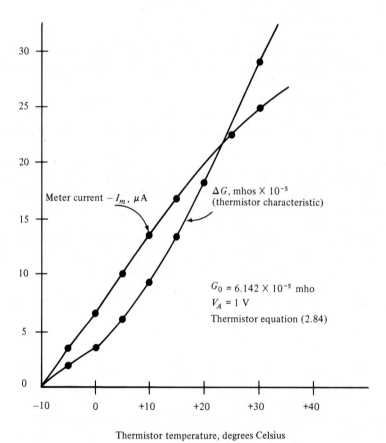

fig. 2.74   *Thermistor characteristic and meter current as a function of thermistor temperature.*

application of the principles of circuit theory in the design of a real device. As you proceed through this text, you will learn to follow the same general procedures (applying Kirchhoff's laws) extended to include capacitance and inductance and some more advanced tools.

characteristics described below were used to perform the mathematical operations of addition, subtraction, differentiation, and integration.

The word *amplifier* as used in the electronics world defines a device whose output voltage, current, or power are enlarged replicas of the input voltage, current, or power. The word *gain* is often used as a synonym for amplification, and both words are used to indicate the *ratio* of the output quantity to the input quantity. The *voltage gain* is usually of the greater interest in an *op amp* than either current gain or power gain, although these gains normally occur simultaneously and are all important. The amplifier is actually a *power converter* that converts the power from a dc power source, such as a battery, into a greatly enlarged replica of the input signal, which may be either ac or dc.

The amplifier is a four-terminal device that may be viewed as a dependent power source, whereas all of the circuit elements considered before were two-terminal devices. We think of the amplifier as having an *input port* with two terminals and an *output port* with two terminals. Quite often both the input signal (or voltage) and the output signal of the amplifier use the same reference potential. In this case, one output terminal may be connected to one input terminal and the amplifier appears as a three-terminal device. This common terminal is connected to the reference potential, considered to be zero potential, or *ground*. The chassis of an electronic amplifier is usually called *ground* although it is rarely connected to the earth or actual ground. The *ground* must be a low-resistance conductor.

The symbol for an *op amp* is shown in Fig. 2.76. The + and − signs at the input terminals indicate the polarity of the output voltage at terminal 3 as compared

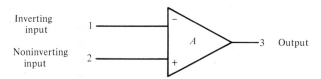

fig. 2.76   *The symbol for an operational amplifier.*

with the input voltage. For example, if the input signal is connected to the inverting input terminal and the noninverting input terminal is connected to the common ground, as shown in Fig. 2.77a, the output voltage has opposite polarity to the input voltage. As the input voltage v_i goes more positive with respect to ground, the output voltage v_o goes more negative with respect to ground. Conversely, as the input becomes more negative, the output becomes more positive. The output voltage may be expressed as

$$v_o = -Av_i \tag{2.93}$$

When the inverting input terminal is at the reference or ground potential and the input signal is applied to the noninverting input terminal as shown in Fig. 2.77b, the input and output voltages either go positive together or negative together with respect to ground. The output voltage may then be expressed as

$$v_o = Av_i \tag{2.94}$$

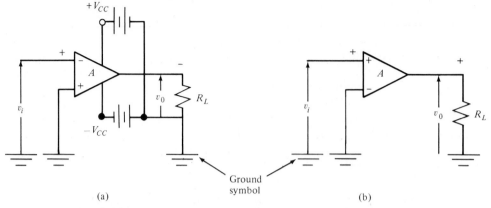

fig. 2.77   Inverting and noninverting connections for the op amp.

The power supply, represented by batteries, is shown connected to the op amp in Fig. 2.77a. Although this power supply must always be connected in order for the amplifier to operate, the actual connections are usually not shown, as indicated in Fig. 2.77b.

The dependent-source model given in Fig. 2.75 may be used for the op amp, as shown in Fig. 2.78, where the element values given are those of a typical op amp. The output resistance is actually nonzero, but it is so small in comparison with the recommended values of load resistance that it is neglected, as indicated in Fig. 2.78.

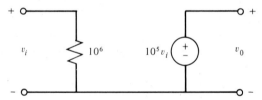

fig. 2.78   A dependent voltage source model for an op amp with typical element values given.

The voltage amplification A provided by an op amp is typically about 100,000, or 10^5. However, the maximum output voltage cannot exceed the power supply voltages, which are typically ± 10 V. Therefore, the maximum usable input voltage $v_i = v_0/A$ is about 10 V/10^5 = 10^{-4} V. This extremely small voltage is an unrealistic limit to place on the input voltage. Therefore, one or two additional resistances are used in addition to the op amp as shown in Fig. 2.79 in order to control the voltage gain of the circuit and thus permit the application of much larger input voltages. These extra resistances shown in Fig. 2.79 provide *negative feedback*, which is a type of gain control. To investigate how this circuit behaves, let us assume that the input voltage v_i is very large in comparison to the amplifier input voltage v_a. This would

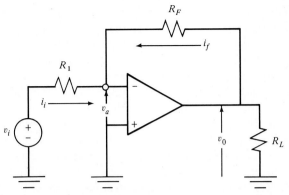

fig. 2.79 An op amp with negative feedback used to control the voltage gain.

be true if $v_i > 0.01$ V. Then we may neglect the voltage v_a and obtain the input current:

$$i_i \approx \frac{v_i}{R_1} \qquad (2.95)$$

Also, the amplifier voltage v_a is negligible in comparison to the output voltage v_o, so we may determine the feedback current i_f from

$$i_f \approx \frac{v_o}{R_F} \qquad (2.96)$$

But the current flowing into the amplifier input $(-)$ terminal is extremely small because the resistance inside the amplifier between the $-$ and $+$ terminals is typically about 1 megohm. Thus the amplifier input signal current is typically $v_a/1\text{M}\Omega = 10^{-4} \text{ V}/10^6 \ \Omega = 10^{-10}$ A. Neglecting this extremely small current and using Kirchhoff's current law $(i_i + i_f = 0)$,

$$i_f = -i_i \qquad (2.97)$$

Using Eqs. (2.95) and (2.96) to replace i_i and i_f in Eq. (2.97),

$$\frac{v_o}{R_F} = -\frac{v_i}{R_1} \qquad (2.98)$$

and

$$v_o = -\frac{R_F}{R_1} v_i \qquad (2.99)$$

Let us use the symbol G_V to designate the voltage gain v_o/v_i of the circuit. Then dividing both sides of (2.99) by v_i,

$$G_V = -\frac{R_F}{R_1} \qquad (2.100)$$

The negative sign appears because of the polarity reversal from the inverting input to the output. Observe from (2.100) that the voltage gain of the circuit is not dependent upon the amplifier voltage gain A, provided that the gain A is very high compared to G_V and the amplifier input voltage is negligible compared to either v_i or v_o. Hence the approximations we made are valid.

The input resistance of the amplifier circuit of Fig. 2.79 is the load resistance on the driving voltage source v_i. This resistance is R_1 since the amplifier end of the resistance is essentially at ground potential. The internal resistance of the driving source v_i may be included as part or all of R_1. The output voltage v_o of the amplifier appears essentially as an ideal voltage source because the resistance of the amplifier is negligible compared with R_L, as long as the current capabilities of the amplifier are not exceeded. The actual output (Thévenin's) resistance of the amplifier circuit is typically a small fraction of an ohm. A model for the op-amp circuit of Fig. 2.79 is given in Fig. 2.80.

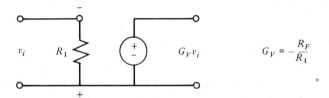

fig. 2.80   A model for the inverting op- amp circuit given in Fig. 2.79.

The voltage gain of the *noninverting* op-amp circuit may also be controlled by negative feedback. The feedback circuit is shown in Fig. 2.81. The feedback resistor R_F must *always* be connected to the *inverting* input terminal, regardless of the input signal connection, in order to obtain the desired gain reduction. Using the voltage divider principle, the feedback voltage v_f applied to the inverting input terminal (Fig. 2.81) is

$$v_f = \frac{R_1}{R_1 + R_f} v_o \qquad (2.101)$$

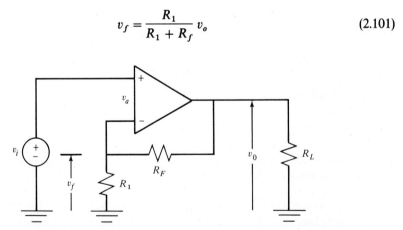

fig. 2.81   Feedback gain control of a noninverting op- amp circuit.

But the amplifier voltage v_a (which is about 10^{-4} V or less, as previously discussed) is negligible compared to either the voltage v_f or v_i. Therefore, $v_i \approx v_f$ and, using (2.101),

$$v_i = \frac{R_1}{R_1 + R_F} v_o \qquad (2.102)$$

Thus the voltage gain v_o/v_i is

$$G_V = \frac{R_1 + R_F}{R_1} = 1 + \frac{R_F}{R_1} \qquad (2.103)$$

With this connection (Fig. 2.81), the load resistance on the driving source voltage v_i is extremely large because the maximum amplifier input current is of the order of 10^{-10} A, even though the input voltage v_i may be of the order of volts. Thus we may assume that the amplifier presents essentially an infinite input resistance to the driving source v_i. An appropriate model for the noninverting op amp with negative feedback is given in Fig. 2.82.

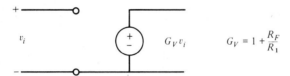

fig. 2.82   A model for the noninverting op amp with feedback.

At this time we need to be aware of one possible problem with op amps. That is, many op amps have very small direct currents, of the order of nanoamperes, that flow into both input terminals. These currents are called *bias currents* and are independent of and in addition to the very small signal currents previously mentioned. When the resistance between either input terminal and ground, such as R_1 in Fig. 2.81 or Fig. 2.79 is less than a few thousand ohms and the voltage gain of the circuit is about 100 or less, the effects of these bias currents will not generally be observed. However, when the resistance from these input terminals to the ground, or power supply terminals, becomes high, the $I_{bias}R$ voltages produced across these resistances act as a dc input signal that is multiplied by the voltage gain G_V of the circuit to produce a dc *offset* voltage in the output. The input voltage that actually operates the op amp is the voltage difference, or differential voltage, between the two input terminals. Therefore, the differential (offset) voltage due to bias currents may be minimized by making the resistances from each of these input terminals to ground (or power supply) approximately equal, as shown in Fig. 2.83. The actual resistance between the inverting input and ground, or power supply, is the parallel combination of R_1 and R_F. Therefore, the compensating resistance should be this same value. These compensating resistances do not otherwise affect the operation of the amplifier because they are normally small in comparison to the input resistance of the op amp.

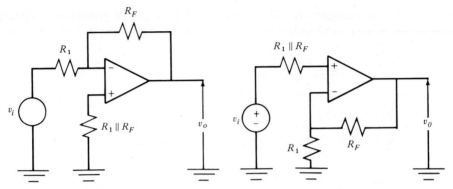

fig. 2.83   Balancing resistors used to minimize dc offset voltage.

Example 2.18

Let us use an op amp in the thermistor thermometer of Section 2.18 to eliminate the need of a costly, sensitive microammeter. The circuit may be as shown in Fig. 2.84. The dots indicate actual circuit connections. The power supply connections to the op amp are shown since a single battery, the same one used to power the bridge, is used to power the op amp instead of the dual supply shown in Fig. 2.77a. The arrangement of Fig. 2.84 is needed because the available indicating meter will in-dicate only unidirectional currents. If the bridge is balanced and the offset voltage is zero, the output voltage (across the meter) is $V_A/2$, providing $R_T = R_2$ and $R_3 = R_1$. We desire that the output voltage will reduce to 0 (ground) when the lowest temper-ature we desire to indicate is reached. This will be accomplished if the op amp we select will allow its output potential to go as low as $-V_{CC}$, which is *ground* in this case. The LM324 is one op amp that will allow this. The other requirement is that the potential, or voltage, at point T decreases by $V_A/2G_V$ as the temperature de-

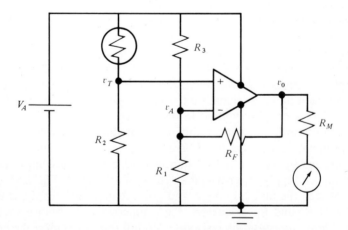

fig. 2.84   The thermistor thermometer with an op amp included to allow the use of an inexpensive, insensitive meter instead of a microammeter.

creases from the temperature required for balance to this minimum temperature. Since the op amp essentially draws no current, the change in voltage at point T may be determined by using the voltage divider technique for the thermistor branch only. The maximum temperature would be the value that causes the potential at point T to rise above $V_A/2$ by the amount that will cause the output voltage to rise to $V_A - 1.5$ V. This is the maximum positive output voltage available from the op amp. If the resistance R_2 is made adjustable over a small range, the temperature at which the meter reads 0 could be adjusted. Also, if R_F is adjustable, then the gain G_V and hence the temperature at which the meter reads maximum could be adjusted.

• • •

PROBLEMS

2.67 A given furnace has an electrically operated valve to control the fuel flow to the burner. This valve requires voltages between 0 and 10 V (either polarity) across its solenoid coil for proper operation. The resistance of the coil is 1 kΩ. The temperature sensor for the heating system produces open-circuit voltages between 0 and 0.1 V and has 5 kΩ internal resistance. Devise a circuit using a typical op amp in the inverting mode that will supply the required voltage to the control valve. How much current must the op amp be able to deliver to the valve? What is the maximum current delivered by the sensor to your circuit?

2.68 Repeat Prob. 2.67 using the op amp in the noninverting mode.

2.69 Design a thermistor thermometer that uses the JA41J1 thermistor, a 10-mA, 50-Ω meter movement, and an LM324 op amp. The thermometer should read the range from $-10°C$ to $30°C$. Use a 9.0-V radio battery for power. Balance the bridge at a temperature that will cause the output voltage to be about halfway between 0 and V_{CC} − 1.5 V. Enough resistance should be added in series with the meter so it will read half-scale at this balance temperature point. Determine the required voltage gain of the op amp by dividing the maximum expected output voltage variation Δv_o by the maximum expected input voltage variation Δv_i applied at the noninverting input terminal with respect to the common mode.

2.70 A potentiometer is a variable-resistance device, as shown in Fig. 2.85. A contact C is made to slide along a wire wound resistor so that the resistance between d and c is proportional to the angle of the rotating shaft that carries the contact c. That is, $R_{dc} = k_1 \times \theta$ where R_{dc} is the resistance between d and c, and θ is the angle of the shaft in the range of $0°$ to $180°$.

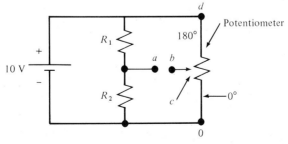

fig. 2.85 Variable V_{ab}.

It is thought that the circuit of Fig. 2.85, using the potentiometer and two fixed resistors R_1 and R_2, can be used to yield a variable voltage V_{ab} as shown in Fig. 2.86. Here V_{ab} as a function of shaft angle θ is a straight line.

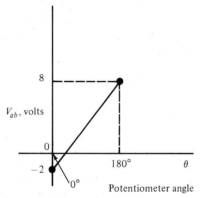

fig. 2.86   Voltage as a function of angle.

a. Assume that the device (not shown here) that uses the voltage V_{ab} draws zero current from the V_{ab} source. Demonstrate that this circuit can (or cannot) in fact be designed to deliver the linear voltage of Fig. 2.86.

b. The device that requires the V_{ab} voltage and will be connected to the a-b terminals presents a resistance at these terminals of 300 kΩ. Select numerical values of R_1, R_2, and potentiometer total resistance such that:
   1. The current drawn from the battery is a minimum and
   2. V_{ab} does not depart from the desired value (Fig. 2.86) by more than 0.2 V at any angle θ.

2.71 a. Show that the circuit of Fig. 2.88 is a model of the circuit of Fig. 2.87 and that this model will give the correct value of V_{ao} and I_a (Fig. 2.87), but does not show explicitly the voltage V_{bo} or the current I_b.

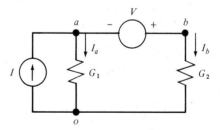

fig. 2.87   Circuit to be modeled.

b. Repeat (a) until you have used each of the following four approaches:
   1. Write the equation for the nodal voltage V_{ao}.
   2. Convert the (V, G_2) combination to a Norton equivalent.
   3. By superposition, add the component of V_{ao} due to I to the component of V_{ao} due to V.
   4. Find the Norton equivalent of the circuit at the a-o terminals, leaving all circuit elements (including G_1) intact. Find the "short circuit current" between a and o and then the resistance between a and o with I and V killed.

2.72 The model of Fig. 2.88 uses a single current source and is valid for the node voltage V_{ao}.

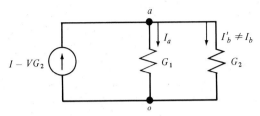

fig. 2.88   A model of the circuit of Fig. 2.87.

   a.   Find for the circuit of Fig. 2.87 a circuit model that is valid for voltage V_{bo} and uses only a single voltage source.

   b.   If you applied Thévenin's theorem in solving (a), apply the nodal approach to the original circuit of Fig. 2.87 for a consistency check.

2.73 We have two automobile batteries A and B. The open-circuit voltages of A and B, respectively, are 12.0 and 12.1 V. When the two are paralleled, the current flow between the two is 1.00 A. Then when the parallel combination of the two batteries is connected to a resistance of 0.11 Ω, the voltage at this resistance is 10.0 V. Find the numerical values in the Thévenin model of each battery.

2.74 Adapt the general nodal equations of (2.68), (2.69), and (2.70) to the circuit of Fig. 2.59 for the purpose of finding the nodal voltages 1, 2, and 3 in respect to o. To do this, merely list the values of G_{ij} coefficients of (2.68) in terms of G_1, G_2, ... of Fig. 2.59 and the I_i coefficients in terms of circuit values given in Fig. 2.59.

2.75 In the circuit of Fig. 2.41a, the following values are given:

$$I_a = 2\text{ A} \qquad R_1 = R_3 = 4\ \Omega$$
$$V_a = 10\text{ V} \qquad R_2 = R_4 = 10\ \Omega$$
$$V_b = 8\text{ V} \qquad R_5 = 2\ \Omega$$

We wish to study V_{ab} as a function of both R and V. You must calculate a few points on the function of V_{ab} as a function R while V is held constant. Then for a new value of V, plot a new function of R. Plot at least two such functions on the coordinate system of Fig. 2.89.

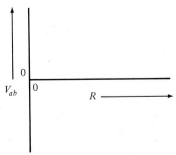

fig. 2.89   Coordinate system for plotting V_{ab} as a function of R for various values of v.

Be thoughtful. Avoid brute force. Select a method for greatest simplicity and maximum accuracy. Anticipate the approximate results before you have made a detailed analysis.

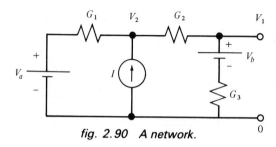

fig. 2.90 A network.

2.76 a. Find the Thévenin model of the network of Fig. 2.90 in respect to the V_{10} terminals by applying the nodal approach to find V_1 in the two-nodal V_1, V_2 situation. This will require the simultaneous solution of two equations.

b. Find V_1 of this network by viewing the circuit as having only one unknown node, V_1. Write the explicit answer for V_1 directly.

Answer

$$R_s = \frac{1}{G_s}$$

$$G_s = G_3 + (G_1 \; s \; G_2)$$

$$V_{oc} = \frac{V_b(G_1 G_3 + G_2 G_3) + I G_2 + V_a G_1 G_2}{G_1 G_2 + G_1 G_3 + G_2 G_3}$$

2.77 In the situation of Fig. 2.90,

$$V_a = 5 \text{ V}$$

$$V_b = 4 \text{ V}$$

$$R_1 = R_2 = 10 \text{ k}\Omega$$

$$R_3 = 20 \text{ k}\Omega$$

$$I = 1 \text{ mA}$$

Determine the numerical values of V_{oc} and R_s of the Thévenin model shown in Fig. 2.91.

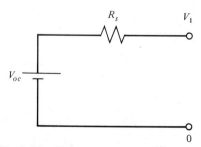

fig. 2.91 Thévenin model of Fig. 2.90.

2.78 Some problems follow that give insight into the limitations and use of dc voltmeters. The D'Arsonval voltmeter, in a sense, is merely a sensitive ammeter with an appropriate series resistor shown as R in Fig. 2.92. The resistance of a voltmeter set on the 5-V range might be as low as 25 kΩ.

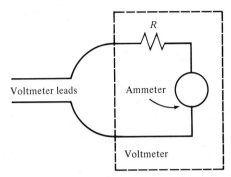

fig. 2.92   Voltmeter.

When a voltmeter is used to measure the voltage at a node in respect to a reference in a network, the voltage at this node is changed because of the insertion of the voltmeter resistance. This is called the voltmeter loading effect, and it may or may not be significant.

A voltmeter having a 5-V range and 25-kΩ resistance is used in attempting to measure the voltage V_1 in the network of Fig. 2.90 having the numerical values given in Prob. 2.77.

a. Calculate the percent loading error where

$$\text{Percent loading error} = \left(\frac{V_1 - V'_1}{V_1}\right) \times 100$$

V_1 is the voltage we wish to measure before the voltmeter is connected and V'_1 is the voltage at the same node when the voltmeter *is* connected.

Note that your solution to Prob. 2.77 brings you very close to the solution to this "loading error" problem.

b. Check your solution by using your value of V_1 and working back to the left through the network to calculate a voltage at the extreme left that is consistent with the given value of $V_a = 5$ V. Likewise, with the resistance of the voltmeter present, start with your value of V'_1 and calculate back to check the applied voltage $V_a = 5$ V.

2.79 Using a microammeter that has a resistance of 5,000 Ω and requires 20 μA for full-scale deflection, design a voltmeter having voltage ranges 5, 25, 125, and 600 V dc. Use a multiposition switch and the appropriate resistors. Show all resistances and the circuit diagram. This voltmeter is said to have a sensitivity of 50,000 Ω/V—that is, 250,000 Ω on a 5-V range or 50,000 Ω on a 1-V range.

2.80 What current in a voltmeter having a sensitivity of 20 kΩ/V is required to give a full-scale deflection?

2.81 Use a 5-V range of a voltmeter having a sensitivity of 50,000 Ω/V to measure V_1 in the network of Figs. 2.90 and 2.91. Determine the percent loading error.

2.82 A voltmeter having a sensitivity of 10 kΩ/V and ranges of 1, 5, and 20 V is available to you. You measure voltages V_{ao}, V_{ab}, and V_{bo} in the circuit of Fig. 2.93. In each measure-

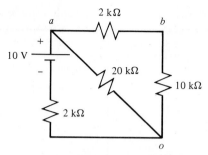

fig. 2.93   Circuit for Prob. 2.82.

ment, you use a range on the voltmeter that gives maximum on-scale indication. For each measurement, calculate the "loading error."

2.83   An ammeter also produces a loading error as it is inserted in some branch to measure the current. The ammeter has a nonzero resistance, which may or may not produce a significant change in the current of the branch in which it is inserted. To find the ammeter "loading error," we find it convenient to use a Thévenin model different from that used to determine voltmeter loading error. For example, in finding the voltmeter loading error in measuring V_{ab} in the circuit of Fig. 2.93, you probably found the Thévenin model at the *a-b* terminals. For this purpose, the Thévenin resistance R_s is given by

$$R_s = 2 \,\|\, (10 + (2 \,\|\, 20)) = 1.711 \text{ k}\Omega$$

The Thévenin voltage is

$$V_{ab_{oc}} = 10 \times \frac{2}{12} \times \frac{(12 \,\|\, 20)}{2 + (12 \,\|\, 20)} = 1.316 \text{ V}$$

This leads to the model of Fig. 2.94, where the voltage we wished to measure is 1.316 V and the measured voltage is v'_{ab}.

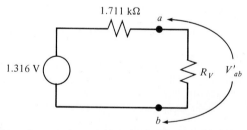

fig. 2.94 V_{ab} is the voltage measured by the voltmeter. The true voltage is 1.316 V.

To measure the current in any branch—the *a-b* branch of the circuit of Fig. 2.93, for example—we open the branch and insert the ammeter with its resistance. To analyze this situation, find the Thévenin model at the *a'-b* terminals of the circuit of Fig. 2.95. Here,

$$R_s = 12 + (2 \,\|\, 20) = 13.82 \text{ k}\Omega$$

$$V_{oc} = 10 \times \frac{20}{22} = 9.091 \text{ V}$$

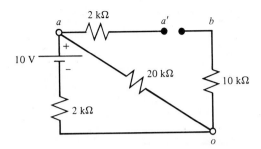

fig. 2.95   Find Thévenin model at the a′ b terminals.

This leads to the circuit of Fig. 2.96, which shows the ammeter resistance in the Thévenin model. This circuit tells us that the current we wished to measure is $9.091/13.82 = 0.6578$ mA and the current in the ammeter when inserted is given by

$$I_m = \frac{9.091}{13.82 + R_A}$$

where R_A is the resistance of the ammeter.

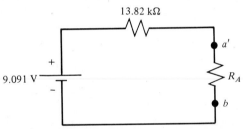

fig. 2.96   Ammeter resistance R_A inserted in Thévenin model.

a. The Thévenin model of Fig. 2.96 leads to the current of 0.6578 mA in the a-b branch. Check this value by some other approach. You may wish to write the nodal voltage V_{ao}, then I_{ab}. Or you may wish to find first the battery current by calculating the total resistance seen by the battery, then applying the current divider concept.

b. If an ammeter having a resistance of 0.30 Ω is used to measure the current in the a-b branch of the circuit of Fig. 2.95, find the loading error given by

$$\text{Percent loading error} = \frac{I - I'}{I} \times 100 \text{ percent}$$

where I and I' are the currents, respectively, before and after the ammeter is inserted.

c. When this same ammeter is used to measure the current in the a-o branch, what is the loading error?

2.84 It is not practical to design the moving coil and the connecting hairspring of a D'Arsonval ammeter to carry more than approximately 1 A. Therefore, many ammeters use a "shunt," as shown in Fig. 2.97. Most of the current I to be measured is shunted around the moving coil of the sensitive instrument.

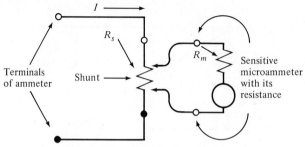

fig. 2.97   Shunt in an ammeter.

a. Design a 10-A ammeter around the microammeter of Prob. 2.79 and a shunt whose resistance you must select. Show numerical values of all currents, voltages, and resistances on a diagram when the current to be measured is 10 A.

b. What is the resistance of this ammeter?

2.85 A voltmeter having a sensitivity of 10,000 Ω/V has 10- and 25-V ranges. The voltage between two nodes in a network is measured, first using the 10-V range, then using the 25-V range of this voltmeter. The respective voltages read on the voltmeter were 9.00 and 9.50 V. Assuming there are no errors in these voltmeter readings, determine the true voltage between these nodes with the voltmeter removed.

Chapter 3

Capacitance and Inductance

Introduction In the preceding chapter your attention was directed to resistance, which is a circuit element that dissipates energy (converts electric energy to thermal energy) but does not store electric energy. In this chapter we expand our development to include the other two passive circuit elements, capacitance and inductance. Both capacitance and inductance store, but do not dissipate, electric energy that can be returned to the source or other parts of the total network, and consequently they possess interesting properties that are not characteristic of resistance.

In this chapter, resistance and capacitance are first considered in combination—the so-called *RC* circuit. Then resistance and inductance are considered in combination—the so-called *RL* circuit. Circuits containing both inductance and capacitance are considered in later chapters.

3.1 SOME FUNDAMENTALS OF CAPACITANCE

Definition of Capacitance

Capacitance means capacity for storing electric charge. Suppose a battery is connected to two conducting bodies separated by an insulating medium (Fig. 3.1). The battery will cause charge to be transferred from one body to the other, leaving one body with a positive charge and the other body with a negative charge of equal magnitude. The amount of charge transferred from one body to the other is determined by the battery voltage, the size and shape of the bodies, the distance between the bodies, and the *permittivity* of the medium separating the bodies. (Permittivity is a property of materials that you will learn more about in courses on electromagnetic field theory.) For example, as the distance between the bodies is decreased, the amount of charge that is transferred increases. Increasing the battery voltage also

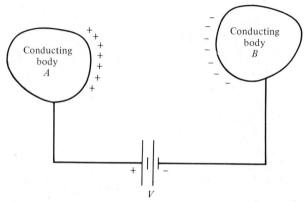

fig. 3.1   A battery connected to two conducting bodies separated by an insulating medium.

increases the amount of charge that is transferred. More specifically, if q is the amount of charge in coulombs transferred from body A to body B, then

$$q = Cv \tag{3.1}$$

where v is the potential difference of body A with respect to body B in volts and C is a constant called the *capacitance* in farads (F). Capacitance is a positive quantity that is a function of the surface areas of the bodies, the permittivity of the medium, and the separation of the bodies. The capacitance increases as the separation of the bodies decreases, and the capacitance increases as the surface area of the bodies increases. An increase of the permittivity of the medium separating the bodies also causes an increase in capacitance.

Although we have used a static case for illustration, (3.1) is true when q and v are time-varying, as indicated by our use of lowercase letters. From (3.1), we can write

$$C = \frac{q}{v} \tag{3.2}$$

from which we see that one farad is equivalent to one coulomb per volt and also that capacitance can be interpreted as the amount of stored charge per volt. Equation (3.1) is a fundamental generalization that has been verified experimentally. We shall accept it as an axiom that defines capacitance.

A *capacitor* is a device that is useful because of its significant capacitance. Capacitors are constructed in a variety of forms, some of which are shown in Fig. 3.2. For example, tuning capacitors for radio receivers are made to have variable capacitance by allowing one set of plates to rotate to a variable degree into another set of intermeshed plates. *Electrolytic* capacitors consist of sets of aluminum foil plates with a conducting semiliquid chemical between them. When a dc voltage is applied between the plates, a very thin dielectric film is formed on one set of plates, thus providing the insulating medium for the capacitor. Since the film is very thin, the capacitance is high.

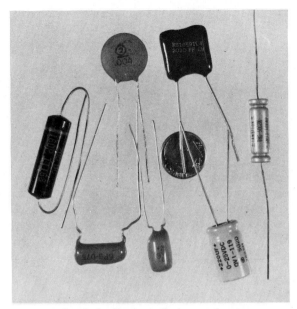

fig. 3.2 Some typical capacitors.

Other typical capacitors are mica capacitors, paper capacitors, and ceramic capacitors, so-called because their insulating medium is, respectively, mica, paper, and ceramic.

You are reminded of the importance of drawing the distinction between the meaning of the word *capacitor* and the word *capacitance*. The *capacitance* property, defined by (3.2), is purposely built into the *capacitor* device. In a rigorous sense, the capacitor unavoidably always includes some resistance and inductance properties. In many situations, however, both resistance and inductance properties of the capacitor are negligible.

Models of Capacitors

The symbol for capacitance is shown in Fig. 3.3, and where the properties of resistance and inductance are negligible, this figure also portrays one *model* of a capacitor. Another important model of a capacitor is shown in Fig. 3.4. In this model, C and R_C represent the properties of capacitance and resistance, respectively. In the property of capacitance, no electric energy is lost in the sense that electric energy is

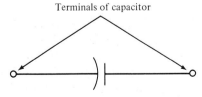

fig. 3.3 One model of a capacitor.

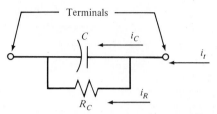

fig. 3.4   *A second model of a capacitor.*

converted to thermal energy; resistance converts energy from electric to thermal. Because the insulating medium between the plates of a capacitor is imperfect, its precise model must include resistance in addition to capacitance. Figure 3.4 shows such a model where resistance is placed in parallel with capacitance. Notice that the capacitor current is given by $i_t = i_C + i_R$. In many situations, R is so large that i_R is negligibly small. In such cases, the resistance is not shown and the capacitor is represented by its capacitance only. This is sometimes called an *ideal* capacitor. In the work that follows, we shall nearly always think of the model of Fig. 3.3 as being appropriate.

You may have expected the precise capacitor model to be in the form of resistance and capacitance in series rather than in parallel, as in Fig. 3.4. The advantages of the parallel RC model will be discussed later.

The resistance R_C in Fig. 3.4 is said to represent "loss" in the capacitor because energy is dissipated (converted to heat and thus "lost") in resistance, but energy is only stored in capacitance, never dissipated.

Any particular model is never an *exact* representation. In practice, we use the simplest model that is adequate for the particular purpose. Therefore the practical value of R_C (Fig. 3.4) for one situation might be very much different from the appropriate value for another situation. Consider the following case. If a battery were connected across the capacitor represented in Fig. 3.4 and then disconnected after a suitable time, charge would be stored in C, positive charge on one plate and negative on the other. However, the charge would flow from one plate to another through the insulating medium until the capacitor became "discharged," having zero charge on each plate. If R_C of the model of Fig. 3.4 were infinitely large, the capacitor would remain charged forever, a physical impossibility. This indicates that R_C cannot be infinite in an exact sense. Since R_C would allow charge to "leak" from one plate to another, the current that would flow through R_C in this special situation is a representation of the *leakage* current that flows through the insulating medium. In another situation where the voltage across the capacitor is varying with respect to time, another loss of power, *polarization loss*, occurs in the insulating medium. This requires the R_C of the model to be much lower than for the case where voltage is steady or varying slowly with respect to time. Thus one model that is adequate for one situation is not necessarily adequate for another.

For alternating voltage at sufficiently high frequencies, the model of a capacitor shown in Fig. 3.4 is inadequate because the inductance of the capacitor also becomes important. Then the model of the capacitor must consist of capacitance, resistance, and inductance. At extremely high frequencies (for example, above 1 GHz), not only is the capacitance-resistance-inductance model of a capacitor invalid, but circuit theory itself is inadequate, and electromagnetic field theory (or "microwave" theory, as it is called at those frequencies) must be used. At this point, you must be wondering how to tell when circuit theory is valid and when it is not and how to decide which model of a capacitor is inadequate. There is no easy answer to this question. The final criterion is always experimental results. However, as you gain experience, you will learn to estimate the range of validity of the various models and the validity of circuit theory. We shall try to do so as we proceed through this book. At this point you should be sure to understand that capacitance is a *property* and a capacitor is a *device*. The usefulness of capacitors is based primarily on their capacitance, although capacitors also have resistance and inductance, as explained above.

Example 3.1
Figure 3.5 describes an experiment that might be conducted in order to find the value of R_C in a model similar to that of Fig. 3.4, where we wish to describe the performance of the capacitor if the applied voltage is constant or changes slowly. In other words, we seek the "leakage" resistance of the capacitor.

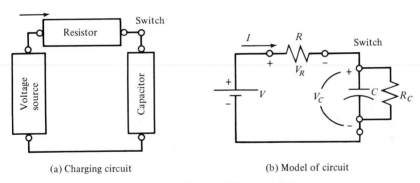

(a) Charging circuit   (b) Model of circuit

fig. 3.5 Circuit of Example 3.1.

We use an appropriate source to apply a steady dc voltage to the series combination of a selected resistor R and the capacitor whose R_C we seek. See the network of Fig. 3.6. The switch has been closed long enough such that the current I is steady

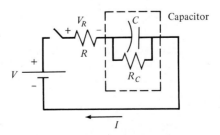

fig. 3.6 To find leakage resistance.

(C is fully charged). We measure V_R and find it to be 12.2 mV when $V = 120$ V and $R = 12.7$ kΩ. Then

$$I = \frac{V_R}{R} = \frac{12.2 \times 10^{-3}}{12.7 \times 10^3} = 0.9606 \ \mu\text{A}$$

Then

$$R_C = \frac{V}{I} - R = \frac{120}{0.9606 \times 10^{-6}} - 12{,}700 = 124.9 \ \text{M}\Omega$$

For a consistency check on some of our procedures, we note that R_C is very large compared to R. It follows that IR_C should be close to 120. Then

$$V = 0.9606 \times 10^{-6} \times 124.9 \times 10^6 = 119.98 \ \text{V}$$

• • •

Current - Voltage Relationship

From (3.2),

$$v = \frac{q}{C}$$

and differentiating, we get

$$\frac{dv}{dt} = \frac{1}{C}\frac{dq}{dt}$$

Since $i = dq/dt$, we can write

$$i = C\frac{dv}{dt} \tag{3.3}$$

where i is in amperes, C in farads, v in volts, and t in seconds. Equation (3.3) applies when current i and voltage v have the senses or directions as shown in Fig. 3.7. If

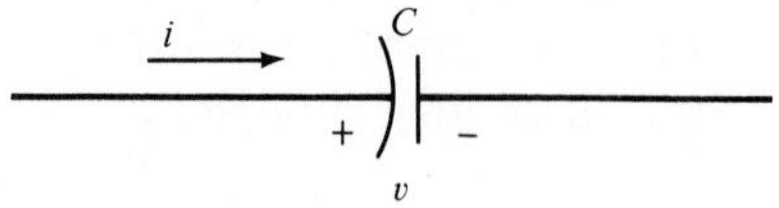

fig. 3.7   Voltage v across a capacitance C and current i through the capacitance.

either the sense of v or i is reversed from those shown here, then $i = -C(dv/dt)$. Equation (3.3) is explicit in i. Where v is known, one can immediately solve for i by taking the derivative of v as indicated. On the other hand, suppose i is known and we wish to find v. Then we write

$$dv = \frac{1}{C} i \ dt \qquad \text{or} \qquad \int_{v_0}^{v} dv = \frac{1}{C} \int_{t_0}^{t} i \ dt \tag{3.4}$$

where v_0 is the value v at t_0 and v is the value of v at t. Equation (3.4) may be rewritten as

$$v = \frac{1}{C} \int_{t_0}^{t} i \, dt + v_0 \tag{3.5}$$

Equation (3.5) is explicit in v and is convenient in finding v, the voltage across capacitance, given the current i in that capacitance. The voltage v_0 is called the initial value of capacitance voltage. Notice that the definite integral has the meaning of the charge accumulation during the time interval $(t - t_0)$. And, of course, the charge accumulation divided by C as shown in (3.5) yields the change of capacitance voltage in that same time interval.

Example 3.2
The v across the C as a function of time in Fig. 3.7 is shown in Fig. 3.8, and $C = 1 \, \mu\text{F}$. Graphs of i and q as functions of time are desired. Using (3.3) and

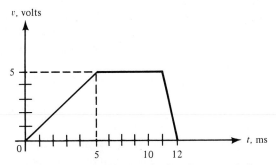

fig. 3.8   Voltage across C in Example 3.2.

graphical differentiation, we can easily find i, as shown in Fig. 3.9. Note that dv/dt is the slope of v. Compare Fig. 3.9 with Fig. 3.8. For $0 \leq t \leq 5$ ms, the slope is 5 V/5 ms $= 1 \times 10^3$ V/s. Multiplying by $C = 10^{-6}$ F gives $i = 1$ mA for $0 < t < 5$ ms. The

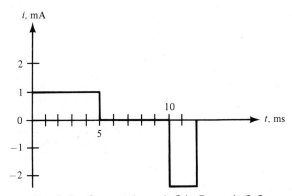

fig. 3.9   Current through C in Example 3.2.

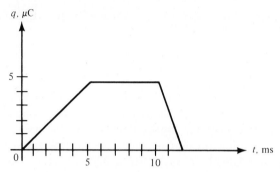

fig. 3.10   Charge stored on the left plate of the capacitance in Example 3.2.

remainder of the curve in Fig. 3.9 can be found in the same manner. Since $q = Cv$, the graph of q is even easier to find, as shown in Fig. 3.10.

The plot of i of Fig. 3.9 was found by the direct application of (3.3) to the known v of Fig. 3.8. Now let us determine whether our solution of i is consistent with (3.5) by performing the indicated integration graphically. The integral of i has the graphical meaning of the area under an appropriate part of the curve for i. The integral of i from 0 to 5 ms is $(1 \times 10^{-3}) \times (5 \times 10^{-3}) = 5 \times 10^{-6}$ A·s $= 5$ μC. When we divide this by $C = 1 \times 10^{-6}$ F, we get a voltage change of 5 V, which is consistent with Fig. 3.8. A little additional thought leads one from the constant value of i between $t = 0$ and $t = 5$ ms to the straight line for v in this same time interval, as shown in Fig. 3.8. You should check other points on the v function by graphically integrating the i function.

Some people resist these graphical interpretations of differentiation and integration. If you are one of these, note the simplicity and directness of these graphical approaches in this particular situation. Graphical interpretations wedded to the familiar formal, symbolic manipulations can give rapid and powerful engineering insights. You will hear much about graphical representations as you proceed through this text.

• • •

PROBLEMS

3.1 In the situation of Example 3.2 and when 5 ms $< t <$ 10 ms, there is a positive voltage applied to C; yet we have concluded that the capacitance current i is zero. Furthermore, for 10 ms $< t <$ 12 ms, the current is negative while the voltage is positive. Justify these conditions to a high school student in shop electricity assuming that the student has no calculus background.

3.2 a. If the voltage v across C in Fig. 3.7 is given by the graph in Fig. 3.11, find the plots of i and q if $C = 2$ μF.

   b. To verify your work, apply (3.5) to your solution for i to find v as shown in Fig. 3.11.

3.3 Think of the plot of Fig. 3.11 as the current in this same capacitance where the maximum value of the current wave is 10 A and the initial value of voltage v_0 is -20 V.

   a. Employ graphical integrations to find the value of v at 3, 5, 10, 11, 13, and 15 μs. Then sketch the plot of v through these points without writing equations for i and the v. The plot of v in each region will not necessarily be a straight line.

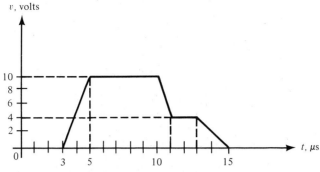

fig. 3.11 Voltage across C in Prob. 3.2.

b. Run a consistency check on your voltage wave by relating the relative slopes of the v wave to the original i wave. For example, the v wave should have zero slope at 3 μs, and constant slope for $5 < t < 10$, and so on.

3.4 For the situation of Prob. 3.3, find the equation of the current in the region where $3 < t < 5$ μs, and then through a standard integration, find the equation for v in this region. Note that you can simplify this procedure by shifting the time axis so that i is zero at $t = 0$ and, of course, i will be 10 A at $t = 2$ μs, and so on. Now compare this result with your solution to Prob. 3.3a.

3.5 If measured values of i and v in Fig. 3.7 are

$$i = 10 \sin \omega t \text{ mA}$$

$$v = -5 \cos \omega t \text{ V}$$

and

$$\omega = 2\pi \times 10^6 \text{ rad/s (radians per second)}$$

find the value of C.

It is important to note that the voltage-current relationships for capacitance are quite different from those for Ohm's law because, for capacitance, the equations involve dv/dt or $\int i\, dt$. A qualitative interpretation of (3.3) is that $C\, dv/dt$ is equal to the amount of current produced by a changing voltage across C. For a given dv/dt, a large capacitance produces a large current and a smaller capacitance produces a smaller current.

There is one more important concept that we should get from (3.3) before we proceed with the development of RC circuit relations. Since the current i is proportional to dv/dt, the voltage across a capacitance cannot change *instantaneously*; that is, there cannot be a finite change in voltage in zero time. We can see that this must be true because an instantaneous change in v would mean that $dv/dt \rightarrow \infty$, and from (3.3), this would require infinite current. Since infinite current is not physically possible, we must conclude that the voltage across a capacitance cannot change instantaneously. We shall make frequent use of this result in determining initial conditions (see Section 3.3) during circuit analysis.

PROBLEM

3.6   A 0.25-μF capacitor is charged to a voltage of 10 V and then connected to a resistance by a switch as shown in Fig. 3.12. The switch is closed at time = 0, and the equation of the voltage across the capacitor is given by

$$v = 10e^{-1,000t} \qquad \text{for } t > 0$$

a.   Find the equation of the current.

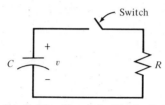

fig. 3.12   Circuit for Prob. 3.6.

b.   Find the necessary numerical value of resistance R.
c.   Are your answers to (a) and (b) consistent with the fact that the voltage across the resistance R must start at 10 V (at $t = 0$) and decrease toward 0 as time increases?
d.   On one set of coordinates, plot v, v_R, i, and q. Is each of these plots consistent with the other plots?

3.2   SOME FUNDAMENTALS OF INDUCTANCE

Definition of Inductance

Consider the simple single-turn loop of wire shown in Fig. 3.13a. The current flowing in the loop produces magnetic flux. The magnetic flux lines are said to "cut" or "link" the loop of wire. The total flux λ passing through the area of the loop is the flux linking this one-turn coil. The *inductance, L,* of the loop is defined by

$$L = \frac{\lambda}{i} \tag{3.6}$$

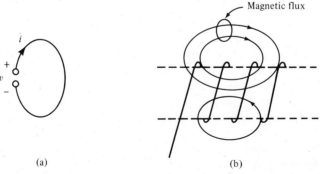

(a)   (b)

fig. 3.13   Simple single - turn loop and a coil used to illustrate the definition of induct-ance.

where L is in henries (abbreviated by H), λ is flux linkages in webers, and i, of course, is current in amperes (abbreviated by A). In the more general case of N turns in a coil, Fig. 3.13b, some of the magnetic flux does not link all of the turns, yet *total flux linkages* λ has meaning and can be used to define inductance. Since each turn in the typical coil nearly closes on itself, one could determine the coil flux linkages by summing up the separate one-turn fluxes.

Inductors are coils specifically designed for their useful inductance. There are many kinds of inductors, some of which are shown in Fig. 3.14. The most common form of inductor is some turns of wire wound in the form of a helix (often with multiple layers). The wire is frequently wound on a core of magnetic material, such as iron, to increase the magnetic flux produced by the current and thus increase the inductance. The wire really need not be wound in the form of a helix, as any collection, or "coil," of turns will serve to enhance the magnetic field and produce significant inductance. For this reason, inductors are often called coils.

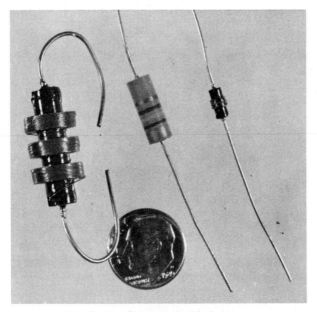

fig. 3.14   Some typical inductors.

The inductance property is present wherever there is electric current because current always produces magnetic flux, which leads to the property of inductance. Of course, the inductive effect, when present, may or may not be large enough to be significant. Coils produce large magnetic fluxes in the magnetic irons of electric motors and generators. Here inductance is not necessarily specifically designed into the machine, but is necessarily highly significant in respect to the performance of the machine. The inductive property is inherent and unavoidable in many situations, especially where currents change very rapidly, such as in computers and high-frequency communication systems.

Models of Inductors

Inductors are represented by models comprised of inductance, resistance, and capacitance. The symbol for inductance is shown in Fig. 3.15. For many purposes, the

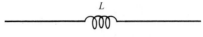

L

fig. 3.15   Symbol for inductance.

precise model of an inductor as shown in Fig. 3.16 must include resistance *R* because of loss in the winding and loss in the medium of the magnetic field. Where there is no solid magnetic material used, the resistance *R* of the inductor model represents the resistance of the conductor or winding. For some inductors, *R* is very low because the conductor has low resistance, but it can never be zero (except for the cryogenic case when *R* approaches zero) because no conductors at typical temperatures are perfectly conducting. Where the winding is wrapped around a magnetic steel (or other lossy, magnetic material), then the *R* of the model is selected to allow for this new loss. Just as is true for capacitors, however, *R* can be low enough to be neglected in many circumstances. In that case, the resistance *R* is not shown in the model and the inductor is represented by its inductance alone. Such an inductor is sometimes called an *ideal* inductor.

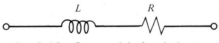

L   *R*

fig. 3.16   One model of an inductor.

As for capacitors, *R* also leads to "loss" in the inductor because energy is dissipated in resistance, but energy is only stored, not dissipated, when we consider only the inductance property.

Where the current is alternating at sufficiently high frequencies, the model in Fig. 3.16 is inadequate because the capacitance between conductors in the inductor becomes important. Then the model of the inductor must include capacitance as well as inductance and resistance. We shall discuss other models of inductors in subsequent chapters. Note especially that inductance is a *property* and an inductor is a *device*.

Current - Voltage Relationship

According to Faraday's law, the potential difference across the terminals of the loop in Fig. 3.13 is given by

$$v = \frac{d\lambda}{dt}$$

which combined with (3.6) gives us

$$v = \frac{d\lambda}{dt} = \frac{d(Li)}{dt} \tag{3.7}$$

When magnetic material is not present, the inductance L is a constant determined only by the size and shape of the coil. Then

$$v = L\frac{di}{dt} \tag{3.8}$$

Equation (3.8) is a general relation that is valid whenever the inductance is constant. The inductance will not be constant when the coil is wound on an iron core, in which case the inductance will be a function of the current. Equation (3.7) is valid in all situations. But in this book, we shall consider only cases where inductance is constant, and therefore (3.8) is valid for all situations to be discussed here. You will learn about the case where inductance is not a constant in courses that include magnetic circuits.

PROBLEM

3.7 Show from (3.8) that one henry is equal to one volt second per ampere. Note that di/dt has units of amperes per second.

Equation (3.8) is an expression of the voltage required to produce a given rate of change (di/dt) of current in inductance L. For a given di/dt, a larger L requires a larger v. We say that inductance opposes a change in current. In this sense, inductance is analogous to mass, which may be thought of as opposing a change in velocity. It is important to note from (3.8) that the current through inductance cannot change instantaneously, as would be expected because inductance opposes a change in current. The kind of argument applies here that we used previously to show that the voltage across a capacitance cannot change instantaneously. If the current through inductance were to change instantaneously, di/dt would be infinitely large, and, according to (3.8), the required voltage would be infinitely large. Since this is not physically possible, we must conclude that the current through inductance cannot change instantaneously.[1] This conclusion is important to the determination of initial conditions, as discussed in following sections of this chapter.

Example 3.3

The current through an inductance of 1 mH is shown in Fig. 3.17. Find the voltage across the inductance. Using graphical differentiation and (3.8), we find the voltage to be as shown in Fig. 3.18. For the interval $0 \le t \le 5$ ms, $di/dt = 5$ mA/5 ms = 1 A/s, as found from the slope of the line in Fig. 3.17. Multiplying di/dt by $L = 1$ mH gives us $v = 1$ mV for $0 \le t \le 5$ ms. Continuing in the same manner gives the complete curve in Fig. 3.18.

---

[1] In a very special hypothetical situation where there are two coils adjacent to each other (magnetic-coupled circuits, treated in Chapter 13), di/dt in a coil can be infinite.

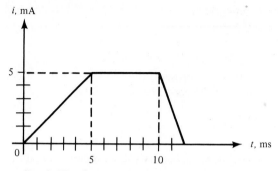

fig. 3.17   Current through L in Example 3.3.

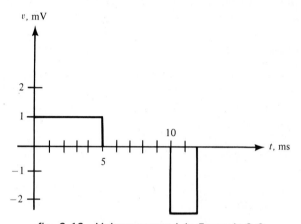

fig. 3.18   Voltage across L in Example 3.3.

• • •

PROBLEMS

3.8 If the current through an inductance of 2 mH is given by the graph in Fig. 3.19, draw a graph of the voltage across the inductance.

3.9 If the measured values of i and v for an inductance L are $i = 10 \sin \omega t$ mA, $v = 5 \cos \omega t$ V, and $\omega = 2\pi \times 10^6$ rad/s, find the value of L.

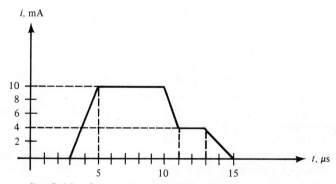

fig. 3.19   Current through the inductance in Prob. 3.8.

By a procedure similar to that used to obtain (3.4) for capacitance, the explicit value of i can be found. From (3.8),

$$di = \frac{v}{L}\, dt \quad \text{and} \quad \int_{i_o}^{i} di = \frac{1}{L} \int_{0}^{t} v\, dt$$

or

$$i = \frac{1}{L} \int_{0}^{t} v\, dt + i_o \tag{3.9}$$

Equation (3.9) is the integral form of (3.8), and so we should be able to recover (3.8) by differentiating (3.9), if we have integrated correctly. Although (3.8) and (3.9) are equivalent relations, we use (3.8) to find v when i is known, and (3.9) to find i when v is known.

PROBLEMS

3.10 Differentiate both sides of (3.9) and show that the result is equivalent to (3.8).

3.11 Run consistency checks on your solutions to Probs. 3.8 and 3.9 by applying (3.9) to the respective solutions to find the current waves of Figs. 3.18 and 3.19, respectively.

3.12 A graph of the voltage across an inductance as a function of time is shown in Fig. 3.20. The current at $t = 0$ is -12.5 A and $L = 2$ mH. Graph the current through the inductance as a function of time, beginning with $t = 0$. After you obtain the result, check it by differentiating to see whether you get the voltage back again.

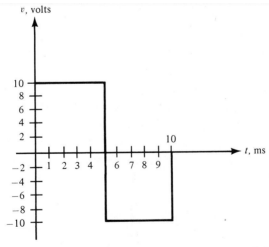

fig. 3.20   Voltage across the inductor in Prob. 3.12.

Inductors are used only sparingly in electronic circuits because they tend to be expensive and because they often require shielding to eliminate stray inductance and unwanted coupling to other circuit elements. However, inductance is very important in tuned circuits and coupled circuits. Furthermore, inductance cannot be avoided in some situations and must be dealt with accordingly.

3.3   SOME RC NETWORKS

So far in this text, we have analyzed networks having:

1.   Resistance only.
2.   Capacitance standing alone.
3.   Inductance standing alone.

Now it is natural to ask about networks made of combinations of resistances, capacitances, and inductances in any series and/or parallel arrangement. Most networks that we face in practice truly are combinations of RC (resistance and capacitance), RL (resistance and inductance), or RLC (resistance, inductance, and capacitance). We take another step in the direction of analyzing these rather general networks by considering another special situation, RC networks. As an example, consider the RC network of Fig. 3.21a. For this network, we will proceed to find v_2

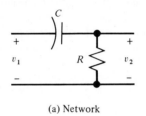

(a) Network

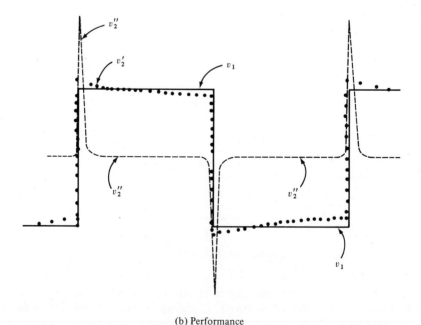

(b) Performance

fig. 3.21   Performance of an RC network

(a function of time) for a particular v_1 where R and C are known. We will analyze many other RC combinations. This particular RC combination and a multitude of other more complex combinations, where there may be more than one R element and more than one C element, are very common in electrical technology. Sometimes the capacitance and/or resistance element is present by design and sometimes present inherently and unavoidably.

The simple RC network of Fig. 3.21a is commonly used in an electronic network to separate a dc voltage from an ac voltage. In this network, if v_1 includes a steady dc component as well as a steady ac component, none of the dc components of voltage can appear in v_2. You may not understand this now, but you will soon learn how to select R and C of this network so that for practical purposes v_2 will be equal to the ac component of v_1. The need for separating the dc and ac components may not necessarily be obvious to you at the moment, but will become apparent as you study electronics.

In many other applications, wave shapes are formed by using RC, RL, and RLC network and electronic components. Figure 3.22 shows some simple wave shapes

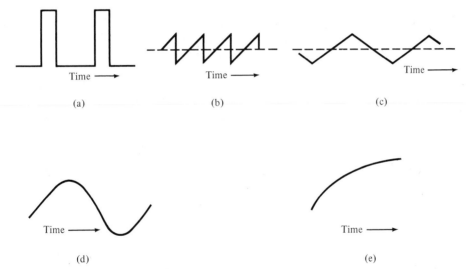

fig. 3.22 Examples of typical voltage and current waveforms used in electrical engineering.

encountered profusely in the technology. We need to learn how to deal with RC networks that are driven by any reasonable function of time such as these. Such RC networks abound in computers, communication systems, control and instrumentation systems, power systems, and, indeed, almost everywhere there are significant electric systems.

To learn how to analyze and design RC networks we will first consider a situation akin to that illustrated in Fig. 3.21. Here the voltage v_1 applied at the left of this network is a so-called "square wave," shown as the solid line of Fig. 3.21. We will soon learn how to select R and C to produce a voltage response like the highly peaked wave v_2''. A different choice of R and C would yield a response v_2' that is a rather close approximation to the applied square wave v_1. Of course, other responses can be generated.

Let us consider a particular application of the RC network of Fig. 3.21a. There are an unlimited number of applications that employ the same fundamental concepts that apply to this isolated case. View the waves v_1 and v_2 of Fig. 3.23 as the

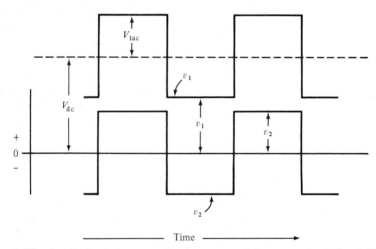

fig. 3.23   Applied voltage v_1 and response v_2 of the network of Fig. 3.21a.

applied and response voltages, respectively, of the RC network of Fig. 3.21a. Notice that v_1 is varying with respect to time, but that it is always positive. On the other hand, v_2 alternates with both positive and negative values as time increases. This applied voltage v_1 has a dc component V_{dc} and an ac component v_{1ac} such that

$$v_1 = V_{dc} + v_{1ac}$$

This situation is inherent in many electronic situations. We desire to use a network that "blocks out" V_{dc} so that the response v_2 is equal to v_{1ac}. We can achieve this goal, not precisely but with acceptable accuracy, by properly selecting R and C of the RC network of Fig. 3.21a. As a step to this end, consider the simpler, but closely related, situation treated in the next section.

Simple RC Performance

In the situation of Fig. 3.24, a switch is closed at time $t = 0$ to apply suddenly a steady voltage V to a series combination of resistance R and capacitance C. We will assume that this circuit had a previous history (not shown here) that left a charge on C. First we will seek the equation for v (across C); later we will find v_2. The application of Kirchhoff's voltage law gives us a start.

$$V = v + v_2 = v + Ri \tag{3.10}$$

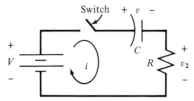

fig. 3.24   Steady voltage applied to series RC.

We desire an equation involving v and no other variable; therefore we eliminate the variable i in favor of v. Because the current i flows through the C, as well as R, we can use $i = C\,dv/dt$; see (3.3). Then it follows that

$$RC\frac{dv}{dt} + v = V$$

which can be rearranged to

$$\frac{dv}{dt} + \frac{1}{RC}v = \frac{V}{RC} \tag{3.11}$$

This is the differential equation for the voltage across the capacitor. It is a first-order, linear, nonhomogeneous differential equation with constant coefficients. At this point, you may want to review methods of solution of differential equations in your favorite book on differential equations.

With (3.11) arranged so that all of the terms *not* involving v are on the right-hand side of the equal sign, the equation is said to be *homogeneous* if the right-hand side is zero, and *nonhomogeneous* if the right-hand side is nonzero, as it is here.

From your study of differential equations, you learned that the general solution to a differential equation must both satisfy the equation and must also contain arbitrary constants, the number of arbitrary constants being equal to the order of the differential equation. The general solution to (3.11) must satisfy (3.11) and must also contain one arbitrary constant because (3.11) is first order. To find the general solution to the differential equation itself, we find first the general solution to the homogeneous part of the equation. We call this the *complementary function* and label it here as v_h (general solution to the homogeneous part). So in this case, the v_h must satisfy

$$\frac{dv}{dt} + \frac{1}{RC}v = 0 \tag{3.12}$$

and also must contain an arbitrary constant. Of course, v_h does not satisfy the original nonhomogeneous equation itself. But the sum of v_h and another term that satisfies the original differential equation, but does not contain an arbitrary constant, will become the general solution that we seek. This second term v_p we call the *particular solution*, sometimes called the *particular integral*. So, the general solution is given by

$$v = v_h + v_p$$

PROBLEM

3.13 The general solution v to (3.11) must have these two properties: (1) It must satisfy (3.11). (2) It must contain one arbitrary constant.

    a. First, for v_h, then for v_p, write a similar sentence defining the property or properties similar to the properties of v as shown above.

    b. Now write statements that prove that $v = v_h + v_p$.

For linear differential equations with constant coefficients (which are the kind describing linear circuits), the complementary function can always be found by use of a trial solution of the form

$$v_h = Ae^{st} \tag{3.13}$$

where A is the necessary arbitrary constant and s is a function of the network parameters, R and C in this case. Substituting (3.13) into the homogeneous differential equation of (3.12) gives us

$$sAe^{st} + \frac{1}{RC} Ae^{st} = 0$$

or

$$\left(s + \frac{1}{RC}\right)Ae^{st} = 0 \tag{3.14}$$

We seek s that makes (3.14) hold for all values of t, and where A may have any arbitrary values. We must now find values of A and s that satisfy (3.14) for all values of t. We have three choices: $A = 0$, $s = -\infty$, or $s = -1/RC$. The first two are not meaningful in this case because each would result in $v_h = 0$ for all time, which is indeed a solution to the homogeneous equation, but not a very useful one. Consequently, we must choose

$$\left(s + \frac{1}{RC}\right) = 0 \quad \text{or} \quad s = -\frac{1}{RC} \tag{3.15}$$

and with this choice, we note that A can have any finite value. Equation (3.15) is called the *characteristic equation*, and for linear circuit equations the degree of the characteristic equation is equal to the order of the differential equation. The complementary function from (3.13) and (3.15) is

$$v_h = Ae^{-t/RC} \tag{3.16}$$

where A is called an *arbitrary constant* because it can take on any finite value. For linear circuit equations, the number of arbitrary constants in the complementary function will always be equal to the order of the differential equation. The complementary function v_h is called the *natural response* because it is determined by the circuit only, not the forcing function. Remember that v_h is the general solution to the homogeneous equation.

There is no all-inclusive method for finding a particular solution to the original equation, although there are standard methods used for certain classes of forcing functions. The particular solution v_p is *any* function that satisfies the original differ-ential equation. Notice that for this case where the right side of the equation is a constant, v_p = a constant satisfies the equation. Let us substitute $v_p = B$ (a constant) into (3.11). We get

$$0 + \frac{B}{RC} = \frac{V}{RC}$$

and

$$B = V = v_p \tag{3.17}$$

The particular solution v_p is called the *forced response* because it results from the forcing function. In this case, the forced response is equal to the dc voltage of the source.

The *general solution* to (3.11) is $v = v_h + v_p$ or

$$v = Ae^{-t/RC} + V \tag{3.18}$$

Our task of finding v is not yet complete because we have not required our solution to satisfy the *initial conditions*, that is, the value of v at the instant s is closed. Satisfying the initial conditions will require A to be a specific value, thus completing the determination of our solution. The initial condition information is not contained in the differential equation itself. The initial voltage on a capacitor (or initial current in an inductor) is determined from the history of the circuit prior to $t = 0$.

To finish the solution, we will make use of the relationship developed in Section 3.1 that states that the voltage across a capacitor cannot change instantaneously. We shall adopt the notation $v(0^-)$ to mean the voltage in the limit as t approaches zero from the left (from the negative side). Thus $v(0^-)$ can be thought of as the voltage at an infinitesimal time before the switch is closed. Similarly, by $v(0^+)$, we mean the voltage in the limit as t approaches zero from the right. Therefore $v(0^+)$ can be thought of as the voltage at an infinitesimal time after the switch is closed. Since the voltage across a capacitor cannot change instantaneously, for the circuit of Fig. 3.24,

$$v(0^+) = v(0^-) \tag{3.19}$$

It is important to note that while, in this particular situation, (3.19) is true, $i(0^-) \neq i(0^+)$.

Example 3.4

Find the value of $i(0^-)$ and $i(0^+)$ in this situation of Fig. 3.24. The current for $t < 0$ is 0 because the switch has not been closed. So $i(0^-) = 0$. On the other hand, for $t = 0^+$, immediately after the switch is closed, $v = v(0^+) + Ri(0^+)$ or $i(0^+) = (V - v(0^+))/R$. We see that, in this case, the current suddenly jumps from 0 at $t = 0^-$ to $(V - v(0^+))/R = (V - v(0^-))/R$ at $t = 0^+$. Notice again that $v(0^+) = v(0^-)$ and $i(0^+) \neq i(0^-)$.

● ● ●

PROBLEM
3.14 a.   In this same situation of Fig. 3.24, is $v_2(0^+) = v_2(0^-)$?

b.   What are these two values? Find an equation for $v_2(0^+)$ in terms of V and $v(0^-)$ by applying Kirchhoff's voltage law at $t = 0^+$.

Now to return to our problem of finding the value of the arbitrary constant A in (3.18), let $t = 0^+$ in this equation; that is,

$$v(0^+) = A + V \qquad \text{or} \qquad A = v(0^+) - V$$

There is a subtle but important point in the fact that in order to find A, we let $t = 0^+$ rather than $t = 0$ or $t = 0^-$ in (3.18). In some situations (not v in this case), the variable that we seek "jumps" at $t = 0$. So we evaluate the arbitrary constants at $t = 0^+$; then our final solution is valid for $t > 0$, but not necessarily at $t = 0$.

In the situation at hand (Fig. 3.24), $v(0^+) = v(0^-)$, so $A = v(0^-) - V$ and the final solution is

$$v = (v(0^-) - V)e^{-t/RC} + V \qquad (3.20)$$

In the special case where the initial capacitance voltage $v(0^-) = 0$,

$$v = V(1 - e^{-t/RC}) \qquad (3.21)$$

where $v(0^-) = 0$.

Since the procedure that we have followed to get (3.20) is the same one that we would follow with more complicated circuits having higher-order equations, let us stop for a moment and summarize. We applied Kirchhoff's laws to the circuit to obtain a differential equation— (3.11) in this case. First we solved the differential equation by solving the homogeneous equation to obtain the natural response, and second we found the forced response (any solution to the nonhomogeneous equation). The natural response always contains one or more arbitrary constants. We evaluated the arbitrary constant—the A in (3.16) —by requiring the general solution (sum of natural and forced response) to satisfy the initial condition. The initial condition is part of the statement of the problem and can never be found from the differential equation itself. In more complicated circuits, the differential equation will be of higher order, and although the mathematics is more involved, exactly the same procedure would be followed in this method, which is called the classical method. Later on, we introduce Laplace transform methods for solving the equations. The classical method is used first to help you gain a better physical understanding of circuits. The Laplace transform method is more abstract.

Before proceeding to the interpretation of (3.20) and (3.21), we should make some consistency checks, as indicated in the problems below, to give us confidence that (3.20) and (3.21) are valid expressions.

PROBLEMS
3.15 Show that the exponents in (3.20) and (3.21) are dimensionless.

3.16 Show that the right-hand side of (3.20) has the dimension of volts.

3.17 Show by direct substitution of (3.20) and (3.21) into the original equation (3.11) that these expressions do satisfy the equation.

3.18 Show that (3.20) satisfies the initial conditions.

Having already the equation for v, we can readily write the equation for v_2 in the situation of Fig. 3.24.

$$v_2 = V - v = V - (v(0^-) - V)e^{-t/RC} - V$$

or

$$v_2 = (V - v(0^-))e^{-t/RC} \tag{3.22}$$

and also

$$i = \frac{v_2}{R} = \left[\frac{V - v(0^-)}{R}\right]e^{-t/RC} \tag{3.23}$$

Example 3.5
Assume that we had *not* already found the solutions for v, v_2, and i for the circuit of Fig. 3.24, and we desire the solution for i only. We decide to find i directly by solving the appropriate differential equation in i.

First we find the differential equation. Then we apply Kirchhoff's law.

$$V = v_2 + v = Ri + \frac{1}{C}\int_0^t i\, dt + v(0^+) \tag{3.24}$$

To eliminate the integral, we differentiate

$$R\frac{di}{dt} + \frac{i}{C} = 0 \quad \text{or} \quad \frac{di}{dt} + \frac{i}{RC} = 0 \tag{3.25}$$

The complementary function or natural solution is

$$i_h = Ae^{st} \tag{3.26}$$

When (3.26) is substituted in (3.25),

$$A\left(s + \frac{1}{RC}\right)e^{st} = 0 \tag{3.27}$$

This tells us, as before, that $s = -1/RC$ and

$$i_h = Ae^{-t/RC} \tag{3.28}$$

The particular solution is zero; therefore,

$$i = i_h + i_p = Ae^{-t/RC} \tag{3.29}$$

To find A, we find $i(0^+)$:

$$i(0^+) = \frac{V - v(0^+)}{R} = \frac{V - v(0^-)}{R}$$

Therefore

$$A = \frac{V - v(0^-)}{R}$$

and

$$i = \left(\frac{V - v(0^-)}{R}\right)e^{-t/RC} \tag{3.30}$$

The solution (3.30) is identical to (3.23).

• • •

PROBLEMS

3.19 Does (3.30) yield the correct value of the initial current $i(0^+)$?

3.20 Does the solution of (3.30) satisfy the differential equation (3.25)?

3.21 Assume that we have not already found solutions for each of v, v_2, and i. Find the differential equation for v_2 of Fig. 3.24 and proceed to find the solution for v_2. Of course, this should be identical to (3.22).

For specific values of V, R, and C, you could now use (3.21), (3.22), and (3.23) to find numerical values for v, v_2, and i for any specified time, and it is important that you be able to do so. But it is even more important that you be able to make a *qualitative interpretation* of these variables. Let us illustrate how to do this in detail so that you can learn how to do it for yourself. First, look at i, since it is mathematically simpler. Since a plot is a powerful method to get an approximate view, choose $V = 10$ V, $R = 10$ kΩ, $C = 100$ μF, and then plot i where $v(0^-) = 0$. Then without changing V, let $R = 20$ kΩ, $C = 200$ μF, and plot i on the same axes to see how the parameters affect the curve. You can easily punch the numbers out on your calculator; the results are shown in Fig. 3.25. We can see that i has a maximum at $t = 0^+$, and decreases (because of the exponential term) thereafter toward zero. We note from (3.23) and the graph that the initial current $i(0^+)$ is the current that would be present if the capacitor were replaced by a steady voltage $v(0^-) = v(0^+)$. Also, since $i(\infty) = 0$, we note that the final current is the current that would be present if the capacitor were an open circuit. Thus we can say that, in this circuit, C initially acts

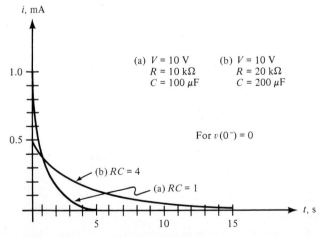

fig. 3.25   The current in Fig. 3.24 for two sets of parameters.

as a battery having voltage $v(0^-)$, and finally it appears to be an open circuit. Of course, at intermediate times, the capacitor voltage varies. This qualitative interpretation is not limited to this circuit, but applies more generally, as we can show from (3.3) and (3.5). From (3.3), we see that when v is not changing, that $i = 0$ because $dv/dt = 0$ and C acts as an open circuit.

PROBLEMS

3.22 a. For this situation of Fig. 3.24 and on one set of coordinates, plot i, v_2, v, and V all against t as the abscissa where $R = 10$ kΩ, $C = 100$ μF, $V = 10$ V, and $v(0^-) = -5$ V. Calculate approximately three important points; then sketch the salient features of each function.

b. Do your plots agree with Kirchhoff's voltage law at any time you wish to check; that is, does $V = v + v_2$?

c. At one midrange point on the plot of i, graphically approximate dv/dt and show that $C\, dv/dt$ is approximately equal to the current at this same value of time.

d. For the value of time t_1 selected in (c), estimate from a graphical view the value of the integral

$$\int_{0+}^{t_1} i\, dt$$

then compute $v(t_1)$ from this. Do not forget $v(0^+)$. Does this result agree reasonably well with your plotted value of $v(t_1)$?

3.23 For the circuit of Fig. 3.24, let $R = 20$ kΩ and $C = 200$ μF. Make the plots of Prob. 3.22 represent the plots of i, v_2, and v for these new (as well as the former) values of R and C by preparing an additional abscissa on the same coordinate paper. In this way, the one family of plots will represent the two different sets of parameters, depending on which abscissa scale is used.

Problems 3.22 and 3.23 and Fig. 3.25 demonstrate the rate that the currents and voltages change. The larger the value of RC, the more slowly these change. This makes physical sense because a larger R reduces the current and thus slows the capacitor voltage change. On the other hand, a larger capacitor can hold more charge, and so the charging current flows for a longer time to achieve a given voltage change.

Time Constant

In each of the equations for i, v, and v_2 for the circuit of Fig. 3.24, the exponential term $e^{-t/RC}$ is present. See (3.21), (3.22), and (3.23). This term decreases in value from one to zero as time increases from zero to infinity. In our particular situation, any one of these three dependent variables (i, v, v_2) may increase or decrease with time, depending on the value of $v(0^-)$. This is illustrated in the solution to Probs. 3.24 and 3.25.

PROBLEM

3.24 For the situation described in Prob. 3.22, select a new value of $v(0^-)$ such that i decreases 20 mA as time changes from zero to ∞. Plot i, v, and v_2 on one set of coordinates. Are these three plots consistent with each other?

In this circuit and many others, the exponential term $e^{-t/\tau}$ describes how the current and voltages increase and decrease with respect to time. The τ is known as the *time constant*. In our particular situation, $\tau = RC$. As we will see presently, τ in other circuit situations will be some other function of the new circuit parameters. Notice that when time t is equal to the time constant, $e^{-t/\tau} = e^{-1} = 0.3679$. So the time constant has the meaning of the time in seconds for the exponential term to decrease from 1 to 0.3679. This means that the current or voltage in question will experience 63.21 percent of its total change in a time equal to one time constant. Since $e^{-3} = 0.04979$, the current decays to about 5 percent of its original value after three time constants have elapsed. The time constant is a very valuable concept because it helps us to quickly approximate a current or voltage as a function of time. See the plot of i in Fig. 3.26. This graph applies for all combinations of $v(0^-)$, V, R, and C; different combinations of these parameters determine $i(0)$ and τ, but not the general characteristic of the curve.

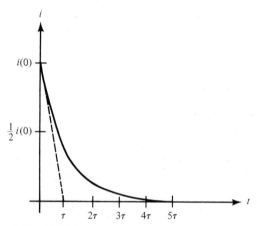

fig. 3.26   Current of Fig. 3.24 in terms of the initial current and the time constant τ. The dotted line is the initial slope of the current.

Also shown in Fig. 3.26 is a broken line that is the slope of $i(t)$ at $t = 0$. Since the broken line crosses zero at $t = \tau$, we can also say that the time constant τ is the time that would be required for i to reach zero if it continued to decrease at its initial rate where $t = 0$.

Now let us make a similar qualitative interpretation of the voltage across the capacitor. A plot of v from (3.20) for the same set of parameters is shown in Fig. 3.27. Here, of course, the capacitance voltage increases as the current decreases. It is important to note that the time constant for the current i is the same as that for the capacitance voltage v. Also note that the voltage v, as $t \to \infty$, approaches the voltage that would exist across C if it were an open circuit. Thus we say that the capacitor charges toward the voltage that would exist across it if it were an open circuit. This concept will be very valuable to you later in analyzing more complicated RC circuits.

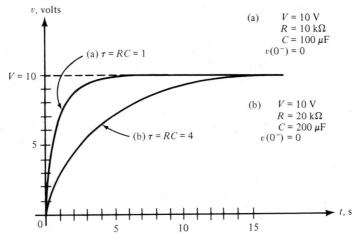

fig. 3.27 The voltage across C in Fig. 3.24 for two sets of parameters.

For the voltage v, the time constant τ is the time required for the voltage to reach approximately 63 percent of its final value. A plot of v in terms of $v(\infty)$ and τ is shown in Fig. 3.28. In a sense, this is a "normalized" curve because it is valid for any combination of V, R, and C. Note also that v approaches the forced response as the natural response dies out.

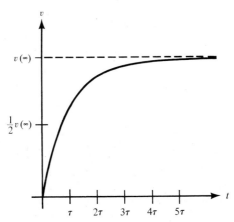

fig. 3.28 Voltage across C in Fig. 3.24 in terms of the final voltage $v(\infty)$ and the time constant ∞.

PROBLEMS

3.25 Draw a broken line in Fig. 3.28 that corresponds to the initial slope of the voltage and give an alternate interpretation for the time constant like the one given for the current based on the broken line in Fig. 3.26.

3.26 Using the circuit of Fig. 3.24 to charge a 10-μF capacitor with a 6-V battery, find the value of R required to charge the capacitor to 95 percent of the battery voltage in 30 ms.

3.27 If $V = 100$ V and $v(0^-) = -20$ V in the circuit of Fig. 3.24, find a value of R and a value of C such that $v(t_1) = 77.69$ V and $i(t_1) = 22.31$ mA, where $t_1 = 2$ ns. If there is more than one combination of R and C that will produce the desired result, discuss the trade-offs involved in choosing the best combination.

3.28 Choose values of R and C in the circuit of Fig. 3.24 that will result in the current being equal to approximately $0.0067i(0)$ at 5 ms, where $v(0^-)$ is $+10$ V. If there is more than one combination of R and C that will work, explain what other factors would be important in choosing the best combination and what constraints might be placed on the system that would require a unique combination of R and C.

Circuit Response by Inspection

We have hassled with one particular RC circuit (Fig. 3.24) at some length. To find a particular current or voltage (a particular dependent variable) we first wrote the differential equation for that variable, and then we solved that differential equation. There are many situations where we do not need to resort to the differential equation to find a response. The following pages treat an approach that can be less time-consuming and at the same time give more direct insight.

It is important to keep in mind the fact that this chapter is constrained to RC and RL situations that are described by first-order differential equations. This particular discussion is also constrained to situations where there is only one constant forcing function—either a current source or a voltage source. See Fig. 3.29 for some examples. In each of these cases a switch is closed or opened at $t = 0$, producing natural responses in currents and voltages.

In these cases where the system is first-order, a response of interest can be written directly without resorting to the techniques of differential equations. One can determine whether a particular RC network is first-order or not without writing the pertinent differential equation. First, keep in mind that the order of the describing equation is fixed exclusively by the network and is therefore independent of the forcing function. Consider the equation

$$\frac{dy}{dt} + ay = b \tag{3.31}$$

which describes any response voltage or current y in a first-order system where a is determined exclusively by the network and is independent of the forcing function, whether voltage or current. As an example, see (3.11), where $a = 1/RC$, $y = i$, and $b = V/RC$. In Fig. 3.29, the fact that some of the networks are first-order and some are second-order is unrelated to whether the forcing function is constant or is some more general function of time.

To get insight into a technique for determining if a particular network is first-order or not, consider the two networks of Fig. 3.30. You will prove in solving Prob. 3.29 that both of these are first-order.

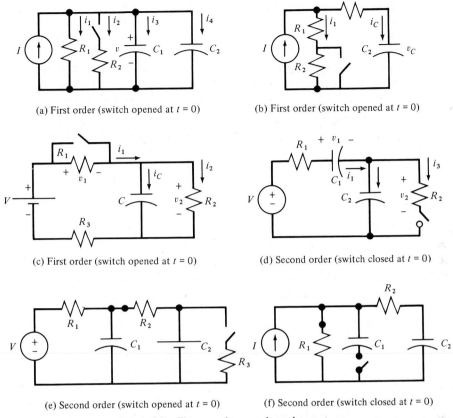

(a) First order (switch opened at $t = 0$)

(b) First order (switch opened at $t = 0$)

(c) First order (switch opened at $t = 0$)

(d) Second order (switch closed at $t = 0$)

(e) Second order (switch opened at $t = 0$)

(f) Second order (switch closed at $t = 0$)

fig. 3.29   First - and second - order systems.

PROBLEM

3.29   The three independent variables of the network of Fig. 3.30a are i_R, i_C, and v.

    a. Find the differential equation for each of these variables in the network of Fig. 3.30a. Arrange the terms of each equation to match the format of (3.31). The three left-hand sides of these three equations, in this format, are identical. The right-hand sides will differ, of course.

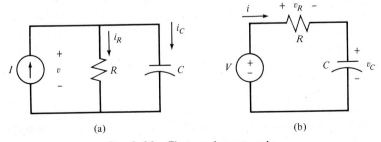

(a)

(b)

fig. 3.30   First - order networks.

b. For each of the three independent variables of Fig. 3.30b, find the differential equation. Place each equation in the format of (3.31).

c. How do the time constants of these two networks compare? Does this seem surprising to you?

The fact that the two networks of Fig. 3.30 are first-order helps us to determine whether more complicated networks similar to those of Fig. 3.29 are first-order or not. If we can reduce a particular network so that we can see that it is equivalent to either of the simple networks of Fig. 3.30, then we can say that the particular network is also first-order. Consider the network of Fig. 3.29a. We can combine R_1 and R_2 into a single equivalent resistance $(R_1 \| R_2)$, and combine C_1 and C_2 to the single capacitance $(C_1 + C_2)$. To make sure that you know how to combine capacitances in parallel and in series, solve Prob. 3.30.

PROBLEMS
3.30 Prove the following:

$$C_s = \frac{C_1 C_2}{C_1 + C_2} \quad \text{and} \quad C_p = C_1 + C_2$$

where C_s and C_p are the series and parallel equivalent capacitances of the series and parallel combinations, respectively, of capacitances C_1 and C_2. Base these proofs on $C = q/v$.

3.31 a. Reduce the network of Fig. 3.29a to the simpler form of Fig. 3.30.

b. From this reduction, show that the time constant for this network is

$$\tau = \frac{R_1 R_2 (C_1 + C_2)}{R_1 + R_2} \tag{3.32}$$

c. Show that this equation is correct dimensionally. In other words, does the right-hand side of this equation have the dimensions of seconds?

In solving Probs. 3.29, 3.30, and 3.31, we have shown that the differential equation for any dependent variable i_1, i_2, i_3, i_4, or v of Fig. 3.29a is first-order.

PROBLEM
3.32 To demonstrate the validity of the technique of Prob. 3.31 for finding the time constant, find the differential equation for i_1. Do this without using the value of the time constant as determined in Prob. 3.31. In other words, begin your proof by applying Kirchhoff's current law. Write $I = i_1 + i_2 + i_3 + i_4$; then proceed toward the differential equation.

At first glance, it may not be obvious that the network of Fig. 3.29b is also a first-order network. In any network, changing the value of the forcing function cannot change the order of the network. So if we let $I \to 0$, then in the limit the current in C_2 is the same as the current in R_1, R_2, and R_3. Think of the situation where $v_C(0^-) \neq 0$ and $I \to 0$. Then it is obvious that the differential equation for any dependent current or voltage is first order because R_1, R_2, R_3, and C_2 are in series for these special conditions. For other situations where $I \neq 0$, the differential equation is still first-order because the order of the equation is independent of the magnitude of the forcing function, which is I in this case. Furthermore, the homogeneous part of the differential equation (corresponding to the left-hand side of (3.31)) is independent of the magnitude of the forcing function. Of course, on the other hand, the right side of the differential equation is very much a function of the magnitude of the forcing function.

A different point of view that might be more convincing is as follows: Replace the combination of the current source I and the shunt path $R_1 + R_2$ with a voltage source as shown in Fig. 3.31. These two networks yield identically the same current in the R_3C_2 branch of the two different networks. However, as a caution, remember that the current i_1 of Fig. 3.29b is lost in the Thévenin conversion. The two networks are equivalent to the right of the a-b terminals of Fig. 3.31. If the network is first-order in one dependent variable, it must be first-order in all other dependent variables. So we see that the network of Fig. 3.29b is of first-order because the equivalent network of Fig. 3.31 is obviously of first-order.

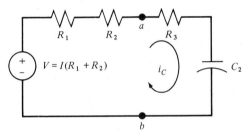

fig. 3.31   Thévenin equivalent of Fig. 3.29b.

PROBLEM

3.33   Without writing the differential equation itself, show that the network of Fig. 3.29c is first-order. You may wish to use the Norton equivalent of the (V, R_1, R_3) combination.

When we examine the separate networks of Figs. 3.29d, 3.29e, and 3.29f, we find no way to reduce each to either of the first-order networks of Fig. 3.30. If you should take the time to write the differential equation for any dependent variable of each of these networks, you will discover that each network is of second order.

Probably you have noticed that in the process of showing that a particular network is first-order, we have simultaneously found the time constant of that network. Remember that the time constant of each of the networks of Fig. 3.30 is given by $\tau = RC$. From this we found the time constant of the network of Fig. 3.29a to be as shown by (3.32). For Fig. 3.29b, $\tau = (R_1 + R_2 + R_3)C$.

PROBLEM

3.34   Show that the time constant of the network of Fig. 3.29c is given by

$$\tau = (R_2\|(R_2 + R_3))C$$

Knowing that the network is first-order and knowing the time constant helps us considerably in writing the response by inspection. Let us consider the next step.

The solution of a first-order differential equation can be written as

$$y = Ae^{-t/\tau} + f \tag{3.33}$$

where y is any dependent current or voltage, τ is the time constant, and f is the forced part of the dependent current or voltage. Even though the forcing function (voltage or current) is a constant, in rather rare situations the forced part of the response can be a function of time. In any event, it will be rather obvious if f is other than a constant.

To find A in (3.33), let $t = 0^+$. It follows that

$$y(0^+) = A + f(0^+) \quad \text{or} \quad A = y(0^+) - f(0^+) \tag{3.34}$$

The technique of "response by inspection" pays off best where f is a constant. Here $f(0^+) = f(\infty)$. And we see from (3.33) that $f(\infty) = y(\infty)$. So where the forced response is constant, we can write the response as

$$y = [y(0^+) - y(\infty)]e^{-t/\tau} + y(\infty) \tag{3.35}$$

Let us review the restrictions on (3.35). This situation applies where the network is first-order and has a time constant of τ, and where the forced response is a constant. This allows us to write the response by inspection if we can find $y(0^+)$, $y(\infty)$, and τ by inspection.

Example 3.6
The circuit of Fig. 3.32 has been at rest and then the switch is closed at $t = 0$.

    a. Find i_2 by inspection.

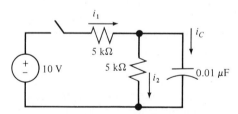

fig. 3.32   Switch closed at $t = 0$.

Solution
Because $v_C(0^+) = 0$, $i_2(0^+) = 0$. Because $i_C(\infty) = 0$, $i_2(\infty) = 10/10{,}000 = 0.001$ A $= 1$ mA. $\tau = 2.5 \times 10^3 \times 10^{-8} = 2.5 \times 10^{-5}$ s. Then

$$i_2 = -e^{-t/2.5 \times 10^{-5}} + 1 \text{ mA}$$

    b. Find i_1 by inspection.

Solution

$$i_1(0^+) = 2 \text{ mA}, \quad i_1(\infty) = 1$$

Then

$$i_1 = e^{-t/2.5 \times 10^{-5}} + 1 \text{ mA}$$

c. Find i_c by inspection.

Solution

$$i_C(0^+) = 2 \text{ mA}, \; i_C(\infty) = 0.$$

Then

$$i_C = 2e^{-t/2.5 \times 10^{-5}} \text{ mA}$$

d. Find v_C by inspection.

Solution

$$v_C(0^+) = 0, \; v_C(\infty) = 5, \; v_C = -5e^{-t/2.5 \times 10^{-5}} + 5 \text{ V}$$

An approximate plot of the responses $i_1, i_2, i_C,$ and v_C are shown in Fig. 3.33.

e. Show a consistency check.

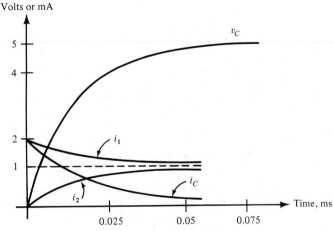

fig. 3.33   *Response in Fig. 3.31.*

Solution 1

$$i_1 = i_2 + i_C$$

$$\underbrace{e^{-t/2.5 \times 10^{-5}} + 1}_{i_1} = \underbrace{-e^{-t/2.5 \times 10^{-5}} + 1}_{i_2} + \underbrace{2e^{-t/2.5 \times 10^{-5}}}_{i_C} \qquad \text{Check}$$

Solution 2

$$v_C = \frac{1}{C} \int_0^t i_C \, dt + v_C(0^+) = \frac{10^{-3}}{10^{-8}} \int_0^t 2e^{-t/2.5 \times 10^{-5}} \, dt + 0$$

$$= -\frac{2 \times 2.5 \times 10^{-8}}{10^{-8}} [e^{-t/2.5 \times 10^{-5}} - 1] = 5[-e^{-t/2.5 \times 10^{-5}} + 1] \qquad \text{Check}$$

Solution 3

As shown in the plots:
- a. v_C should increase from 0 to 5 V.
- b. i_1 should drop from 2 to 1 mA.
- c. i_C should drop from 2 to 0 mA.
- d. i_2 should increase from 0 to 1 mA.

● ● ●

PROBLEMS

3.35   In the circuit of Fig. 3.29a, the switch has been open until the circuit comes to rest. Then the switch is closed at $t = 0$. $R_1 = 1,000$ ohms, $R_2 = 500$ ohms, $I = 0.2$ A, $C_1 = 0.1 \, \mu F$, and $C_2 = 0.2 \, \mu F$.
- a.   For $t > 0$, find i_1, i_2, i_3, and i_4 by inspection.
- b.   Are your solutions consistent with $i_1 + i_2 + i_3 + i_4 = 0.2$?
- c.   Sketch all of these currents on one set of coordinates for $-0.1\tau < t < 2\tau$, where τ is the time constant.

3.36   In the situation of Prob. 3.35, the switch is closed at $t = 0$, but then it is open again at $t = 20 \, \mu s$. By inspection, find the equations for i_1, i_2, i_3, and i_4 after the switch has been opened at $t = 20 \, \mu s$. You may find it convenient to measure time for this problem with the new $t = 0$ at the instant the switch is opened after having been closed for the 20 μs.

Square-Wave Response

Let us apply the background we acquired earlier to the problem where the forcing function of Fig. 3.23 had both dc and ac components. This forcing function v_1 is applied to the circuit of Fig. 3.21a, reproduced here for convenience. Remember that

$$v_1 = V_{dc} + v_{1_{ac}}$$

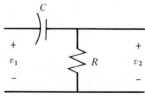

fig. 3.21a   RC network.

This circuit is used commonly in electronics to simultaneously block out the dc component and also make $v_2 \approx v_{1_{ac}}$. The dc component V_{dc} has an important impact on v_2 if we were interested in the performance of the circuit immediately after the voltage v_1 was first applied. But we have an interest in the performance only after certain response terms have essentially decayed to zero. Let us use the principle of superposition to simplify our approach to the problem. The dc component of v_1 will produce an exponentially decaying term in v_2. This component will have no practical significance after perhaps five time constants. So we will ignore the effect of V_{dc} upon v_2, but we will remember that our solution is not valid immediately after v_1 is first applied. So under these conditions, v_1 is a square wave, as shown in Fig. 3.34. Our problem then is to select R and C such that, for practical purposes, v_2 is also the same square wave.

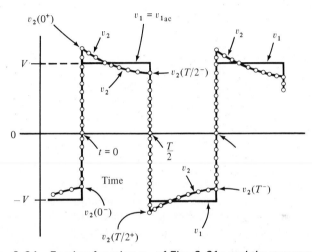

fig. 3.34   Forcing function v_1 of Fig. 3.21a and the response v_2.

To begin the analysis, we arbitrarily select $t = 0$ as shown in Fig. 3.34. We desire to find v_2 for $t > 0$, but unfortunately we do not immediately recognize the value of $v_2(0^+)$. Its value is dependent upon the previous history of the circuit. We can write, however,

$$v_2 = [v_2(0^+) - v_2(\infty)]e^{-t/\tau} + v_2(\infty) \qquad \text{for } 0 < t < T/2 \qquad (3.36)$$

where $v_2(\infty)$ has the special meaning of the value of v_2 at $t = \infty$ if v_1 remained at V for all $t > 0$. We can see that $v_2(\infty) = 0$. But, of course, (3.36) is valid only for $0 < t < T/2$. So it follows that

$$v_2 = v_2(0^+)e^{-t/\tau} \qquad (3.37)$$

where $\tau = RC$. The value of $v_2(0^+)$ is still unknown, but let us proceed to find v_2 for $T/2 < t < T$. From (3.37),

$$v_2(T/2^-) = v_2(0^+)e^{-T/2\tau} \qquad (3.38)$$

After working Prob. 3.37, you will see that

$$v_2(0^+) = v_2(0^-) + 2V \quad \text{and} \quad v_2(T/2^+) = v_2(T/2^-) - 2V \tag{3.39}$$

PROBLEM

3.37 Prove the relationship of (3.39). Keep in mind that at $t = 0$, $T/2$, T, $3T/2$, and so forth, v_1 "jumps," whereas v_C cannot "jump."

This is a steady-state condition, the circuit is bilateral, and the applied voltage is such that v_1 for $0 < t < T/2$ is equal to $-v_1$ for $T/2 < t < T$. Then it follows that $v_2(t) = -v_2(T/2 + t)$ and, in particular,

$$v_2(0^+) = -v_2(T/2^+) \tag{3.40}$$

But from (3.39) and (3.38),

$$v_2(T/2^+) = v_2(T/2^-) - 2V = v_2(0^+)e^{-T/2\tau} - 2V \tag{3.41}$$

Eliminating $v_2(T/2^+)$ from (3.40) and (3.41),

$$v_2(0^+) = -v_2(0^+)e^{-T/2\tau} + 2V \quad \text{or} \quad v_2(0^+) = \frac{2V}{1 + e^{-T/2\tau}} \tag{3.42}$$

So it follows that

$$v_2 = \left(\frac{2V}{1 + e^{-T/2\tau}}\right)e^{-t/\tau} \quad \text{for } 0 < t < T/2 \tag{3.43}$$

Equation (3.43) describes the desired response v_2 of Fig. 3.34.

We see that if τ is made sufficiently large compared to T, then from this equation we see that $v_2 \rightarrow V$ all through the interval $0 < t < T/2$. On the other hand, if τ is sufficiently small compared to T, then the $v_{2_{max}} \rightarrow 2V$ and $v_{2_{min}} \rightarrow 0$.

PROBLEMS

3.38 a. Now it is your turn. We started out to design this RC network so that v_2 was a close approximation to v_1. Finish the problem under the following constraints. The frequency of v_1 is 1.0 MHz. From considerations not treated here, R must have a value of 3,000 Ω. Select a value of C such that the maximum departure of v_2 from v_1 is 1 percent of V.

   b. For these conditions and from $0^- < t < T^+$, plot on one set of coordinates v_1, v_2, and the voltage across the capacitance C. Check the consistency of your plot against the following facts:

   1. v_c cannot "jump."
   2. Whenever v_1 "jumps," v_2 must "jump" exactly the same amount.
   3. At all times, $v_1 = v_2 + v_C$.
   4. The average value of each of v_1, v_2, and v_C over one period is equal to zero.
   5. The maximum and minimum values of v_C are 0.01 V and -0.01 V, respectively.

3.39 a. We wish to use the same circuit configuration of Fig. 3.21a to generate a highly peaked wave much like v_2'' of Fig. 3.21b. The square wave and value of R of Prob. 3.38 are to be used. Find a new value of C such that $v_{2_{max}} = 1.95 \times V$.

   b. For these conditions of Prob. 3.38, plot on one set of coordinates v_1, v_2, and v_C.

   c. For each of the five "facts" stated in Prob. 3.38b, write the same fact or similar fact that applies to the conditions of Prob. 3.39.

3.4 ENERGY STORAGE IN CAPACITANCE

A capacitance stores charge, the charge produces an electric field, and energy is stored in the electric field. You may have seen a dramatic manifestation of the discharge of the energy stored in a capacitance when someone short-circuited the terminals of a large capacitance that was charged by a high voltage, producing a noise like a rifle shot. Some very unfortunate manifestations of the energy stored in capacitance have occurred in the electrocution of maintenance personnel who have unintentionally touched the terminals of a charged capacitance in a high-voltage power supply or other device, such as a Q-switched laser.

An expression for the total energy stored in a capacitance can be obtained by integrating the power delivered to a capacitance. From Chapter 1, the fundamental definition of power is $p = vi$. The energy stored in a capacitance is the integral over time of the power delivered to the capacitance. That is

$$w = \int_0^t vi \, dt + w(0)$$

where w is the total energy stored in the capacitance up to a time t, and v and i are as shown in Fig. 3.5. By making use of (3.3), we can write

$$w = C \int_0^t v \frac{dv}{dt} \, dt + w(0)$$

and changing the variable of integration from t to v, we get

$$W = C \int_0^v v \, dv \tag{3.44}$$

assuming that the voltage across C was zero at $t = 0$, and is v at time t. Equation (3.44) can be easily integrated; the result is

$$W = \frac{1}{2} Cv^2 \tag{3.45}$$

which is the expression for energy stored in a capacitance charged by a voltage v.

PROBLEMS

3.40 Show that (3.45) has proper dimensions. You may want to refer back to Chapter 2.

3.41 State whether, according to (3.45), W increases or decreases as the capacitance is increased; whether W increases or decreases as the voltage is increased. Explain why you would expect W to increase or decrease, whichever you stated, with C and with v.

3.42 Find the energy stored in C in Fig. 3.35, as $t \to \infty$, in terms of significant variables such as R_1, R_2, C, and I. Note the equation for i.

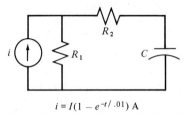

$$i = I(1 - e^{-t/.01}) \text{ A}$$

fig. 3.35 A current source charging a capacitor through a resistor.

It is important for you to understand the difference between energy *stored* and energy *dissipated*. The previous chapters include discussions of energy dissipation, but not energy storage, since this is the first chapter in which an element capable of storing energy is introduced. Energy is stored in capacitance but not dissipated in capacitance. Any dissipation of energy by a capacitor (not capacitance) is accounted for by the parallel resistance in the model of a capacitor (see Fig. 3.4). Energy is dissipated in resistance but not stored in resistance. By dissipation of energy, we will usually mean conversion of electric energy to heat in an irreversible way.

As an example, let us calculate the energy dissipated in the resistance in the circuit of Fig. 3.36 as the capacitance discharges after the switch is closed at $t = 0$. Let the initial capacitance voltage be V_0.

$$W_R = \int_0^\infty v_R i_R \, dt = \int_0^\infty \frac{v_R^2}{R} \, dt \tag{3.46}$$

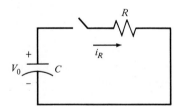

fig. 3.36 C discharges through R.

By inspection,

$$v_R = [v_R(0^+) - v_R(\infty)]e^{-t/RC} + v_R(\infty)$$

or

$$v_R = V_0 e^{-t/RC} \tag{3.47}$$

Eliminating v_R between (3.46) and (3.47), it follows that

$$W = \int_0^\infty \frac{V_0^2}{R} e^{-2t/RC} \, dt = \tfrac{1}{2}CV_0^2$$

This result shows us that the energy dissipated in the resistor as the capacitance discharges is equal to the energy that was initially stored in the capacitance. This, of course, is consistent with the principle of conservation of energy.

PROBLEM

3.43 Calculate the total energy dissipated in the circuit of Fig. 3.24, assuming that the switch remains closed forever.

Example 3.7

Before the switch of Fig. 3.37 is closed, capacitances C_1 and C_2 are charged to 10 and 20 V, respectively. After the switch is closed, the circuit at $t = \infty$ reaches a new steady-state condition.

    a.   Find the total stored energy prior to the time when the switch is closed.
    b.   Find the total stored energy at $t = \infty$.
    c.   Find the equation of the current after the switch is closed.
    d.   Find the energy dissipated in R in terms of the current.
    e.   For consistency, show that the initial stored energy is the sum of the final stored energy and the energy dissipated by the resistance R.

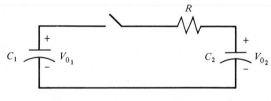

$$C_1 = 1\ \mu F; C_2 = 2\ \mu F$$

fig. 3.37 C_1 and C_2 initially charged to V_{0_1} and V_{0_2}, respectively.

Solution

    a.   Initial stored energy $= \tfrac{1}{2}(C_1 V_{0_1}^2 + C_2 V_{0_2}^2)$

$$W_i = \frac{10^{-6}}{2} (1 \times 10^2 + 2 \times 20^2) = 450 \times 10^{-6} \text{ joules (J)}$$

    b.   Total final charge (both capacitances) must be equal to the total initial charge. Furthermore, in the final state, the voltage across one capacitance comes to the voltage of the other.

$$C_1 V_{0_1} + C_2 V_{0_2} = V_f(C_1 + C_2) \qquad\qquad (3.48)$$

                 ↑   ↑
            Total   Total
        initial charge   final charge

where V_f = final voltage on both capacitances. From (3.48),

$$V_f = \frac{C_1 V_{0_1} + C_2 V_{0_2}}{C_1 + C_2} = \frac{10^{-6}(1 \times 10 + 2 \times 20)}{10^{-6}(1 + 2)} = 16.67 \text{ V}$$

c. From inspection,

$$i = [i(0^+) - i(\infty)]e^{-t/\tau} = \frac{V_{0_2} - V_{0_1}}{R} - e^{-t/\tau}$$

where $\tau = \dfrac{RC_1 C_2}{C_1 + C_2}$

d. Energy dissipated:

$$W_d = \int_0^\infty i^2 R \, dt = \frac{(V_{0_2} - V_{0_1})^2 R}{R^2} \int_0^\infty e^{-2t/\tau} \, dt$$

$$W_d = \frac{(V_{0_2} - V_{0_1})^2 R C_1 C_2}{2R(C_1 + C_2)} = \frac{(20 - 10)^2 \times 1 \times 2 \times 10^{-6}}{2(1 + 2)}$$

$$= 33.33 \times 10^{-6} \text{ J}$$

$$W_f = \tfrac{1}{2} V_f^2 (C_1 + C_2) = \frac{16.67^2 \times 3 \times 10^{-6}}{2} = 416.8 \times 10^{-6} \text{ J}$$

$$W_f + W_d = (416.8 + 33.3) \times 10^{-6} = 450.1 \times 10^{-6} \text{ J}$$

which checks against $W_i = 450 \times 10^{-6}$ J.

● ● ●

PROBLEM

3.44 In Example 3.7 the two capacitors were initially charged with the polarities shown in Fig. 3.37. Find the solutions called for in Example 3.7, but where the polarity of the initial voltage on C_2 is reversed from that shown in Fig. 3.37.

3.5   INTEGRATOR CIRCUIT

For many different voltages v, the circuit of Fig. 3.38a can be adjusted such that v_C is very nearly, but not exactly, equal to the integral of v. The integral of a voltage is useful in certain automatic control, instrumentation, and other situations.

For the integrator circuit of Fig. 3.38a, assume that R and C are selected such that the effect of C upon i is negligible. In other words, v_C is so small compared to v that, for practical purposes, i is determined by v and R. Yet at the same time, it is practical, in many situations, to select R and C such that v_C is large enough to be useful.

Under the assumptions stated above, it follows that

$$v_C = \frac{1}{RC} \int_0^t v \, dt \tag{3.49}$$

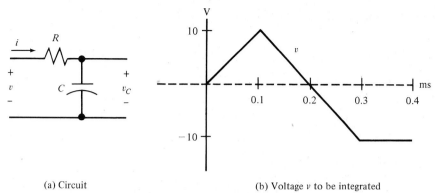

(a) Circuit (b) Voltage v to be integrated

fig. 3.38 An integrator circuit.

because $i = v/R$, where $v_C = 0$. Equation (3.49) is a useful approximation where v_C is small compared to v_R but still large enough to be useful.

$$\int_0^t v \, dt = v_C RC \tag{3.50}$$

From (3.50), we see that the simple RC series circuit of Fig. 3.37 can be designed so that it can be used as a practical integrator having an error that in some circumstances is negligible.

PROBLEM

3.45 We wish to integrate the voltage v of Fig. 3.38b using our RC integrator circuit. Assume that, for reasons not given here, R is selected to be 100 kΩ. Make the additional assumption that our circuit will have satisfactory performance if v_C at $t = 0.1$ ms is 1 percent of the maximum value of v. Now determine the necessary value of C. *Suggestions :* Evaluate the integral of (3.50) graphically. Then assume that (3.50) is sufficiently accurate for the purposes of finding the desired value of C. Solve Prob. 3.45 for a precise determination of the error produced by the integrator circuit that we have selected.

The performance of the integrator circuit of Fig. 3.38a may be remarkably improved if an *op amp* is included in the circuit, as illustrated in Fig. 3.39. Since the amplifier input voltage v_a is so very small due to the large voltage amplification A of the op amp, as discussed in Chapter 2, essentially all of the input voltage v_i appears

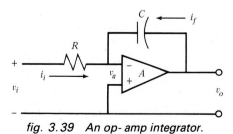

fig. 3.39 An op-amp integrator.

across R. Therefore, the input current is essentially

$$i_i = \frac{v_i}{R} \tag{3.51}$$

Similarly, the output voltage v_o appears across C, so

$$i_f = C \frac{dv_o}{dt} \tag{3.52}$$

However, $i_f = -i_i$, essentially, since the op-amp input current is extremely small. Therefore, equating (3.52) to (3.51)

$$C \frac{dv_o}{dt} = -\frac{v_i}{R} \tag{3.53}$$

We may now obtain v_o by cross multiplying and then integrating both sides of (3.53) to obtain

$$v_o = -\frac{1}{RC} \int_0^t v_i \, dt + V_C(0) \tag{3.54}$$

The factor $1/RC$ is known as the *gain* of the integrator. Observe that the op-amp integrator also inverts, or changes the sign. If this is an undesirable feature, an additional op amp may be used in the inverting mode to reverse the sign and provide additional voltage gain if desired.

PROBLEM
3.46   Use the op-amp circuit of Fig. 3.39 to integrate the waveform of Fig. 3.38b. Use $R = 10^4 \, \Omega$ and $C = 10 \, \mu F$. Sketch the output voltage and estimate the maximum error if the op amp has voltage gain $A = 10^5$ and input resistance $R_{i_a} = 10 \, M\Omega$.

Differentiator Circuit

The R and C circuit elements can be arranged to give a response voltage that is approximately proportional to the derivative of an input voltage. For reasons that are not treated here, the differentiator circuit is not as useful as the integrator circuit previously discussed.

PROBLEMS
3.47   Show how a series RC circuit can produce a response voltage that is approximately proportional to the derivative of some driving voltage. Discuss the design constraint and design equation.

3.48   Show how the R and C elements of the op-amp integrator can be exchanged to produce a differentiating circuit. Does the op-amp circuit have the same limitations as the simple RC circuit? Explain.

3.6   SUMMARY OF BASIC RC BEHAVIOR

In this chapter we have developed the relations for first-order basic RC circuits, and we have discussed the interpretation of these results. In complex analyses it will be important for you to have a good understanding of the fundamental relations of this chapter as a basis for building further understanding. It is very important for you to extract all the understanding that you can of the qualitative and approximate behavior of the basic *RC* circuits treated in this chapter, as well as a good under-standing of the application of the principles and the analysis in a more mathemat-ical sense. The understanding of the approximate behavior can be very valuable in guiding you in the analysis and understanding of more complex circuits. Conse-quently, we have summarized some important points below.

1. Capacitance is the ability to store charge, and capacitance can be thought of as the charge stored per unit volt.
2. $i = C \, dv/dt$ is the fundamental voltage-current relation for capacitance. This relation can be thought of as the current produced by a given rate of change of voltage across the capacitance.
3. From the relation in (2) above, it can be seen that the voltage across C cannot change instantaneously because, in the real world, currents must be finite. The current in a capacitance, however, *can* "jump" or change instantaneously.
4. The relationship

$$v_C = \frac{1}{C} \int_0^t i \, dt + v_C(0)$$

   tells us that direct current (i in one direction only) cannot flow for an infinite time. Otherwise, v_C would approach infinity. In more common language, we say a capacitance cannot pass a steady direct current for an indefinite period of time.
5. In any first-order *RC* circuit where the forcing function is steady, any response quantity of y (voltage or current) is given by $y = [y(0^+) - y(\infty)]e^{-t/\tau} + y(\infty)$, where $y(0^+)$ and $y(\infty)$ are the initial and final values of the response, respec-tively.
6. Since v cannot change instantaneously, and uncharged capacitance C initially looks like a short circuit, and since C will not pass a direct current, C finally looks like an open circuit for the transient or natural response. The time con-stant in going from one condition to the other for a simple series *RC* circuit is $\tau = RC$. "Initially" and "finally" refer to the beginning and end of a transient response. If a capacitor has an initial charge, then the capacitor can be represented by a voltage source in series with C, the voltage source being equal to the initial voltage across the capacitor.
7. A capacitor is modeled by a capacitance in parallel with resistance. Capacitance alone dissipates no energy and stores energy equal to $\frac{1}{2}Cv^2$.

3.7 ANALYSIS OF RL CIRCUITS

As the first step in our development of *RL* circuit relations, we shall consider the circuit shown in Fig. 3.40. We can analyze this circuit by applying Kirchhoff's laws and (3.7). The procedure is straightforward, and with the experience that you have

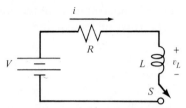

fig. 3.40   Simplest RL circuit with a source. The switch S is closed at $t = 0$.

gained to this point in your study of this text, you should be able to write the following differential equation:

$$V = Ri + L\frac{di}{dt}$$

from Kirchhoff's voltage law. Rearranging this equation in the standard form gives us

$$\frac{di}{dt} + \left(\frac{R}{L}\right)i = \frac{V}{L} \tag{3.55}$$

This is a first-order, linear, constant-coefficient, nonhomogeneous differential equation with a time-invariant forcing function, which is similar in form to (3.12), the differential equation for the voltage across a capacitance in the RC circuit described in Section 3.3. Later we will write the response i and also v_L for this RL circuit by inspection. First, however, let us review the classical differential equation approach by solving (3.55). As explained in Section 3.3, the general solution to (3.55) consists of the sum of the complementary function and a particular solution. Since the explanation of the procedure for solving this kind of differential equation given in Section 3.3 applies here, we shall not repeat that explanation but merely follow through the steps to obtain the solution. If you have not studied Section 3.3, you should do so before proceeding on in this section.

The complementary function to (3.55) is

$$i_h = Ae^{-(R/L)t}$$

and a particular solution is

$$i_p = \frac{V}{R}$$

Remember that the complementary function is the natural response of the system and the particular solution is the forced response, the forcing function in this case being the voltage V. The general solution is

$$i = Ae^{-(R/L)t} + \frac{V}{R} \tag{3.56}$$

with the arbitrary constant A to be found from the initial conditions.

Since S is open for $t < 0$,

$$i(0^-) = 0$$

and since we know from Section 3.2 that the current through an inductance cannot change instantaneously, we know that the initial condition is

$$i(0^+) = 0 \tag{3.57}$$

Requiring (3.56) to satisfy (3.57) gives us

$$A + \frac{V}{R} = 0$$

Solving for A and substituting into (3.56) gives us the final solution

$$i = \frac{V}{R}(1 - e^{-(R/L)t}) \tag{3.58}$$

The voltage across L, as found by using (3.8), is

$$v_L = Ve^{-(R/L)t} \tag{3.59}$$

PROBLEM

3.49 Make consistency checks on (3.58) and (3.59).
  a. Show that the dimensions are proper.
  b. Show that (3.58) satisfies (3.55) by direct substitution.
  c. Show by substituting $t = 0$ into (3.58) that the initial condition $i(0^+) = 0$ is satisfied.
  d. $L = 0$ means that the inductor becomes a short circuit. Show that $\lim_{L \to 0} i$ and $\lim_{L \to 0} v_L$ from (3.58) and (3.59), respectively, give the values that you would get from the circuit in Fig. 3.40 if the inductor were replaced by a short circuit.

Now that we have obtained the mathematical expressions, let us look at the physical interpretation. As we have pointed out previously (particularly in Section 3.4), plots are a good way to get an approximate feel for characteristic behavior. Figures 3.41 and 3.42 show plots for two sets of parameters, as indicated.

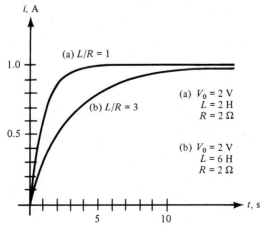

fig. 3.41 The current in Fig. 3.40 for two sets of parameters.

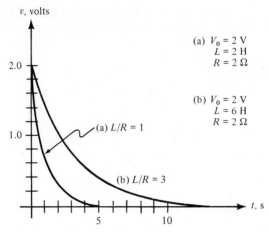

fig. 3.42   The voltage across L in Fig. 3.40 for two sets of parameters.

PROBLEM

3.50   By inspection, write responses i and v_L for the RL circuit of Fig. 3.40. These, of course, should agree respectively with (3.58) and (3.59).

The time constant in an RL, first-order circuit like that of the RC circuit is the time required for the exponent of the e^{-at} to become (-1). Or τ, the time constant, is $= 1/a$ when the decaying term is written as e^{-at}. We see that the time constant of the network of Fig. 3.40 is L/R. Like the RC first-order circuit, the time constant of the RL first-order circuit usually can be written by inspection.

PROBLEM

3.51   After the switch is closed, is this circuit of Fig. 3.43 first order? If so, what is the numerical value of the time constant?

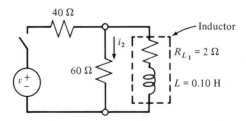

fig. 3.43   Is this circuit first - order with switch closed?

3.52   Is this circuit of Fig. 3.44 first order? If so, what is the numerical value of the time constant?

3.53   In the circuit of Fig. 3.43, the switch has been opened such that all currents are zero, then it is closed at $t = 0$.
    a.   Find the equation where $t > 0$ for i_2.
    b.   Plot i_2 from $t = 0^-$ to $t = 3\tau$; $v = 100$ V.

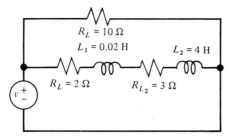

fig. 3.44   Is this circuit first - order?

Let us consider an example where the initial current in the inductance is non-zero.

Example 3.8

Consider the circuit of Fig. 3.45. An inductor having resistance of 10 Ω and inductance of 1 H is connected as shown. The current reaches its steady value with the

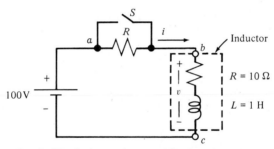

fig. 3.45   Inductor b - c and R ohms in series.

switch closed, and then the switch is opened at $t = 0$. For $t > 0$, find the equation of v_{bc}, the voltage across the inductor where $R = 30 \, \Omega$.

$$\tau = \frac{1}{40} = 0.025 \text{ s}$$

$$i(0^+) = i(0^-) = \frac{100}{10} = 10 \text{ A} \qquad \text{(current cannot “jump” in } L)$$

$$v(0^+) = v_{bc}(0^+) = 100 - (30 \times 10) = -200 \text{ V}$$

$$v(\infty) = \frac{10}{40} \times 100 = 25 \text{ V} \qquad \text{(voltage divider)}$$

$$v = -225e^{-t/0.025} + 25$$

Figure 3.46 shows a plot of the inductor voltage after the switch has been opened. One could be surprised in finding that this voltage starts at -200 V and then passes through zero. At the time the switch is opened, the inductor current remains at its

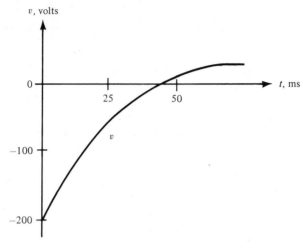

fig. 3.46   Inductor voltage v.

former value of 10 A, producing a 300-V drop across the 30-Ω resistance. So at this instant, the inductor voltage of 200 V adds to the source voltage of 100 V to yield the necessary 300 V across the 30-Ω resistance. Furthermore, the voltage generated in the inductance (exclusive of the resistance in the inductor) is 300 V at this instant. The current is decreasing at the rate of 300 A/s in 1 H of inductance at this instant to generate the necessary 300 V.

● ● ●

PROBLEMS

3.54   By inspection, solve for v_R in Fig. 3.45 (across $R = 30\ \Omega$). For a consistency check, plot v and v_R on the same coordinate system. Show that the sum of these two is the source voltage of 100 V.

3.55   In the circuit of Fig. 3.45, the switch had been closed sufficiently long for the current to reach a steady value. Then the switch is opened at $t = 0$, then closed again at $t = 20$ ms, and finally opened again at $t = 30$ ms. At 40 ms, find the value of v and i. Plot the history of both v and i from 0^- to 40^+ ms.

3.8   ENERGY STORAGE IN INDUCTANCE

Energy is stored in the magnetic field produced by the current flowing in an inductance, in contrast to capacitance, in which energy is stored in the electric field produced by the charge on the plates of the capacitor. The circuit of Fig. 3.45 is a reasonable model of a voltage source, a switch, and an inductor in series. Here R represents the resistance through the air path as the switch is opened. In many situations, we think of a switch having infinite resistance when it is open and zero resistance when closed. But the current in an inductor cannot, of course, be interrupted instantaneously. So as the contacts open in this situation, the current must continue to flow in the air path as an arc. The arc will have significant resistance, and this will be very much a function of the current. However, let us think of this resistance being constant leading to a simple but crude analysis.

PROBLEM

3.56   In the circuit of Fig. 3.45, the switch resistance R is 500 Ω. Show that the voltage across the contacts of the real switch is 5,000 V at the instant the switch contacts separate. Assume that the circuit reached a steady condition prior to the time the switch is opened.

The arc at a switch that interrupts an inductive current is a manifestation of the stored energy in the inductor. The inductive switching voltage may be very high as suggested by Prob. 3.56. In reducing the inductive current to zero, the stored energy must leave the inductor. In the situation of Fig. 3.45, the stored energy leaving the inductor is dissipated in the combination of the switch resistance and the inductor resistance.

The energy stored in an inductance is the time integral of the power $p = vi$ delivered to the inductance. Thus

$$w = \int_0^t vi \, dt + W(0)$$

where v and i are, respectively, the voltage across the inductance and the current through the inductance. Using (3.8), we get

$$w = L \int_0^t \frac{di}{dt} i \, dt + W(0)$$

which becomes, with a change of variable,

$$W = L \int_{I_1}^I i \, di + W(I_1)$$

Where $I_1 = 0$,

$$W = L \int_0^I i \, di = \frac{1}{2} LI^2 \tag{3.60}$$

which is the classic expression for the energy stored in an ideal inductor with current I through it.

PROBLEMS

3.57   Show that (3.60) has proper dimensions.

3.58   State whether, according to (3.60), W increases or decreases as the inductance is increased; whether W increases or decreases as the current is increased. Is this behavior consistent with what you would expect from your physical understanding? Explain.

3.59   Find the energy stored in L in Fig. 3.47 as $t \to \infty$ in terms of R_1, R_2, and L.

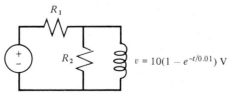

fig. 3.47   A voltage source producing current through an inductor (see Prob. 3.56).

3.60   In this circuit of Fig. 3.48, the switch has been closed long enough for the circuit to reach the steady state. We do not need a parallel resistance in the model of a switch in this situation because opening the switch does not immediately interrupt the inductor current. The switch is opened at $t = 0$.

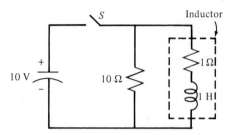

fig. 3.48   *Switch opened at $t = 0$.*

  a. Calculate the stored energy in the inductor at $t = 0$.
  b. Calculate the energy dissipated in the 10-Ω resistance and also in the resistance of the inductor on the time interval $0 < t < \infty$. Show that the stored energy is equal to the sum of these two dissipated energies.

3.61   For the conditions of Prob. 3.60, find the equation for the voltage across the switch for $t > 0$.

3.9   SUMMARY OF BASIC RL BEHAVIOR

In the previous chapter and this one, we have developed and discussed the basic RC and RL relations. We have tried to help you gain good qualitative understanding and physical insight as well as the ability to apply the mathematical relations. It is very important that you learn to use your qualitative understanding and physical insight as a guide in mathematical analysis, in measurement and diagnostics, and interpreting results. Some of the important points that were developed in this chapter are summarized below.

1. The fundamental voltage-current relationship for inductors is $v = L\,di/dt$. This relationship can be interpreted as the voltage required to produce a given rate of change of current, and inductance opposes a change in current.
2. From the relation in (1) above, it can be seen that the current through an inductor cannot change instantaneously, because that would require infinite voltage.
3. Since i in inductance cannot change instantaneously, an inductance with zero initial current initially appears as an open circuit since $i = 0$ and, in situations where inductor current is steady, the inductance appears to be a short circuit.

  PROBLEMS

3.62   By inspection, find i_2 for $t > 0$ in the circuit of Fig. 3.49 where the switch is closed at $t = 0$ after it has been open for infinite time. *Suggestion:*   To find $i_2(0^+)$, use the nodal method to find $v_a(0^+)$. At this instant of time, $v_C(0^+)$ can be treated as an impressed voltage.

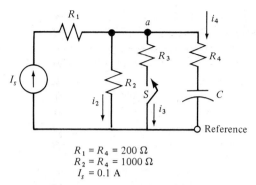

$R_1 = R_4 = 200\ \Omega$
$R_2 = R_4 = 1000\ \Omega$
$I_s = 0.1$ A

fig. 3.49   τ and current source.

3.63 a. For the situation of Fig. 3.49, also find i_3 and i_4 for $t > 0$.

b. On one set of coordinates, plot I_s, i_2, i_3, and i_4. Are these plots consistent with Kirchhoff's current law?

3.64 a. In order to find i_2 in Fig. 3.49 by the classical differential equation approach, find the differential equation in i_2. Of course this differential equation *cannot* involve other currents or voltages except I_s. You face a moderately difficult task here. You may wish to first find a differential equation for the voltage v_a, then eliminate v_a with $i_2 R_2$.

b. Solve this differential equation for i_2 to check against your solution of Prob. 3.63.

3.65 In the situation of Fig. 3.50, the switch S is opened at $t = 0$. By your best technique, find v_2 for $t > 0$.

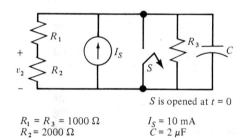

S is opened at $t = 0$

$R_1 = R_3 = 1000\ \Omega$ $I_S = 10$ mA
$R_2 = 2000\ \Omega$ $C = 2\ \mu$F

fig. 3.50   Circuit for Prob. 3.65.

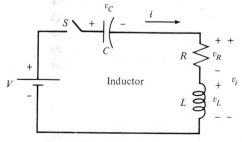

fig. 3.51   Series RLC circuit.

3.66   In Fig. 3.51, switch S is closed at $t = 0$ under the conditions that $v_C(0^-) = V/4$ and $i(0^-) = 0$. In terms of V, R, L, and C, find the following:

$$i(0^+) = \qquad \left.\frac{di}{dt}\right|_{t=0+} = \qquad \left.\frac{d^2i}{dt^2}\right|_{t=0+} =$$

$$v_R(0^+) = \qquad \left.\frac{dv_R}{dt}\right|_{t=0+} = \qquad \left.\frac{d^2v_R}{dt^2}\right|_{t=0+} =$$

$$v_L(0^+) = \qquad \left.\frac{dv_L}{dt}\right|_{t=0+} = \qquad \left.\frac{d^2v_L}{dt^2}\right|_{t=0+} =$$

$$v_C(0^+) = \qquad \left.\frac{dv_C}{dt}\right|_{t=0+} = \qquad \left.\frac{d^2v_C}{dt^2}\right|_{t=0+} =$$

Suggestions: Notice that you have not been asked to find i, v_C, v_L, and so on, as functions of time. This circuit is second-order and will be analyzed in a later chapter. Obviously, if you knew i for all values of $t > 0$, then you could very easily evaluate $i(0^+)$, $[di/dt]_{t=0+}$, and $[d^2i/dt^2]_{t=0+}$. But this is not the way to go. To solve this problem, you must search through your bag of tricks such as: Kirchhoff's voltage law, $v_L = L\, di/dt$, and so forth. Remember also that we may take the derivative of both sides of an equation such as $V = v_C + v_R + v_L$. *Caution:* Knowing that $i(0^+) = 0$ tells you nothing directly about $[di/dt]_{t=0+}$. The value of a function at a point is completely independent of the derivative of that function at the same point.

3.67   The precise model of a capacitor includes the shunt resistance R_C as shown in Fig. 3.52. In an approximate model, the value of this shunt resistance is assumed to be so high that its effect can be neglected. In this problem, we will investigate the effect of R_C upon a particular circuit response.

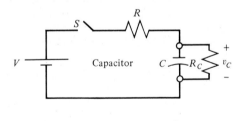

$V = 10$ V, $R = 10{,}000$, $C = 10^{-7}$ F,
$R_C = 10$ MΩ, $v_C(0^-) = 3$ V

fig. 3.52   Importance of R_c.

Switch S has been closed and then opened at some $t < 0$ such that $v_C(0^-) = 2$ V. Switch S is closed at $t = 0$, then opened again at $t = 2$ ms.
  a.   With $R_C = 10$ MΩ (the precise model) as shown, find v_C for $0 < t < 2$ ms and also for $2 < t < 1{,}000$ ms.
  b.   Now find v_C, where R_C is infinite. This, of course, is the performance of the approximate model.
  c.   Compare the performances of the two models in both time regions.

3.68   The precise model of an inductor includes series resistance R_L as shown in Fig. 3.53. The simpler but less precise model does not include this resistance.

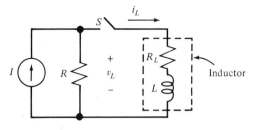

$R = 10\,\Omega, R_L = 1\,\Omega, L = 0.1\,H, I = 10\,A$

fig. 3.53   Circuit for Prob. 3.68.

This circuit is at rest when switch S is closed at $t = 0$. Find the equation for i_L for both the precise and the approximate models. Compare the two performances.

3.69   In the circuit of Fig. 3.54, the switch S is closed at $t = 0$. Also, $v_C(0^-) = 0$.

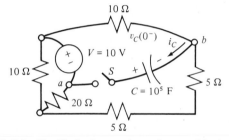

fig. 3.54   Find i_c by inspection where $v_c(0^-) = 0$.

a.   By inspection, find the equation for i_C. Apply Thévenin's theorem to this circuit to reduce the network to a simpler configuration. Consider the Thévenin model of Fig. 3.55. This particular model is formed by extracting the a-b branch including S and C, then finding the Thévenin voltage V_t and the Thévenin resistance R_i. From this Thévenin model and by inspection, write the equation for i_C.

b.   To test the consistency of your solution to (a), find the equation for i_C without the use of Thévenin's theorem.

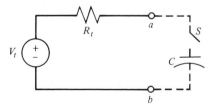

fig. 3.55   Thévenin model of Fig. 3.54.

3.70   Find i_C in the network of Fig. 3.54 where $v_C(0^-)$ is 5 V rather than zero. Does it occur to you that your Thévenin model of Prob. 3.69a brings you very close to a solution to this problem?

3.71 In the circuit of Fig. 3.56a, the op amp has very high-voltage amplification A, so v_a is very nearly 0. Also, the input, or driving point, resistance to the op amp is very high, so essentially zero current enters the input terminals.

a.   If the integrator input voltage v_i is as shown in Fig. 3.56b, the value of $R = 1\ \mathrm{M\Omega}$ and $C = 1\ \mu\mathrm{F}$, sketch the output voltage v_o as a function of time.

b.   Repeat (a) with C changed to $0.1\ \mu\mathrm{F}$.

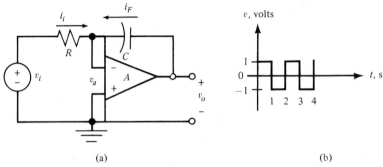

(a)   (b)

fig. 3.56   An op - amp integrator.

3.72 In order to determine the accuracy of the approximate integration performed by the circuit of Fig. 3.38:

a.   First, determine the precise value of the integral of v at the three points $t = 0.1, 0.2$, and 0.3 ms, where v is as shown in Fig. 3.38b. This can be done very quickly by a graphical view of the pertinent areas.

b.   Second, calculate the *exact* value of the voltage across C at these three points. Call this voltage v_C'. To find v_C, the method of "response by inspection" is not necessarily helpful, because the forcing function V is not a constant. Perhaps it will be best if we write the pertinent differential equation and solve this equation along the formal lines treated early in this chapter to find v_C. Now this exact value of v_C multiplied by RC yields the "measured value" of the integral that we desire. Calculate the error of our circuit at these three points by taking the difference between the "measured value" and the exact value of the integral. We have reason to believe that the maximum error will not greatly exceed 1 percent.

Chapter 4

Sinusoidal Steady-State Forcing Functions

Introduction In this chapter we investigate a special technique, the phasor transform, for finding the forced response of sinusoidal forcing functions. This technique is very important because of the widespread use of systems with sinusoidal forcing functions. Power systems everywhere produce sinusoidal waveforms. Electric clocks, refrigerators, washing machines, electric machines in industry, and countless other devices are designed to be used with sinusoidal voltage sources. The space around us is constantly filled with electromagnetic waves that are sinusoids or are analyzed as sinusoids, launched from broadcast radio stations, television stations, amateur radio stations, citizen band transmitters, aircraft control towers, and many other kinds of transmitters. Other devices, such as lasers and ultrasound generators, emit sinusoidal waves. The natural response of steel beams and drum heads to mechanical forcing functions is described in terms of a series of sinusoidal waves, called a Fourier series. In fact, Fourier series are widely used in engineering and science for analyzing a wide variety of systems.

In this chapter we shall develop the basic concepts of the phasor transform, a powerful and widely used tool for finding sinusoidal forced responses, and in other chapters we shall extend the application of the phasor transform to more complicated circuits. The phasor transform is a method for finding the particular

151

solution (forced response) to a linear differential equation with constant coefficients when the forcing function is a sinusoid. For sinusoidal forcing functions, the forced response is also called the *steady-state response* because the forced response is all that remains of the total response after the natural response has decayed away— that is, when the steady-state condition has been reached. Remember (see Chapter 3) that the total response is the sum of the natural response (solution to the homogeneous differential equation) and the forced response (particular solution to the nonhomogeneous equation). Thus the phasor transform is a method for finding the *sinusoidal steady-state response*. Although the phasor transform is restricted to finding the steady-state response, it is used widely because the steady-state response is all that is desired in many applications, in fact, in most of those described above. A principal advantage of the phasor transform is that it can be used to find the steady-state response without writing and solving the differential equations inherent in the classical solution. Furthermore, the use of the phasor transform leads to the definition of *impedance*, which provides powerful conceptual advantages.

As an illustration of phasor transform methods, we shall consider a specific example, a circuit that produces a rotating magnetic field. Such circuits are used in a number of important areas, for example in electric motors. Another interesting use of rotating magnetic fields is in magnetic bubble logic devices, which are being developed to produce extremely high densities of information storage. The rotating magnetic field is used to move the bubbles from place to place in the magnetic films in which they are produced.

The rotating-field principle, here used to illustrate the usefulness of the phasor approach, involves a special rotating field that is circular in the sense that the magnetic field vector is constant in magnitude and has a constant angular velocity. To produce this circular field in the magnetic bubble application, two identical coils may be mounted orthogonally as shown in Fig. 4.1. Then the two currents are made to be equal in magnitude and are made to differ in phase by 90°. You are not necessarily expected to understand at this time how this arrangement produces the so-called circular field. This situation merely illustrates that the phasor concept treated in the coming pages is a powerful weapon for solving many problem situations involving sinusoidal currents and voltages.

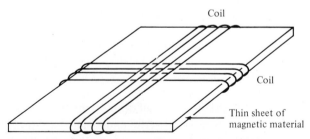

fig. 4.1   Two coils used to produce a rotating magnetic field in the plane of the film. The sinusoidal current in one coil differs in phase by 90° from the sinusoidal current in the other coil.

Figure 4.2 illustrates the use of a circular field in electric motors in terms of what might be called an elementary motor. It consists simply of two identical coils mounted orthogonally around an ordinary magnetic compass. The needle of the

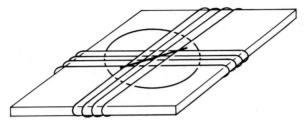

fig. 4.2 *An elementary motor consisting of two orthogonal coils of wire wound around a magnetic compass.*

compass lines up with the magnetic field produced by the two coils of wire and therefore rotates if the magnetic field rotates. Thus the needle forms the rotor of the elementary motor, and the coils are the field windings. The elementary motor is easy to construct and makes an interesting demonstration, but some important under-standing is needed to produce currents in the coils that are equal in magnitude and 90° out of phase. This can be done as shown in Fig. 4.3a.

We shall use the circuit in Fig. 4.3 as an example for developing the basic techniques of sinusoidal steady-state analysis and design. For simplicity, only one branch of the circuit is considered first, with the design of the entire circuit com-pleted at the end of the chapter, where we have the full power of the phasor transform method at our disposal.

The circuit of Fig. 4.3 works well at higher frequencies, but it must be modified for use at the low frequencies needed for the elementary motor. Some modifications are discussed at the end of the chapter.

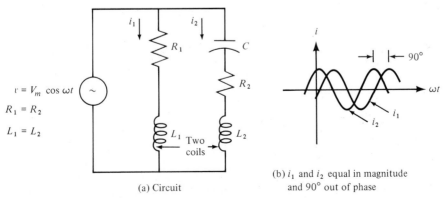

(a) Circuit

(b) i_1 and i_2 equal in magnitude and 90° out of phase

fig. 4.3 *A circuit for producing 90° phase shift between i_1 and i_2.*

4.1 SINUSOIDAL FORCING FUNCTIONS

Before we begin the development of the phasor transform, let's review sinusoids and find the complete response of a circuit with a sinusoidal forcing function by classical solution of the differential equation, just to make sure that we get the phasor transform in proper perspective when we develop it.

Review of Sinusoids

The two most familiar forms of sinusoids are probably

$$f = F_m \sin \omega t$$

$$g = G_m \cos \omega t$$

Plots of these functions are shown in Fig. 4.4, with the horizontal axes labeled in terms of both ωt and t. F_m and G_m are called *amplitudes*, or *peak values*, ω is the *radian frequency*, ωt is the *argument*, and t is time. A sinusoid is not necessarily a function of time, but we will be concerned here only with sinusoidal functions of time. The time period of the sinusoid is T, so defined because the sinusoid repeats

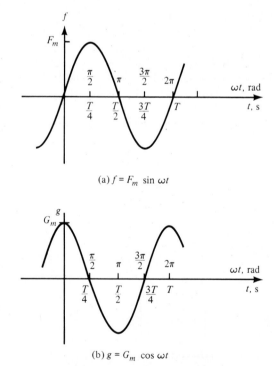

(a) $f = F_m \sin \omega t$

(b) $g = G_m \cos \omega t$

fig. 4.4   Graphs of f and g.

itself every T seconds. The radian frequency is related to the time period by

$$\omega = \frac{2\pi}{T}$$

and the frequency is defined by

$$f = \frac{1}{T}$$

so that

$$\omega = 2\pi f$$

In terms of its argument, a sinusoid repeats itself every 2π radians. The units of ω are radians/second (rad/s) and the units of f are hertz (abbreviated Hz).

We often make use of shifted sinusoids such as

$$f_1 = F_m \sin\left(\omega t + \frac{\pi}{6}\right)$$

$$g_1 = G_m \cos\left(\omega t - \frac{\pi}{3}\right)$$

The function f_1 is just the function f shifted to the left by $\pi/6$ radians, and the function g_1 is just the function g shifted to the right by $\pi/3$ radians. The function f_1 is said to *lead* f, and g_1 is said to *lag* g. We generally speak of all functions of the form $\sin(\omega t + \theta)$ and $\cos(\omega t + \phi)$ as sinusoids because the cosine function can always be written as a shifted sine function through the relation

$$\sin\left(\omega t + \phi + \frac{\pi}{2}\right) = \cos(\omega t + \phi)$$

In the expressions $\sin(\omega t + \theta)$ and $\cos(\omega t + \phi)$, θ and ϕ are called *phase angles*, and they may be expressed either in radians or degrees (2π radians $= 360°$).

PROBLEMS
4.1 Write an expression for a sinusoid that:
   a. Has an amplitude of 2 units.
   b. Has a period of 0.1 second.
   c. Has a value of 2 when $\omega t = 5\pi/3$.
   Plot the sinusoid versus t and versus ωt.
4.2 $v = \cos(\omega t + \pi/4)$. Plot v versus ωt and versus t. The frequency is 100 Hz.

Classical Analysis of Sinusoidal Forced Response

Let's proceed now to find the current in one leg of the circuit of Fig. 4.3, the $R_1 L_1$ leg shown in Fig. 4.5, where R_1 and L_1 have been renamed R and L, respectively.

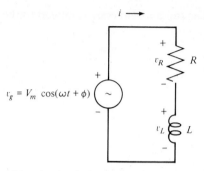

fig. 4.5 An RL circuit driven by a sinusoidal voltage generator (one branch of the circuit of Fig. 4.3).

From the circuit laws, we find that the differential equation describing the current i is

$$L\frac{di}{dt} + Ri = V_m \cos(\omega t + \phi) \qquad (4.1)$$

Equation (4.1) is like (3.55), but with a different forcing function. The complete solution to (4.1) is the sum of the natural response (complementary function) and the forced response (particular solution). The natural response is the same as the natural response obtained from (3.55), since the homogeneous equation is the same in each case. Consequently

$$i_h = Ae^{-(R/L)t}$$

A standard method for finding a particular solution is to substitute the trial solution

$$i_p = B_1 \cos(\omega t + \phi) + B_2 \sin(\omega t + \phi)$$

into (4.1) and find values of B_1 and B_2 that satisfy the equation. Making the substitution gives us

$$-\omega L B_1 \sin(\omega t + \phi) + \omega L B_2 \cos(\omega t + \phi)$$
$$+ R B_1 \cos(\omega t + \phi) + R B_2 \sin(\omega t + \phi) = V_m \cos(\omega t + \phi)$$

Collecting terms results in

$$(-\omega L B_1 + R B_2)\sin(\omega t + \phi) + (\omega L B_2 + R B_1 - V_m)\cos(\omega t + \phi) = 0$$

This equation will be satisfied for all values of time if

$$-\omega L B_1 + R B_2 = 0$$
$$R B_1 + \omega L B_2 = V_m$$

Simultaneous solution of these last two equations gives us

$$B_1 = \frac{R}{R^2 + \omega^2 L^2} V_m$$

$$B_2 = \frac{\omega L}{R^2 + \omega^2 L^2} V_m$$

Hence for $i = i_h + i_p$ we get

$$i = Ae^{-(R/L)t} + \frac{R}{R^2 + \omega^2 L^2} V_m \cos(\omega t + \phi) + \frac{\omega L}{R^2 + \omega^2 L^2} V_m \sin(\omega t + \phi)$$

If the initial condition is that $i(0^+) = 0$, the value of A is

$$A = -V_m \frac{R \cos \phi + \omega L \sin \phi}{R^2 + \omega^2 L^2}$$

and the final expression for i is

$$i = -V_m \frac{R \cos \phi + \omega L \sin \phi}{R^2 + \omega^2 L^2} e^{-(R/L)t} + \frac{R}{R^2 + \omega^2 L^2} V_m \cos(\omega t + \phi)$$

$$+ \frac{\omega L}{R^2 + \omega^2 L^2} V_m \sin(\omega t + \phi) \qquad (4.2)$$

The first term in (4.2) decays to zero as t gets large, as the natural response should. The other terms remain steady as t gets large, and are called the *steady-state response*, which is synonymous to forced response. Since the forcing function in this case is a sinusoidal voltage, we call the last two terms in (4.2) the *sinusoidal steady-state response*. In all the sinusoidal steady-state analysis and design that follows, we automatically assume that the natural response has decayed away to a negligible value, leaving only the forced response.

You can begin to get some insight into the nature of sinusoidal forced response by interpreting the last two terms in (4-2). The sum of these terms (steady-state response) decreases with R and with ωL; why? Because R is resistance to current and because the inductance L is a measure of the resistance to a change in the current, and ω tells how fast the current is changing. Remember that the current is constantly changing because of the sinusoidal forcing function. Thus ωL is a measure of the resistance to change, the resistance increasing with both inductance and the frequency of change. Consequently, the forced amplitude goes down as ωL goes up.

It is probably not too hard for you to imagine how complex the mathematical relations in this method might become for a more complicated circuit. The one that we analyzed is perhaps the simplest meaningful circuit. Fortunately, there is a much better method of analysis, the phasor transform, which is described in the next section.

PROBLEMS

4.3 Find an expression for the steady-state current in the circuit of Fig. 4.6.

4.4 Find an expression for the complete response of the voltage across the capacitance in Fig. 4.6. The voltage across the capacitance at $t = 0$ is zero. Plot the voltage across the capacitance versus ωt.

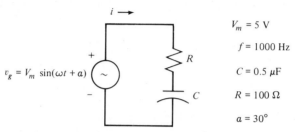

$V_m = 5$ V

$f = 1000$ Hz

$C = 0.5 \ \mu$F

$R = 100 \ \Omega$

$a = 30°$

fig. 4.6. An RC circuit driven by a sinusoidal generator (see Problem 4.3).

4.2 DERIVATION OF THE PHASOR TRANSFORM

As we pointed out in the introduction to this chapter, the phasor transform is a method for finding a particular solution to a linear differential equation with constant coefficients when the forcing function is a sinusoid. As you will see from the derivations that follow, the phasor transform offers more advantages than just a way to find a particular solution, one notable advantage being the definition of impedance.

Particular Solution by Phasor Transform

In developing the phasor transform, let's use the *RL* circuit of Fig. 4.5, which we have already analyzed by the classical solution of the differential equation, as a simple illustration of the techniques.

Definition of a Phasor. We begin the derivation by defining a new quantity **I**, such that

$$i = \text{Re}[\mathbf{I}e^{j\omega t}] \tag{4.3}$$

where i is the current in Fig. 4.5, **I** is called a *phasor*, $j = \sqrt{-1}$, ω is the radian frequency of the voltage generator, t is time, and $e^{j\omega t}$ is a complex number. Phasors are always designated by boldface type. "Re" means the real part of the quantity in the brackets; for example,

$$\text{Re}[2 + j3] = 2$$

$$\text{Re}[e^{j\pi/6}] = \cos \frac{\pi}{6}$$

Since the definition of a phasor involves complex numbers, you may wish to review the algebra of complex numbers in a math book, or study Appendix A, which summarizes what you need to know about complex numbers in studying this book. The last relation above is easily understood from *Euler's relation*:

$$e^{jx} = \cos x + j \sin x \qquad (4.4)$$

which we will be using often in the work that follows. Two other important relations are

$$\cos x = \frac{e^{jx} + e^{-jx}}{2} \qquad (4.5)$$

$$\sin x = \frac{e^{jx} - e^{-jx}}{2j} \qquad (4.6)$$

At this point, **I** is an unknown complex quantity which we shall require to be independent of time. Our next step is to substitute (4.3) into the differential equation (4.1), and then find an equation that **I** must satisfy to make i satisfy (4.1). We will find that it is easier to find **I** first and then to find i from (4.3) than to solve the differential equation directly.

Transformation of the Differential Equation. Substituting (4.3) into (4.1) gives us

$$L \frac{d}{dt} (\text{Re}[\mathbf{I}e^{j\omega t}]) + R \, \text{Re}[\mathbf{I}e^{j\omega t}] = V_m \cos(\omega t + \phi) \qquad (4.7)$$

Now we need to use the relation that

$$\text{Re}[F] = \frac{F + F^*}{2} \qquad (4.8)$$

where F is any complex number and F^* means the complex conjugate of F. For example, if $F = 4 + j5$, then

$$\frac{F + F^*}{2} = \frac{(4 + j5) + (4 - j5)}{2} = 4$$

Consequently,

$$\text{Re}(\mathbf{I}e^{j\omega t}) = \frac{\mathbf{I}e^{j\omega t} + \mathbf{I}^* e^{-j\omega t}}{2}$$

and we can write (4.7) in the form

$$L \frac{d}{dt} \left(\frac{\mathbf{I}e^{j\omega t} + \mathbf{I}^* e^{-j\omega t}}{2} \right) + R \left(\frac{\mathbf{I}e^{j\omega t} + \mathbf{I}^* e^{-j\omega t}}{2} \right) = \frac{V_m}{2} [e^{j(\omega t + \phi)} + e^{-j(\omega t + \phi)}] \qquad (4.9)$$

where we have also used (4.5) in the right-hand term.

Since we have restricted **I** to be a complex constant (that is, not a function of time),

$$\frac{d}{dt} (\mathbf{I}e^{j\omega t}) = j\omega \mathbf{I}e^{j\omega t}$$

Taking the derivatives in (4.9) and collecting terms then gives us

$$(j\omega L\mathbf{I} + R\mathbf{I} - V_m e^{j\phi})e^{j\omega t} + (-j\omega L\mathbf{I}^* + R\mathbf{I}^* - V_m e^{-j\phi})e^{-j\omega t} = 0 \qquad (4.10)$$

Since we want (4.3) to be true for all values of t, then (4.10) must be true for all values of t. The only way (4.10) can be true for all values of t is for the following to be true:

$$j\omega L\mathbf{I} + R\mathbf{I} - V_m e^{j\phi} = 0 \qquad (4.11)$$

$$-j\omega L\mathbf{I}^* + R\mathbf{I}^* - V_m e^{-j\phi} = 0 \qquad (4.12)$$

You can see that (4.11) and (4.12) must be true by trying two values of t, for example, $t = 0$, and $t = \pi/2\omega$. Since the coefficient of $e^{j\omega t}$ and the coefficient of $e^{-j\omega t}$ must each vanish separately for (4.10) to be true, $e^{j\omega t}$ and $e^{-j\omega t}$ are said to be *linearly independent*.[1] Both (4.11) and (4.12) contain the same information, since (4.12) can be obtained by taking the complex conjugate of (4.11).

Equation (4.11) is the equation that $\mathbf{I}$ must satisfy if i satisfies (4.1). Note that (4.11) is an algebraic equation; that is, it contains no derivatives. We speak of (4.11) as being the *transform* of (4.1), and we say that (4.11) is (4.1) transformed into the frequency domain. The inherent advantage of the phasor transform is that the differential equation in the time domain is transformed into an algebraic equation in the frequency domain, and algebraic equations are much easier to solve than differential equations. Equation (4.11) can be obtained directly from (4.1) by noting that d/dt "transforms" as $j\omega$, and $V_m \cos(\omega t + \phi)$ "transforms" as $V_m e^{j\phi}$, so (4.11) can be obtained without going through all the algebraic steps that we did in the derivation. Furthermore, as shown in the next section, $d^n i/dt^n$ transforms into $(j\omega)^n\mathbf{I}$, which allows us easily to transform any differential equation. However, as we shall soon see, we will shortly develop methods based on impedance that will allow us to analyze and design circuits without writing the differential equations at all.

Equation (4.1) is a *time-domain equation* because it contains functions of time, that is, $i(t)$. Equation (4.11) is a *frequency-domain equation* because it contains functions of ω, that is, $\mathbf{I}(\omega)$. We can see that $\mathbf{I}$ must be a function of ω by solving (4.11) for $\mathbf{I}$ to get

$$\mathbf{I} = \frac{V_m e^{j\phi}}{j\omega L + R} \qquad (4.13)$$

The phasor $\mathbf{I}$ is a complex quantity that is a function of ω; $\mathbf{I}$ is not a function of t. Now let's find i from $\mathbf{I}$ by using (4.3) and then review what we have done.

The Inverse Transform.   Since we have an expression for $\mathbf{I}$, we can find i from (4.3) by direct substitution. The result is

$$i = \mathrm{Re}\left[\frac{V_m e^{j\phi}}{j\omega L + R}\, e^{j\omega t}\right] \qquad (4.14)$$

---

[1] C. R. Wylie, *Advanced Engineering Mathematics*, 3rd ed., New York: McGraw-Hill, 1966, p. 444.

Rationalizing and then combining the exponentials gives us

$$i = \text{Re}\left[\frac{V_m(R - j\omega L)e^{j(\omega t + \phi)}}{R^2 + \omega^2 L^2}\right]$$

Using the relationship in (4.4) for $e^{j(\omega t + \phi)}$, multiplying the numerator out, and taking the real part, results in

$$i = \frac{R}{R^2 + \omega^2 L^2}\, V_m \cos(\omega t + \phi) + \frac{\omega L}{R^2 + \omega^2 L^2}\, V_m \sin(\omega t + \phi) \qquad (4.15)$$

which is the same as the forced response of (4.2), as obtained by classical methods.

An equivalent form for (4.15) can be obtained by returning to (4.14) and writing the denominator in polar form:

$$R + j\omega L = \sqrt{R^2 + \omega^2 L^2}\,e^{j\psi}$$

where $\psi = \tan^{-1}\dfrac{\omega L}{R}$. Then we get

$$i = \text{Re}\left[\frac{V_m e^{j\phi}e^{j\omega t}}{\sqrt{R^2 + \omega^2 L^2}\,e^{j\psi}}\right] = \text{Re}\left[\frac{V_m e^{j(\omega t + \phi - \psi)}}{\sqrt{R^2 + \omega^2 L^2}}\right]$$

$$i = \frac{V_m \cos(\omega t + \phi - \psi)}{\sqrt{R^2 + \omega^2 L^2}} \qquad (4.16)$$

This form, which is equivalent to (4.15), is more convenient than (4.15) for many purposes, such as plotting i versus t. Equation (4.16) can also be obtained from (4.15) by using trigonometric identities.

Both (4.16) and (4.15) are called the *inverse transform* of **I**, and it is the inverse transform that gives us i, the time-domain representation of the current. In summary, our method has been:

1. To define the phasor **I** by (4.3).
2. To transform the differential equation (4.1) to an algebraic equation in the frequency domain by substituting (4.3) into (4.1), expanding, and using linear independence.
3. To solve for **I** from the transformed equation.
4. To take the inverse transform of **I** to obtain i.

At this point, we have shown how the phasor transformation works for one specific case. The next subsection shows formally that the phasor transformation is valid in the general case. If you are not interested in the details of the derivation, you can skip the next section without loss of continuity. (Sections considered optional are identified with asterisks throughout this book.)

Generalized Phasor Transformation of Linear Differential Equations. Without dwelling on details of the proof, we will indicate how the phasor transformation may be used to obtain a particular solution of any linear differential equation with constant coefficients when the forcing function is a sinusoid. Consider the following nth-order differential equation,

$$a_n \frac{d^n f}{dt^n} + a_{n-1} \frac{d^{n-1} f}{dt^{n-1}} + \cdots + a_0 f = C \cos(\omega t + \phi) \tag{4.17}$$

where f is a function of time, but the a's are not functions of time. The phasor transformation is

$$f = \text{Re}[F e^{j\omega t}] \tag{4.18}$$

Substituting (4.18) into (4.17), using the relations in (4.8) and (4.5), taking the derivatives, and collecting terms, results in

$$[a_n(j\omega)^n F + a_{n-1}(j\omega)^{n-1} F + \cdots + a_0 F - C e^{j\phi}] e^{j\omega t}$$
$$+ [a_n(-j\omega)^n F^* + a_{n-1}(-j\omega)^{n-1} F^* + \cdots + a_0 F^* - C e^{-j\phi}] e^{-j\omega t} = 0 \tag{4.19}$$

Since $e^{j\omega t}$ and $e^{-j\omega t}$ are linear independent, (4.19) can be satisfied for all t only if

$$a_n(j\omega)^n F + a_{n-1}(j\omega)^{n-1} F + \cdots + a_0 F = C e^{j\phi} \tag{4.20}$$

and

$$a_n(-j\omega)^n F^* + a_{n-1}(-j\omega)^{n-1} F^* + \cdots + a_0 F^* = C e^{-j\phi} \tag{4.21}$$

Since (4.21) is the complex conjugate of (4.20), it contains no new information and is not needed. Equation (4.20) is said to be the phasor transform of (4.17). Since any constant-coefficient, linear differential equation can be put in the form of (4.17), (4.20) is a result that holds for any such differential equation, and we can therefore find a particular solution by solving for F from (4.20), substituting into (4.18), and finding f.

However, we need not go through all the algebra associated with (4.19) each time we wish to find a particular solution. You have probably already noticed that we can get (4.20) from (4.17) by replacing $d^n f/dt^n$ by $(j\omega)^n F$, $d^{n-1} f/dt^{n-1}$ by $(j\omega)^{n-1} F$, and so on, and $C \cos(\omega t + \phi)$ by $C e^{j\phi}$. We can formalize this process a little more by stating that we can transform the differential equation term by term into the frequency domain and that

$$\mathscr{P}[f] = F \tag{4.22}$$

$$\mathscr{P}[af] = aF \tag{4.23}$$

$$\mathscr{P}\left[\frac{d^n f}{dt^n}\right] = (j\omega)^n F \tag{4.24}$$

$$\mathscr{P}[\cos(\omega t + \phi)] = e^{j\phi} \tag{4.25}$$

where the script $\mathscr{P}$ means "the phasor transform of." Equation (4.22) is the basic definition of the phasor transform, (4.23) states that the transform of a constant times a function is equal to the constant times the transform of the function, (4.24) is a relation for the transform of a derivative, and (4.25) is the relation for the transform of a sinusoid. With these transform relations, we can quickly and easily transform a differential equation to the frequency domain, solve for the unknown phasor, transform back to the time domain, and thus obtain the particular solution to the differential equation.

For example, let's find a particular solution to

$$\frac{d^3f}{dt^3} + 3\frac{d^2f}{dt^2} + 50\frac{df}{dt} - 60f = 5\cos\left(10t + \frac{\pi}{3}\right) \qquad (4.26)$$

Using (4.22) to (4.25) to transform the equation term by term, we get

$$(j10)^3\mathbf{F} + 3(j10)^2\mathbf{F} + 50(j10)\mathbf{F} - 60\mathbf{F} = 5e^{j\pi/3}$$

From which

$$\mathbf{F} = \frac{5e^{j\pi/3}}{-j1000 - 300 + j500 - 60}$$

$$\mathbf{F} = \frac{5e^{j\pi/3}}{-360 - j500} = \frac{-5e^{j\pi/3}}{616.12e^{j54.25°}}$$

$$\mathbf{F} = -8.12(10^{-3})e^{j5.75°}$$

Taking the inverse gives us

$$f = \text{Re}[\mathbf{F}e^{j\omega t}] = -\text{Re}[8.12(10^{-3})e^{j5.75°}e^{j10t}]$$

$$f = -8.12(10^{-3})\cos(10t + 5.75°)$$

You should find the particular solution to (4.26) by the classical method to appreciate the differences between the two methods (Prob. 4.9). However, we shall see shortly that the phasor transform method has even greater advantage in circuit theory because we can write the circuit equations directly in the frequency domain without even writing the differential equations.

Now that we have seen how the phasor transform works, let's extend the formalism slightly by defining transform pairs. The particular solution to (4.17) will always be a sinusoid, since the derivative of the sine is the cosine and the derivative of the cosine is the negative of the sine and since the sum of sinusoids is also a sinusoid. Thus the particular solution can always be written in the form

$$f = F_m \cos(\omega t + \alpha) \qquad (4.27)$$

where F_m is the amplitude and α is the phase angle. Using our original definition of a phasor in (4.3), we can therefore write

$$f = \text{Re}[\mathbf{F}e^{j\omega t}]$$

and

$$F_m \cos(\omega t + \alpha) = \text{Re}[\mathbf{F}e^{j\omega t}]$$

Expanding the left-hand side according to (4.5) and the right-hand side according to (4.8), we get

$$\frac{F_m e^{j\alpha} e^{j\omega t}}{2} + \frac{F_m e^{-j\alpha} e^{-j\omega t}}{2} = \frac{\mathbf{F} e^{j\omega t}}{2} + \frac{\mathbf{F}^* e^{-j\omega t}}{2}$$

And because of the linear independence of $e^{j\omega t}$ and $e^{-j\omega t}$, we have

$$F_m e^{j\alpha} = \mathbf{F}$$

We now have what is called a *transform pair*:

$$f = \text{Re}[\mathbf{F} e^{j\omega t}] \tag{4.28}$$

$$\mathbf{F} = F_m e^{j\alpha} \tag{4.29}$$

Equation (4.28) tells us how to get f when we know $\mathbf{F}$, and (4.29) tells us how to get $\mathbf{F}$ when we know f. f is said to be in the time domain and $\mathbf{F}$ in the frequency domain. We also say that

$$\mathbf{F} = \mathscr{P}[f] \tag{4.30}$$

$$f = \mathscr{P}^{-1}(\mathbf{F}) \tag{4.31}$$

where (4.30) states that $\mathbf{F}$ is the *phasor transform* of f and (4.31) states that f is the *inverse phasor transform* of $\mathbf{F}$. Thus the formalism of the phasor transform is complete. We can transform from the time domain to the frequency domain by (4.30) and from the frequency domain to the time domain by (4.31).

Review of the Phasor Transform for Differential Equations. Now let's stand back and review the process of finding a particular solution using phasor transforms. Figure 4.7 illustrates the general process, with the differential equation for the current in Fig. 4.5 as an example. The differential equation is transformed to the frequency domain resulting in an *algebraic* equation in the frequency domain. The algebraic equation is solved in the frequency domain to give an expression for the phasor. Then the inverse transform, (4.18), is used to transform back to the time domain, giving the desired particular solution to the differential equation. The main advantage of the phasor transform is that it requires only the solution of an algebraic equation in the frequency domain instead of the solution of a differential equation as required in the time-domain analysis. The algebraic equation results because the time derivative of $e^{j\omega t}$ is just a constant times itself, which relation is the whole reason for making the phasor transformation.

This method is a general method, not limited to circuit theory. If the algebra of complex numbers does not cause you great difficulty, you will find the phasor transform faster and easier than the classical method. Next we will show how the phasor transform can be extended in circuit theory so that the differential equations need not even be written, thus shortening the analysis and making the phasor transform even more powerful.

PROBLEMS

4.5 Evaluate the following:

    a. $\text{Re}\left[\dfrac{1}{3 + j2}\right]$ b. $\text{Re}[(2 + j5)e^{j\pi/4}]$

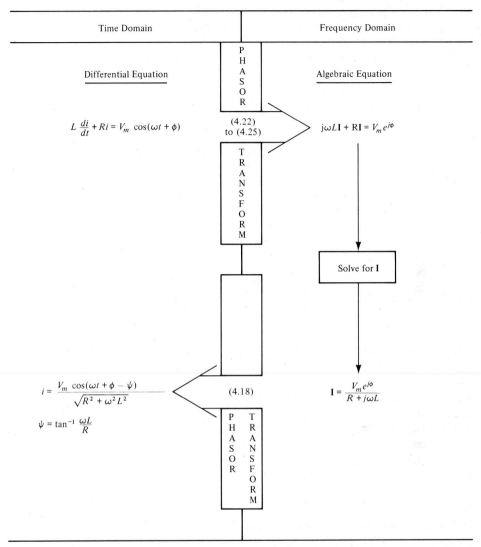

Time Domain		Frequency Domain

Differential Equation

Algebraic Equation

P H A S O R

$$L\frac{di}{dt} + Ri = V_m \cos(\omega t + \phi)$$

(4.22) to (4.25)

$$j\omega L\mathbf{I} + R\mathbf{I} = V_m e^{j\phi}$$

T R A N S F O R M

Solve for **I**

$$i = \frac{V_m \cos(\omega t + \phi - \psi)}{\sqrt{R^2 + \omega^2 L^2}}$$

$$\psi = \tan^{-1}\frac{\omega L}{R}$$

(4.18)

$$\mathbf{I} = \frac{V_m e^{j\phi}}{R + j\omega L}$$

P H A S O R   T R A N S F O R M

fig. 4.7   *How the phasor transform is used to find a particular solution to a constant-coefficient, linear differential equation with a sinusoidal driving function.*

c. $\left(\dfrac{e^{j\pi/5}}{4 + j3}\right)^*$ d. $\dfrac{d}{dx}(e^{j3x})$

e. $\text{Re}\left[\dfrac{e^{j\pi/3}}{j3 + 2}\right]$

4.6   Write $4/(5 + j7)$ in polar form.

4.7   Write $3.5e^{j35°}$ in rectangular form.

4.8   Write $\cos(2\pi t + 20°)$ as the sum of two exponentials.

4.9 Find a particular solution to (4.26) by substituting the trial solution $f = A \cos(10t + \pi/3) + B \sin(10t + \pi/3)$ into (4.26) and finding the values of A and B that satisfy the equation.

4.10 Transform the following differential equation to the frequency domain:

$$5\frac{d^4v}{dt^4} + 2\frac{d^3v}{dt^3} + 10\frac{d^2v}{dt^2} + 4\frac{dv}{dt} + 2v = 3\cos(t+1)$$

4.11 Find a particular solution to

$$3\frac{d^3i}{dt^3} + \frac{d^2i}{dt^2} + i = 5\sin(2t+3)$$

and show by direct substitution that it does satisfy the equation.

4.12 a. If $f = 3\sin \omega t$, find $\mathscr{P}[f]$.
b. If $\mathbf{F} = (2 + j5)$, find $f = \mathscr{P}^{-1}[\mathbf{F}]$.

4.13 Find values of a, b, and c that make $-\cos(10t + 30°)$ a particular solution to

$$a\frac{d^2f}{dt^2} + b\frac{df}{dt} + cf = 100\cos(10t + 30°)$$

4.3 IMPEDANCE AND CIRCUIT LAWS IN THE FREQUENCY DOMAIN

The real power of the phasor transform in circuit theory comes from being able to apply Kirchhoff's laws in the frequency-domain representation and from being able to define impedance in the frequency domain.

Kirchhoff's Laws in the Frequency Domain. Let's use the simple circuit of Fig. 4.5 as an example to show how Kirchhoff's laws apply in the frequency domain. For that case, Kirchhoff's voltage law gives

$$v_g - v_R - v_L = 0 \tag{4.32}$$

which we have purposely written in the form of the general relation

$$\sum_n v_n = 0 \tag{4.33}$$

We can transform (4.32) into the frequency domain by defining the phasors according to

$$v_g = \text{Re}[\mathbf{V}_g e^{j\omega t}]$$
$$v_R = \text{Re}[\mathbf{V}_R e^{j\omega t}]$$
$$v_L = \text{Re}[\mathbf{V}_L e^{j\omega t}]$$

substituting these relations into (4.32), using (4.8) and linear independence, just as we did to get (4.11), and getting

$$\mathbf{V}_g - \mathbf{V}_R - \mathbf{V}_L = 0 \tag{4.34}$$

We call (4.34) Kirchhoff's voltage law in the frequency domain for the circuit of Fig. 4.5.

The same procedure can be generalized to show that (4.33) can be transformed into the phasor domain as

$$\sum_n \mathbf{V}_n = 0 \tag{4.35}$$

Likewise, Kirchhoff's current law in the time domain

$$\sum_n i_n = 0$$

can be transformed to the phasor domain as

$$\sum_n \mathbf{I}_n = 0 \tag{4.36}$$

If you studied the subsection on the generalized phasor transform (identified with an asterisk), you will want to know that we can use these results to add the following relation to the transform relations in (4.22) to (4.25), and thus extend the formalism:

$$\mathscr{P}[f_1 + f_2] = \mathscr{P}[f_1] + \mathscr{P}[f_2] = \mathbf{F}_1 + \mathbf{F}_2 \tag{4.37}$$

Equation (4.37) tells us that we can transform a relation like (4.33) directly into (4.35).

Impedance and Admittance. Now that we can write Kirchhoff's laws directly in terms of phasors, that is, in the frequency domain, we can avoid writing differential equations in the time domain if we can write voltage-current relations like $v_L = L\, di_L/dt$ directly in the frequency domain. All we need do is transform the following three voltage-current relations for inductance, capacitance, and resistance, respectively, into the frequency domain:

$$v_L = L\frac{di_L}{dt}$$

$$i_C = C\frac{dv_C}{dt}$$

$$v_R = Ri_R$$

With (4.24), the transformation is easy, and we get

$$\mathbf{V}_L = j\omega L \mathbf{I}_L \tag{4.38}$$

$$\mathbf{I}_C = j\omega C \mathbf{V}_C \tag{4.39}$$

$$\mathbf{V}_R = R\mathbf{I}_R \tag{4.40}$$

These relations allow us to define *impedance*, a very important and useful concept. It is important that you realize that impedance is defined only in the *frequency* domain and cannot be defined in the time domain. Impedance is defined simply as the ratio of phasor voltage to phasor current. Thus

$$\mathbf{Z}_L = \frac{\mathbf{V}_L}{\mathbf{I}_L} = j\omega L \tag{4.41}$$

is called the impedance of inductance, and

$$\mathbf{Z}_C = \frac{\mathbf{V}_C}{\mathbf{I}_C} = \frac{1}{j\omega C} = \frac{-j}{\omega C} \qquad (4.42)$$

$$\mathbf{Z}_R = \frac{\mathbf{V}_R}{\mathbf{I}_R} = R \qquad (4.43)$$

are called the impedances of capacitance and resistance, respectively. Impedance in the frequency domain is similar in some ways to resistance in the time domain. As a matter of fact, you must have noticed that the impedance of resistance is equal to its resistance. The unit of impedance is the ohm. In general, impedance is a complex number because it is the quotient of two complex numbers; that is, $\mathbf{V}/\mathbf{I}$. The impedance of either inductance or capacitance is imaginary; see (4.41) and (4.42). The impedance of resistance, of course, is real, whereas in most models or networks the impedance has both real and imaginary parts, as you will see presently. Because impedance $\mathbf{Z}$ is complex, it is presented, like the phasors $\mathbf{V}$ and $\mathbf{I}$, in boldface type. The magnitude of impedance of an inductance is specifically called *inductive reactance* X_L ($\mathbf{Z}_L = j\omega L = jX_L$), and that of a capacitance is called *capacitive reactance* X_C ($\mathbf{Z}_C = -j/\omega C = -jX_C$).

By considering the voltage-current relations for combinations of elements as was done for resistances in Chapter 2, we can show that impedances combine like resistances. Impedances in series add as resistors in series:

$$\mathbf{Z}_s = \mathbf{Z}_1 + \mathbf{Z}_2 + \mathbf{Z}_3 + \cdots + \mathbf{Z}_n$$

where $\mathbf{Z}_s$ is the equivalent impedance of the n impedances in series. Similarly,

$$\frac{1}{\mathbf{Z}_p} = \frac{1}{\mathbf{Z}_1} + \frac{1}{\mathbf{Z}_2} + \cdots + \frac{1}{\mathbf{Z}_n}$$

where $\mathbf{Z}_p$ is the equivalent impedance of n impedances in parallel.

As you might expect, we also define admittances (similar to conductance of resistance) in the frequency domain according to

$$\mathbf{Y} = \frac{1}{\mathbf{Z}}$$

Admittances may be combined as conductances are combined.

Neither impedance nor admittance is ever transformed to the time domain. Both are strictly frequency-domain quantities, with no meaning in the time domain.

In rectangular form, impedance can always be written as the sum of a real part and an imaginary part. The customary notation is to write

$$\mathbf{Z} = R + jX$$

where R is called the *resistive* or *real component* of $\mathbf{Z}$, or the *resistance*, and X is called the *imaginary* or *reactive component* of $\mathbf{Z}$, or the *reactance*. Similarly, admittance can be written as

$$\mathbf{Y} = G + jB$$

where G is called the *conductive* or *real component* of $\mathbf{Y}$, or the *conductance*, and B is called the *susceptive* or *imaginary component* of $\mathbf{Y}$, or the *susceptance*. The susceptance of an inductance is specifically called *inductive susceptance*, B_L, whereas that of a capacitance is called *capacitive susceptance*, B_C.

Frequency-Domain Representation of Circuits.   We can use the impedance and admittance relations to define a *frequency-domain representation* of circuit elements, as shown in Fig. 4.8. Each element is represented by its impedance, with the impedance

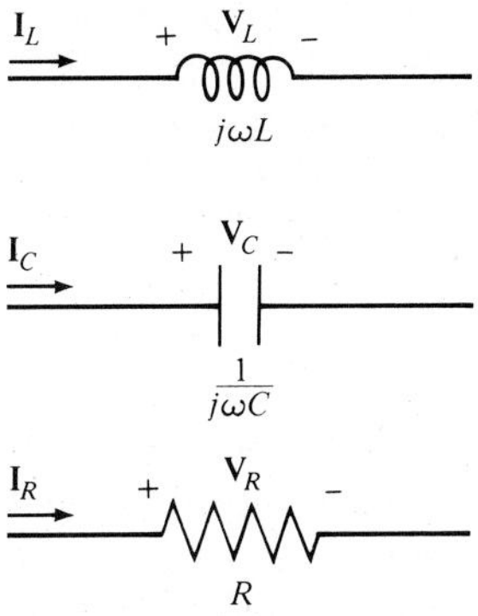

fig. 4.8   Frequency - domain representation of circuit elements.

relations as given in (4.41) to (4.43). For example, the frequency-domain representation of the *RL* circuit in Fig. 4.5 is shown in Fig. 4.9. The voltage generator is represented by a phasor voltage $\mathbf{V}_g$.

Analysis of the circuit in Fig. 4.9 is easy in the frequency domain. From Kirchhoff's laws and the impedance relations, we can write directly

$$\mathbf{V}_g = R\mathbf{I} + j\omega L\mathbf{I} \qquad (4.44)$$

and hence

$$\mathbf{I} = \frac{\mathbf{V}_g}{R + j\omega L} \qquad (4.45)$$

This is the same relation that we obtained previously in (4.13); the inverse transform is obtained the same way and is given in (4.16).

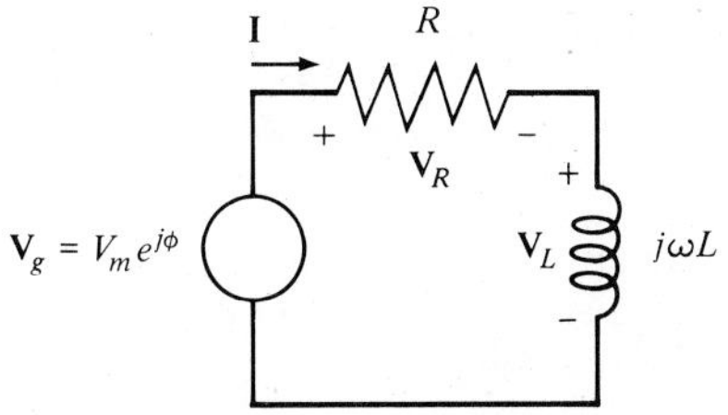

fig. 4.9   The frequency- domain representation of the circuit in Fig. 4.5.

However, (4.45) can be obtained even without writing (4.44), because impedances combine just like resistances. Thus the impedances R and $j\omega L$ in series can be combined as the sum of the two, and a circuit equivalent to the one in Fig. 4.9 is shown in Fig. 4.10, from which (4.45) can be written directly.

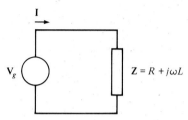

fig. 4.10   Circuit equivalent to the one in Fig. 4.9.

We now have a very powerful method of analysis of sinusoidal forced response, because all of the methods for resistive circuits can be applied to the frequency-domain representation of sinusoidal circuits. That is, loop and nodal methods, parallel and series equivalent impedances and admittances, Thévenin's theorem, Norton's theorem, superposition, and impedance matrices (as described in the following chapters) can all be applied to frequency-domain circuit representations. You may not appreciate the full meaning of the last two statements if you have not had some experience in circuit analysis, but as you gain experience, you will come to realize that the phasor transformation is of tremendous importance in circuit theory. The following example should help you see how phasor theory is used, and we will extend the application into more complex situations in the remainder of this chapter and in following chapters.

An Example of Circuit Analysis in the Frequency Domain. Consider the circuit shown in Fig. 4.11a, and let's see how to find the current through the resistance. The frequency-domain representation is shown in Fig. 4.11b and an equivalent circuit in Fig. 4.11c, where the impedances of the capacitance and inductance have been combined into one equivalent impedance like two parallel resistances would be combined; that is,

$$\mathbf{Z} = \frac{\mathbf{Z}_C \mathbf{Z}_L}{\mathbf{Z}_C + \mathbf{Z}_L}$$

From Fig. 4.11c, it is easy to write

$$\mathbf{I} = \frac{\mathbf{V}_g}{R + \mathbf{Z}}$$

Putting in numbers, we get

$$\mathbf{Z} = \frac{(-j159.1)(j628.3)}{-j159.1 + j628.3} = -j213.0$$

$$\mathbf{I} = \frac{10e^{j30°}}{220 - j213.0} = \frac{10e^{j30°}}{306.3e^{-j44.08°}} = 0.03265e^{j74.08°}$$

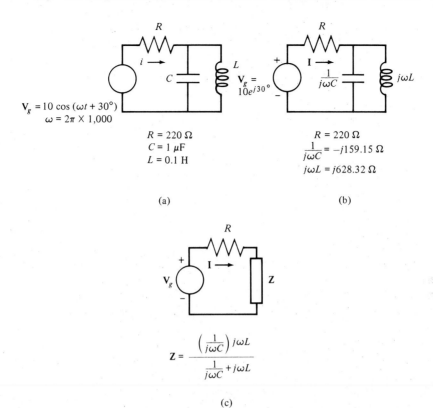

(a)

(b)

(c)

fig. 4.11 (a) Time domain representation of a circuit. (b) Frequency-domain representation of the circuit in (a). (c) Circuit equivalent to the circuit in (b).

Taking the inverse transform of **I** gives us

$$i = \text{Re}[\mathbf{I}e^{j\omega t}] = \text{Re}[0.03265e^{j(\omega t + 74.08°)}] = 0.03265 \cos(\omega t + 74.08°) \text{ A}$$

This is a good place to point out that the magnitude of the phasor is equal to the amplitude of the corresponding sinusoid in the time domain. That is, $|\mathbf{I}| = 0.03265$ A and the amplitude of i is 0.03265 A. Also, the angle of the phasor **I** is 74.08°, and the phase angle of the sinusoid is 74.08°. It is always true that the angle of the phasor is equal to the phase angle of the corresponding sinusoid. These two correspondences are extremely important because they allow us to think about the characteristics of the sinusoids by looking at the properties of the phasor, without formally making the inverse transform.

As far as applying the circuit laws, the circuit in Fig. 4.11a is no more difficult to analyze than the same circuit with C and L replaced by resistors; the added complexity stems from the impedances being complex numbers.

PROBLEMS

4.14 Draw a diagram of the frequency-domain representation of the circuit shown in Fig. 4.12.

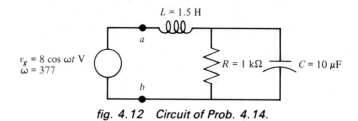

fig. 4.12   Circuit of Prob. 4.14.

4.15 Find the numerical value of the equivalent parallel impedance of the resistance and capacitance in Fig. 4.12.

4.16 Find the numerical value of the equivalent impedance between the points *a* and *b* in Fig. 4.12.

4.17 Find the numerical value of the equivalent admittance of the resistor and capacitor in Fig. 4.12.

4.18 Find the numerical value of the equivalent admittance between the points *a* and *b* in Fig. 4.12.

4.19 Find the time-domain expression for the current through the resistor in Fig. 4.12.

4.20 Find the time-domain expression for the current through the capacitor in Fig. 4.12.

4.21 Choose new values of R and C for the circuit of Fig. 4.6 such that the amplitude of the voltage across C is one-half the amplitude of the generator voltage. Tell whether the voltage across C leads or lags the generator voltage, and by what angle.

4.4   SUMMARY OF METHODS OF PHASOR TRANSFORMATIONS

We began this chapter by introducing a circuit for producing rotating magnetic fields as an example of design and analysis that phasor transforms can be applied to with advantage. Then as we developed the phasor transform methods, we analyzed one branch of the circuit as an example. After completing a rather involved development of phasor transforms, we still have not discussed the design of that circuit, mainly because we did not want to digress into a detailed discussion of the application of phasor transforms at the expense of having a cohesive development of the methods. We shall return to the design of the rotating magnetic field in the next section, after summarizing the basic methods of phasor transforms.

The phasor transform was introduced as a tool for finding the forced response of circuits with sinusoidal generators. We showed that the phasor transform is not limited to circuit theory but can be used to find a particular solution of any constant-coefficient, linear differential equation with a sinusoidal forcing function. Phasor transforms are powerful tools in sinusoidal steady-state circuit theory because they make possible frequency-domain representations of circuits that include impedances and admittances, resulting in circuit techniques that are exactly analogous to those used with resistive circuits. Consequently, the mathematical procedures are simpler, and, more important, the frequency-domain representations offer qualitative insight that would otherwise not be possible.

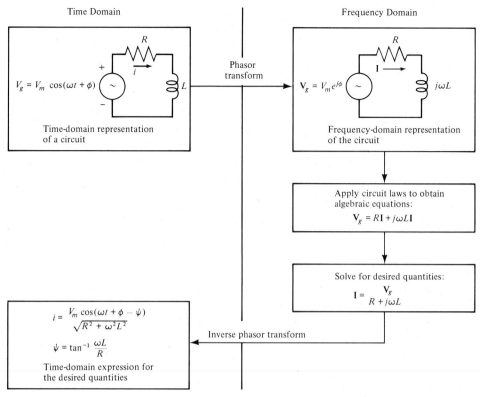

fig. 4.13 Procedure for circuit analysis using phasor transforms.

A summary of the phasor transform procedures is shown in Fig. 4.13. After you have used phasor transforms for a while, you will tend to work mostly in the frequency domain without starting with an explicit time-domain representation of the circuit, and you will make inverse transforms of the phasors infrequently. The reason is that you will think of phasors in terms of what they mean in the time domain, but without writing down the time-domain representations. For instance, it will become automatic for you to think of the magnitude of a phasor as the amplitude of the corresponding sinusoid in the time domain. Figure 4.14 summarizes these useful correspondences.

4.5 CIRCUIT FOR PRODUCING A ROTATING MAGNETIC FIELD

A basic circuit for producing a rotating magnetic field was described in the introduction to this chapter and was used as an example in developing phasor transforms. With the tools developed in the preceding sections, we are now ready to describe the design of the circuit, which is shown in Fig. 4.3.

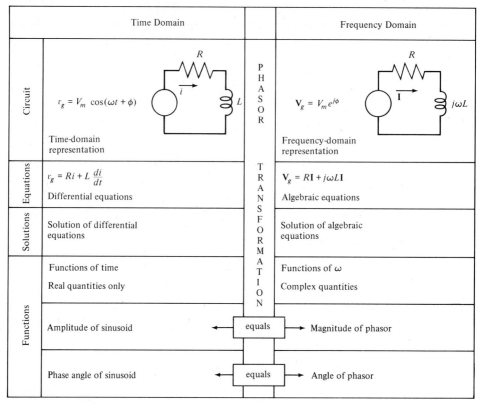

	Time Domain		Frequency Domain
Circuit	$v_g = V_m \cos(\omega t + \phi)$ Time-domain representation	P H A S O R	$\mathbf{V}_g = V_m e^{j\phi}$ Frequency-domain representation
Equations	$v_g = Ri + L \dfrac{di}{dt}$ Differential equations	T R A N S F O R M A T I O N	$\mathbf{V}_g = R\mathbf{I} + j\omega L\mathbf{I}$ Algebraic equations
Solutions	Solution of differential equations		Solution of algebraic equations
Functions	Functions of time Real quantities only		Functions of ω Complex quantities
	Amplitude of sinusoid ← equals → Magnitude of phasor		
	Phase angle of sinusoid ← equals → Angle of phasor		

fig. 4.14 Correspondence between time-domain quantities and frequency-domain quantities.

Design of the Circuit

Frequency-Domain Representation of the Circuit. The first step in the design is to transform the circuit to the frequency domain, as shown in Fig. 4.15. The phasor transform is applied as described in preceding sections, and the circuit elements are represented by their impedances. Note that the internal impedance of the generator is assumed to be negligible.

Equations for Phasor Currents. The expressions for $\mathbf{I}_1$ and $\mathbf{I}_2$ of Fig. 4.15 are easily written in both rectangular and polar form:

$$\mathbf{I}_1 = \frac{\mathbf{V}_g}{R_1 + j\omega L_1} = \frac{V_m e^{j\psi_1}}{\sqrt{R_1^2 + (\omega L_1)^2}} \tag{4.46}$$

$$\psi_1 = -\tan^{-1}\left(\frac{\omega L_1}{R_1}\right) \tag{4.47}$$

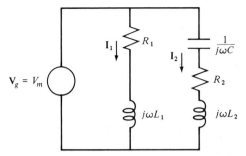

fig. 4.15 Frequency-domain representation of the circuit that produces a rotating magnetic field (Fig. 4.3).

$$\mathbf{I}_2 = \frac{\mathbf{V}_g}{R_2 + \dfrac{1}{j\omega C} + j\omega L_2} = \frac{V_m e^{j\psi_2}}{\sqrt{R_2^2 + \left(\omega L_2 - \dfrac{1}{\omega C}\right)^2}} \qquad (4.48)$$

$$\psi_2 = -\tan^{-1}\left[\left(\omega L_2 - \frac{1}{\omega C}\right)\Big/ R_2\right] \qquad (4.49)$$

As explained in the introduction to this chapter, the magnetic field produced by the two coils will be circular if the currents in the coils have equal amplitudes and differ in phase by 90°. These requirements correspond to the following frequency-domain requirements:

$$|\mathbf{I}_1| = |\mathbf{I}_2| \qquad (4.50)$$

$$\psi_1 - \psi_2 = \pm 90° \qquad (4.51)$$

Therefore, to complete the design, we need to choose the circuit elements so that (4.50) and (4.51) are satisfied. In addition, the coils must satisfy other requirements and the currents must be strong enough to produce a sufficiently strong magnetic field. We will assume that the coils have a specified inductance and resistance as required to produce the desired magnetic field strength. Consequently, R_1 and R_2 will represent the dc resistance of coils 1 and 2, respectively, plus any resistance that might be needed to satisfy (4.50) and (4.51), but the minimum values that R_1 and R_2 can have are the dc resistances of the respective coils.

Phasor Diagrams. Although we could substitute appropriate expressions from (4.46) to (4.49) into (4.50) and (4.51) and try to find suitable values of R_1, R_2, and C, there is another method that is both easier and provides more insight. This method is based on a *phasor diagram*, which is just a graphical representation of the phasors. Since phasors are represented by complex numbers, they can be represented in the complex plane by directed line segments, or vectors, as described in Appendix A. A phasor diagram of $\mathbf{I}_1$ and $\mathbf{I}_2$ is shown in Fig. 4.16, with $\mathbf{I}_1$ and $\mathbf{I}_2$ having equal magnitudes and differing in angle by 90°. Relating the magnitudes of $\mathbf{I}_1$ and $\mathbf{I}_2$ to the circuit components is complicated by the components appearing in the denomi-

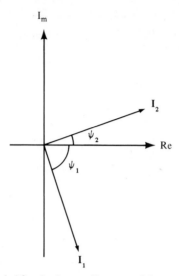

fig. 4.16   A phasor diagram of I_1 and I_2.

nators of (4.46) and (4.47). We can get a simpler diagram by noting that (4.50) and (4.51) will be satisfied if

$$Z_1 = Z_2 \tag{4.52}$$

(note that lightface italic type implies magnitude only) and

$$\phi_1 - \phi_2 = \pm 90° \tag{4.53}$$

where

$$\mathbf{Z}_1 = R_1 + j\omega L_1 \tag{4.54}$$

$$\mathbf{Z}_2 = R_2 + \frac{1}{j\omega C} + j\omega L_2 \tag{4.55}$$

$$\phi_1 = \tan^{-1}\left(\frac{\omega L_1}{R_1}\right) \tag{4.56}$$

$$\phi_2 = \tan^{-1}\left[\left(\omega L_2 - \frac{1}{\omega C}\right) \middle/ R_2\right] \tag{4.57}$$

A diagram of $\mathbf{Z}_1$ and $\mathbf{Z}_2$ is shown in Fig. 4.17, with $\phi_2 - \phi_1 = 90°$ and $Z_1 = Z_2$. Figure 4.18 shows congruent triangles containing $\mathbf{Z}_1$ and $\mathbf{Z}_2$ that can be used to find values of the circuit parameters that will make $\phi_2 - \phi_1 = 90°$ and $|\mathbf{Z}_1| = |\mathbf{Z}_2|$. The sides of the triangle are related to the circuit parameters by (4.54) to (4.57). Triangles OAB and DCO are congruent because $Z_1 = Z_2$ requires that $OB = OD$, and $\phi_2 - \phi_1 = 90°$ requires the angles to be equal, as marked in the diagram. Since the triangles are congruent, it must be true that

$$OA = CD \quad \text{and} \quad OC = AB$$

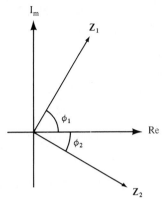

fig. 4.17 Diagram of Z_1 and Z_2 in the complex plane.

which means that

$$R_2 = \omega L_1 \qquad (4.58)$$

$$R_1 = \frac{1}{\omega C} - \omega L_2 \qquad (4.59)$$

Equations (4.58) and (4.59) can now be used to choose R_1, R_2, and C, assuming that L_1 and L_2 are specified by magnetic field requirements. There is no unique solution to (4.58) and (4.59) because many combinations of R_1, R_2, and C will satisfy the equations for given values of L_1 and L_2. However, it is clear that once ωL_1 is specified, R_2 is also fixed. And if ωL_2 is specified, then R_1 and C must be chosen to satisfy (4.59). Thus our task is to choose proper values of R_1 and C.

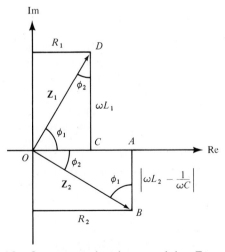

fig. 4.18 Congruent triangles containing Z_1 and Z_2.

The procedure used to obtain (4.58) and (4.59) is a good example of the advantages of phasor diagrams. In comparison to the use of the phasor diagram, substituting appropriate expressions from (4.46) to (4.49) into (4.50) and (4.51) and obtaining (4.58) and (4.59) is a laborious procedure. Not only is the procedure laborious, but the resulting relations are so obscure that you may not actually realize from looking at the expressions that (4.58) and (4.59) must be true. Not only does the use of the phasor diagram reduce the work, but it provides more insight into the nature of the relations; it helps you to visualize how $\psi_2 - \psi_1$ can be equal to 90°, and how the capacitor is required to cause this phase difference.

PROBLEMS

4.22 On one phasor diagram, show the following phasors:

$$\mathbf{I}_1 = 6e^{j30°} \qquad\qquad \mathbf{I}_2 = \frac{1}{R + jX}$$

$$\mathbf{V}_1 = 3 + j2 \qquad\qquad \mathbf{F} = (2 - j3)e^{j60°}$$

4.23 Draw a phasor diagram showing $\mathbf{V}_g$ and $\mathbf{I}$ for the circuit of Fig. 4.11.
4.24 Let $\mathbf{V}_C$ be the phasor voltage across C and $\mathbf{V}_R$ the phasor voltage across R in Fig. 4.6. Make a phasor diagram showing $\mathbf{I}, \mathbf{V}_R$, and $\mathbf{V}_C$.

The Design Parameters. Now let's finalize the design using the following typical values:

$$\omega = 2\pi(15)(10^3) = 94.25(10^3) \tag{4.60}$$

$$L_1 = L_2 = 225 \ \mu\text{H} \tag{4.61}$$

$$\text{Coil dc resistance} = 1.3 \ \Omega \tag{4.62}$$

From (4.58), we must choose

$$R_2 = \omega L_1 = 21.21 \ \Omega \tag{4.63}$$

Since this value includes the dc resistance of coil 2, we must add 18.9 Ω in series with coil 1. Now let's choose R_1 and ωC to satisfy (4.59) and to minimize R_1, hoping that this will minimize the power dissipation in the circuit (power relations are developed subsequently). At this point, we do not really know that minimizing R_1 will minimize the average power dissipation. We suspect it might, because average power is dissipated only by the resistors in the circuit, and not by the inductance or the capacitance. On the other hand, though, the power dissipated is a strong function of the current through the resistors, which is affected by the inductances and capacitances. At any rate, let's choose R_1 to be the minimum possible value, which is the dc resistance of coil 1. Then from (4.59), with

$$R_1 = 1.3 \ \Omega \tag{4.64}$$

we get

$$\frac{1}{\omega C} = 1.3 + 21.21 \ \Omega$$

and

$$C = 0.471 \ \mu F \tag{4.65}$$

Putting these values in (4.46) to (4.49), we get

$$\mathbf{I}_1 = \frac{V_m e^{-j86.49°}}{21.25} \tag{4.66}$$

$$\mathbf{I}_2 = \frac{V_m e^{j3.51°}}{21.25} \tag{4.67}$$

A quick check shows that $\psi_2 - \psi_1 = 90°$ and $|\mathbf{I}_1| = |\mathbf{I}_2|$, showing that the design criteria are satisfied. The next subsection explains how the magnetic field vector rotates and describes a modified circuit for the elementary motor. If you are not interested in that, you can skip to Section 4.6.

* Interpretation of the Circuit Characteristics

How to Produce the Circular Field. With the design parameters that we have just obtained, let's look at how the currents produce the circular field. Doing so will help you connect the phasor-domain characteristics with the time-domain characteristics. Let's suppose that we adjust the generator voltage so that $V_m = 21.25$ V, resulting in a convenient current amplitude of 1 A. Transforming $\mathbf{I}_1$ and $\mathbf{I}_2$ to the time domain gives us

$$i_{1_f} = \cos(\omega t - 86.49°) \tag{4.68}$$

$$i_{2_f} = \cos(\omega t + 3.51°) \tag{4.69}$$

We have called the time-domain currents i_{1_f} and i_{2_f} to emphasize that they are the forced response (steady-state response). The current i_{1_f} in coil 1 will produce a magnetic field vector that lies in the plane of the magnetic sheet (Fig. 4.2) and is perpendicular to the magnetic field vector produced by i_{2_f} in coil 2. To see how the magnetic field rotates, let's say that i_{1_f} produces a magnetic field vector in the x direction, and i_{2_f} produces a magnetic field vector in the y direction, as shown in Fig. 4.19. Expressions for B_x and B_y are

$$B_x = Ki_{1_f} = K \cos(\omega t - 86.49°) \tag{4.70}$$

$$B_y = Ki_{2_f} = K \cos(\omega t + 3.51°) \tag{4.71}$$

where K is a constant determined by the magnetic properties of the coil. The total magnetic field produced by the two currents is the vector sum of B_x and B_y, which can be obtained by graphical vector addition (see Appendix A).

We shall follow the magnetic field through one revolution by summing B_x and B_y vectorially for appropriate values of t. First we choose $t = t_1$, such that

$$\omega t_1 - 86.49° = 0 \qquad \text{or} \qquad \omega t_1 = 86.49°$$

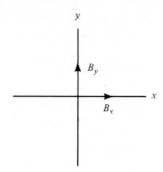

fig. 4.19 The magnetic field vectors B_x and B_y produced by i_{1_f} and i_{2_f}, respectively.

For this value of t,

$$\omega t_1 + 3.51° = 86.49° + 3.51° = 90°$$

Then

$$B_x = K \cos(0°) = K$$

$$B_y = \cos(90°) = 0$$

The vector sum of B_x and B_y is just B_x, since $B_y = 0$. Next we choose $t = t_2$ such that

$$\omega t_2 - 86.49° = 45°$$

which means that

$$\omega t_2 + 3.51° = 135°$$

and

$$B_x = K \cos 45° = 0.707K$$

$$B_y = K \cos 135° = -0.707K$$

The vector sum of B_x and B_y for $t = t_2$ is shown in Fig. 4.20.

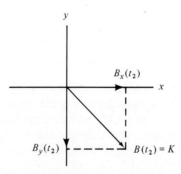

fig. 4.20 Vector sum of B_x and B_y for $t = t_2$.

By continuing this procedure for appropriate values of t, we get the data shown in Table 4.1. The corresponding vector sums are shown in Fig. 4.21, which makes it apparent that the magnetic field B rotates clockwise as time increases. An interesting demonstration of

TABLE 4.1 B_x and B_y for several values of t

t	$\omega t - 86.49°$	$\omega t + 3.51°$	$B_x(t)$	$B_y(t)$
t_1	0	90°	K	0
t_2	45°	135°	$0.707K$	$-0.707K$
t_3	90°	180°	0	$-K$
t_4	135°	225°	$-0.707K$	$-0.707K$
t_5	180°	270°	$-K$	0
t_6	225°	315°	$-0.707K$	$0.707K$
t_7	270°	360°	0	K
t_8	315°	405°	$0.707K$	$0.707K$
t_9	360°	450°	K	0

this rotation can be set up by letting the horizontal displacement of an oscilloscope beam represent the vector B_x and the vertical displacement represent B_y. The electron beam can be made to rotate around a point by connecting a voltage proportional to i_1 to the horizontal input of an oscilloscope and a voltage proportional to i_2 to the vertical input. Have some fun by designing a circuit like the one in Fig. 4.3 and hooking it up to an oscilloscope to produce a circle. The pattern that you see is one form of a *Lissajous pattern*. Other Lissajous patterns can be obtained by varying the frequency and phase of the signals applied to the horizontal and vertical inputs of an oscilloscope.

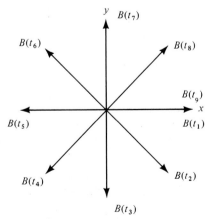

fig. 4.21   Vector sums of B_x and B_y for the values of t shown in Table 4.1.

PROBLEM

4.25 Choose a set of values for R_1 and C other than those used above that satisfy (4.59) and show that (4.50) and (4.51) are satisfied. Draw a phasor diagram showing $\mathbf{I}_1$ and $\mathbf{I}_2$. Use expressions like (4.70) and (4.71) with your values of current and show how the magnetic field rotates.

A Modified Elementary Motor Circuit. The parameters of Fig. 4.3 may not be practical at motor frequencies of 60 Hz or lower if we wish to produce a circular field.[2] The necessary size of the capacitor may be much too large. One can simultaneously decrease the capacitance of the capacitor and increase independently the resistances of R_1 and R_2 and meet the specifications that i_1 and i_2 are equal in magnitude and out of phase by 90°. These changes will result in a decreased magnitude of the currents, so a compromise will be required. See Fig. 4.22.

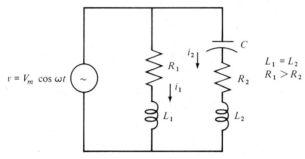

fig. 4.22 Modified parameters for producing 90° phase shift between i_1 and i_2.

PROBLEM

4.26 We wish to produce a circular field using the circuit of Fig. 4.22. The frequency is 60 Hz. The two coils are identical so that $L_1 = L_2 = 0.100$ H. The capacitor is selected such that its reactance has a magnitude four times the reactance of either coil. You may adjust R_1 and R_2 independently to achieve the desired results, but neither of these resistances can be less than coil resistance of 10 Ω. Find the numerical values of R_1, R_2, and C. *Suggestion:* A straight algebraic approach may be rather involved. Note that the impedances of the separate branches must be equal in magnitude, and their angular differences must be 90°. Plot these two impedances, clearly marking the real and imaginary parts.

An Op-Amp Motor Circuit

A circuit that is practical for the elementary motor is shown in Fig. 4.23. In this circuit, an op amp is used to produce a 90° phase shift without using an unreasonably large capacitor. Analysis of this circuit provides a good illustration of the power of the phasor transform techniques. The op-amp circuit involving the resistor R_3, the capacitor C, and the op amp is simply the integrator circuit discussed in

[2] Practical single-phase capacitor motors in starting produce rotating but not circular fields. This compromise can reduce the size and thus the cost of the capacitor.

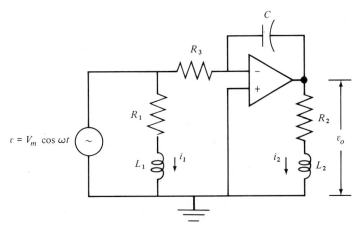

fig. 4.23   *A circuit that can be used for an elemental motor.*

Section 3.5. You may recall that the integration of a sinusoid results in a 90° phase shift (sine function to cosine function, for example). However, we will use the steady-state sinusoidal impedance concept to generalize the basic op-amp voltage-gain equation $G = R_F/R_1$, developed in Section 2.19, to $G = -Z_F/Z_1$. In the circuit of Fig. 4.23, $Z_F = 1/j\omega C$ and $Z_i = R_3$. Then, the voltage gain, or ratio of V_o to V_i is

$$G = -\frac{V_o}{V_i} = -\frac{\dfrac{1}{j\omega C}}{R_3}$$

where V_o and V_i are the phasor values of v_o and v_i, respectively. Then

$$V_o = \frac{j}{\omega C R_3} V_i$$

If C and R_3 are chosen so that $\omega C R_3 = 1$, then V_o and V_i are equal in magnitude but differ in phase by 90° as indicated by the j factor. The capacitor C may be small because R_3 may be very large. If the coils are identical, $R_1 = R_2$ and $L_1 = L_2$; then $|I_1| = |I_2|$ and the angles differ by 90° by simply choosing $\omega R_3 C = 1$. This condition can easily be satisfied at low frequencies allowing easy construction of an elementary motor. Note that the op amp must furnish the current I_2, so this coil current may not exceed the capability of the op amp. Also, the voltages v_i and v_o must not exceed either the op-amp voltage ratings or power supply voltages (not shown).

PROBLEM

4.27   A given op amp has a maximum output current capability of 100 mA. If this op amp is used in the circuit of Fig. 4.23, with $R_1 = 50\ \Omega$, $R_2 = 100\ \Omega$, $R_3 = 10^4\ \Omega$, $L_1 = 0.1$ H, and $L_2 = 0.2$ mH, determine the values of C and V_m required in order to produce the desired rotating field at 60-Hz frequency and have the maximum current through the coils equal to the capability of the op amp. What must be the minimum power supply voltages and voltage ratings for the op amp?

4.6 PERIODIC FUNCTIONS OR WAVES

So far in this chapter, we have gained considerable experience with current and voltage waves that were sinusoidal, as shown in Fig. 4.24a. Electrical engineers must also deal with nonsinusoidal waves such as those shown in Figs. 4.24b, 4.24c, and the *i* wave of Fig. 4.24d. True, the currents and voltages in our power systems (residences, commercial buildings, and industries) are typically sinusoids or near sinusoids. However, others such as communication signals are typically non-sinusoidal, although we often analyze the performance of these systems as though the signals were sinusoids. A "sawtooth" wave like that of Fig. 4.24c is found in television systems and electrical instruments of various kinds. When a sinusoidal voltage *v* such as Fig. 4.24a is applied to an iron-core transformer, the resultant current *i* might be such as shown in Fig. 4.24d. Electrical engineering applications include many other kinds of nonsinusoidal waveforms.

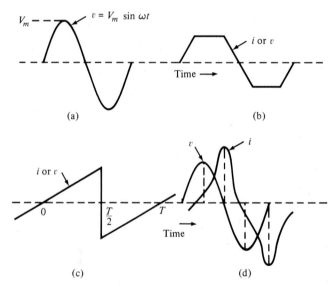

fig. 4.24   Waves or periodic functions in circuits.

So far in this chapter we have learned about how to deal with circuits where currents and voltages are sinusoids, but we must be aware that other periodic functions or waves are important to us and that most of the analysis in this chapter so far is constrained to the steady-state sinusoidal situation. In the remainder of this chapter we will broaden our perspective of the sinusoidal circuit.

In Section 4.7 we will deal with periodic functions or waves, some of which are sinusoidal and others are nonsinusoidal. (See Figs. 4.28, 4.29, and 4.31 as additional examples of nonsinusoidal waves.)

It is also important, before proceeding with Section 4.7, that we understand that a particular wave of applied voltage does not necessarily produce a current wave of the same shape. In Chapter 5 you will see how a sinusoidal voltage applied across any linear *RLC* network produces steady-state or *forced* sinusoidal currents and voltages everywhere in the network. Now it must be clear in your mind that the sinusoidal situation is special, not general. For example, a triangular voltage applied to an inductance does not produce a steady-state current of the same triangular shape. Problems 4.28 and 4.29 make it clear that the sinusoidal situation is special.

PROBLEMS

4.28   If, in Fig. 4.25, the maximum value of the steady-state current is 1 A and the maximum value of the voltage wave is 300 V, find the simplest possible circuit that would have this current wave when the particular voltage wave is applied. The waves are symmetrical as suggested by the figure, and the voltage wave is zero one-half of the time.

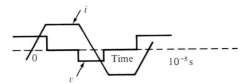

fig. 4.25   Current and voltage waves.

4.29   The sinusoidal voltage wave of Fig. 4.26 is applied to some circuit and the resultant steady-state current *i* is shown in Fig. 4.26. Find the simplest possible circuit that will perform in this manner. Describe the circuit numerically.

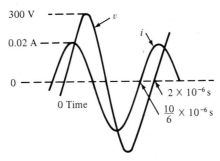

fig. 4.26   Current and voltage waves.

4.7   POWER AS A FUNCTION OF TIME AND AVERAGE POWER

In Chapter 1 we learned that power as a function of time is given by

$$p = iv \tag{4.72}$$

where p is in watts, i is in amperes, and v is in volts. This equation is perfectly general and holds for all situations where p, i, and v are any functions of time.

Where v has the polarity and i the direction indicated in Fig. 4.27, (4.72) gives the power flowing into the "box." "Box" refers to any component, circuit, or system to which electric power is delivered. The box might contain an electric motor, an industrial plant, a city, an integrated circuit, a computer, or whatever. A negative power flowing into the "box" means positive power leaving the box. Consider a case where a motor is in the box. Of the instantaneous power that flows into the box, some might be converted to mechanical power, some converted to thermal power (where the heat is wasted, it is considered to be a loss), and some stored in the magnetic fields or inductances of the motor. If the box contained a water immersion heater having no inductive or capacitive properties, then all of the power into the box would be converted to thermal power to raise the temperature of the water.

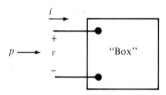

fig. 4.27   p, i, and v.

Where i and v are periodic, p will also be periodic. In a typical situation, i and v will be positive during part of the period and negative during another part. Of course, if i has the opposite sign (by the standard of Fig. 4.27) from v, then p will be negative.

PROBLEM
4.30 For the situation of Prob. 4.28,
    a. Plot power as a function of time on the same coordinate system on which you have plotted both i and v.
    b. By inspection, what is the average value of the power over a full period of the p wave?

Your solution to Prob. 4.30 should demonstrate that for this particular situation the average value of the power over one period is zero. Your plot shows that the power flow into the inductor is positive for a half-period, and then the power *leaving* the inductor (negative power into) is positive for the next half. Or put differently, the inductor unit absorbs energy through one half-period and then gives up that same energy back to the electric circuit during the next half-period. Where the voltage applied to any inductance or any capacitance is periodic (any wave shape), that circuit element absorbs energy through one half-period, then gives up the same energy in the next half-period.

PROBLEMS
4.31 The sinusoidal v wave of Fig. 4.26 is applied to the "box" of Fig. 4.27. The resultant current wave i is also shown in Fig. 4.26.
    a. On one set of coordinates, plot the waves of i, v, and p. Prepare these plots quickly

without detailed calculations. Note that p will be zero whenever either i or v is zero. Note also that p will be negative when i and v have opposite signs.

b. Estimate, do not calculate with precision, the maximum value of power, the average value of power, and the maximum magnitude of the negative power. Notice that none of these are simply related to the product of the maximum current and the maximum voltage.

4.32 On the same set of coordinates used for the solution of Prob. 4.31, prepare an approximate plot of the *energy* absorbed by the "box" over two periods.

4.33 As was stated before, Fig. 4.24d shows a typical v-i relationship of an iron-core transformer. The maximum value of v is 300 V and the maximum value of i is 2 A. Estimate the average power absorbed by this transformer. Do this quickly without detailed calculation.

You should remember that, for the most part in this text, analyses are constrained to linear circuits (R, L, and C independent of current and voltage), with only an occasional side trip into the nonlinear realm, such as in Prob. 4.33. In these linear circuits, a sinusoidal current produces a sinusoidal voltage in the steady-state condition—or a sinusoidal voltage produces a sinusoidal current. Figure 4.26 is an illustration. But we must not draw the conclusion that a periodic wave of non-sinusoidal shape will similarly produce a voltage of the same shape in these linear circuits. Figure 4.25 illustrates this point where we have current and voltage in an inductance. In general, one can expect almost any current wave shape arising from a particular nonsinusoidal voltage wave shape. Furthermore, if one or more elements in a circuit are nonlinear, then a sinusoidal voltage will *not* produce a sinusoidal current. Figure 4.24d illustrates voltage and current in an iron-core transformer where the inductance is nonlinear, or, in other words, where the inductance is a function of current.

4.8 AVERAGE POWER

Power as a function of time is given by (4.72). Except as a means to an end, in most situations we are not interested in how power varies with respect to time. We are very much more interested, typically, in the average power wherever the phenomenon is periodic. For a periodic situation, the average power output of an electronic amplifier, the average power converted to heat, the average power converted to light, and the average power converted to mechanical power are of first interest to us. Where power is periodic, then the average power is given by

$$P = \frac{1}{T} \int_0^T p \, dt = \frac{1}{T} \int_0^T iv \, dt \qquad (4.73)$$

where P is average power and T is period. Notice that this equation can be applied to any periodic situation independent of wave shape. Problem 4.34 illustrates this point.

PROBLEMS
4.34 For the wave shapes of Fig. 4.25, assume that the maximum value of the current is 10 A and the maximum value of the voltage is 20 V. Prepare a graphical plot of p or iv. Be

careful with the signs to make sure that p is positive when both i and v have the same sign, and negative when these two have opposite signs. Now, without writing equations for i or v, approximate the integral of (4.73) to find the average power P. For this situation, P is 50 W. Justify this graphically without writing equations.

4.35 Where the maximum value of the voltage wave is 300 V and the maximum value of the current wave is 2 A in the situation of Fig. 4.24d, approximate the average power P. The voltage wave here is a sinusoid and you can, of course, write its equation; on the other hand, do not attempt to write the equation of the current wave i. To solve this problem, make an approximate sketch of p or iv, and from this sketch estimate the value of P. When you talk to your classmates about the solution to this problem, do not expect a precise comparison between the numerical values of P. Do expect, however, to agree on whether P is zero, positive, or negative. Also, expect to agree on whether P is greater or less than the product of the maximum value of voltage and maximum value of current.

We have a very special interest in the average power where both current and voltage are sinusoids, as indicated in Fig. 4.26. We will come back to this situation shortly.

4.9 EFFECTIVE VALUES

The *effective* value of a current or voltage is very useful when we are interested in *average* power P. This is especially true in the case of sinusoids. Earlier in this chapter we used the symbols I and V (lightface italic type as contrasted with boldface type) to represent the respective maximum values of sinusoids. That was convenient at that time, but now through the remaining part of this text, the effective value will dominate the scene, so I and V will represent effective values and I_m and V_m will represent maximum values. Of course, **I** and **V** will represent phasors. The effective value will dominate many of the situations throughout the remaining part of this text. Its meaning is given by (4.74).

$$P = I^2 R \qquad (4.74)$$

where I is the *effective* value of the current in resistance R producing *average* power P.

Consider the following illustration. A steady current $i = 10$ A flows through an immersion water heater having a resistance of 12 Ω. The power (i^2R) or $10^2 \times 12 = 1.2$ kW is also steady, so the average power P is also 1.2 kW. From (4.74), the effective current I is $\sqrt{P/R} = \sqrt{1,200/12} = 10$ A. So, for this case, the instantaneous value i, the average value I_{av}, and the effective value I are separately equal to 10 A. This is not surprising, but consider the situation where the current in the same 12-Ω resistance is sinusoidal, $i = I_m \sin \omega t$, and I_m is such that the same average $P = 1.2$ kW is in this resistance. Because $p = i^2R$, it is clear that the instantaneous p is periodic having values $0 \le p \le I_m^2 R$. If maximum value I_m were 10 A, then the maximum power would be 1.2 kW, and the average power P would certainly be less than the maximum in this case where instantaneous power varies from its maximum to zero. So for a sinusoid, the maximum value of a current is certainly

greater than its effective value. We shall soon discover how much greater. By definition, the effective value of the current is such that when it is squared and multiplied
by the resistance through which it flows, the result is the average power in that
resistance; see (4.74). Then solving for the square of the effective value of current, it
follows that

$$I^2 = \frac{P}{R} = \frac{1}{RT} \int_0^T iv \, dt = \frac{1}{RT} \int_0^T i \times iR \, dt = \frac{1}{T} \int_0^T i^2 \, dt \tag{4.75}$$

or

$$I = \sqrt{\frac{1}{T} \int_0^T i^2 \, dt} \tag{4.76}$$

The effective value is sometimes called the *rms* value, as indicated by (4.76). In
finding the effective or *rms* value, notice that we take the square root (*r*) or the mean
(*m*) of the squared (*s*) values.

You are reminded of the importance of using technical words and symbols
consistent with their precise meanings. Notice again that *effective* (not average)
value of current leads to *average* (not effective) value of power; see (4.74) again.
Effective power is not defined. Average value of current has some significance, but
not with respect to average power.

Example 4.1

Find the effective value of the current wave of Fig. 4.24c if the peak value is 10 A.
By inspection, one can see that the appropriate integral of i^2 from $T/2$ to T is
identically equal to the corresponding integral from 0 to $T/2$. Therefore,

$$I^2 = \frac{2}{T} \int_0^{T/2} i^2 \, dt \tag{4.77}$$

and for $0 < t < T/2$,

$$i = \frac{10}{T/2} t = \frac{20}{T} t \tag{4.78}$$

It follows that

$$I^2 = \frac{2}{T} \int_0^{T/2} \frac{20^2}{T^2} t^2 \, dt \tag{4.79}$$

It follows after performing the integration that

$$I^2 = \frac{2 \times 20^2}{T^3} \frac{t^3}{3} \bigg|_0^{T/2} = \frac{2 \times 20^2}{2^3 \times 3} = \frac{100}{3} \tag{4.80}$$

and finally we get

$$I = \frac{10}{\sqrt{3}} = 5.773 \text{ A} \tag{4.81}$$

One would certainly expect the effective value to be less than the maximum value (10 A in this case). It is interesting to note that the average value of this wave over a period is zero and its average value between zero and $T/2$ is 5 A. One should carefully draw the clear distinction between average value of a current and its effective value. Except in special cases, we do not try to relate the average value of a current to average value of power, as one might suspect. We will, however, continually relate effective value of current to average power.

• • •

PROBLEMS

4.36 Find the effective value of the rectangular current wave of Fig. 4.28 as a function of k. Do this quickly by performing a graphical integration.

Answer : $I = \sqrt{k}$

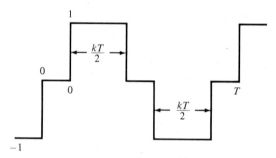

fig. 4.28   Rectangular wave.

4.37 Apply (4.73) to the situation of Fig. 4.28, where $k = 1/2$ and where this current flows through resistance R, in order to find P, the average power in R. Also, compute P from your effective value of Prob. 4.36. Obviously these two values of P should agree.

The effective value of any wave can be found by the application of (4.76).

The effective value of a voltage wave is defined in much the same way that the effective value of current was defined by (4.74). In other words, where $P = V^2/R$, V is the effective value of the voltage.

PROBLEM

4.38 Prove that effective value of voltage is given by (4.82).

$$V = \sqrt{\frac{1}{T} \int_0^T v^2 \, dt} \tag{4.82}$$

The effective value of a sinusoidal current or a sinusoidal voltage is extremely important to much of the remaining content of this textbook. Given

$$i = I_m \sin(\omega t + \alpha)$$

it follows that we can find the effective value squared by

$$I^2 = \frac{1}{T} \int_0^T i^2 \, dt = \frac{I_m^2}{T} \int_0^T \sin^2(\omega t + \alpha) \, dt \qquad (4.83)$$

For the $\sin^2$ term, we substitute an appropriate trigonometric identity so that

$$I^2 = \frac{I_m^2}{T} \int_0^T \left[\frac{1}{2} - \frac{1}{2} \cos 2(\omega t + \alpha) \right] dt \qquad (4.84)$$

PROBLEMS

4.39 Refer to the appropriate trigonometric identity and prove that (4.84) follows from (4.83).

4.40 a. Justify by graphical inspection that

$$\sin^2 \omega t = [\tfrac{1}{2} - \tfrac{1}{2} \cos 2\omega t]$$

Equation (4.84) simplifies considerably in the fact that the integral of the cosine term over one period is zero.

b. Formally perform the indicated integrations to prove that

$$\int_0^T \sin(n\omega t + \alpha) \, dt = \int_0^T \cos(n\omega t + \beta) \, dt = 0 \qquad (4.85)$$

where $\omega T = 2\pi$ and $n = 1, 2, 3, 4, \ldots$.

Notice that in doing so you have proved that the integral of any sine or cosine term over an integral number of full periods is equal to zero.

Then it immediately follows from (4.84) and (4.85) that the effective value of a sinusoidal current is given by

$$I^2 = \frac{I_m^2}{2T} \int_0^T dt = \frac{I_m^2}{2} \qquad (4.86)$$

or

$$I = \frac{I_m}{\sqrt{2}} = 0.7071 I_m \qquad (4.87)$$

and from Prob. 4.38, the effective value of a sinusoidal voltage is

$$V = \frac{V_m}{\sqrt{2}} = 0.7071 V_m \qquad (4.88)$$

Of course (4.87) and (4.88) apply only to sinusoids; however, (4.76) can be applied to any periodic wave to determine its effective value.

When measuring sinusoidal voltages and currents, it is important to understand whether the instrument indication is the maximum, average, or effective value of the voltage or current. The traditional ac voltmeter and ammeter indicate effective values. The dc voltmeter and ammeter typically indicate average values. In the typical ac voltmeter and ammeter, instantaneous force produced on the moving element is proportional to the squared value of the current or voltage being measured. The inertia of the moving element averages this force. Then the scale is adjusted so that the "needle" indicates effective value. In these situations, the instrument is said to have no *waveform* error. In an electronic voltmeter, the current in the moving coil is ofttimes proportional to the maximum value of the voltage or current to be measured. Then its scale might be adjusted to indicate effective value, assuming the wave is sinusoidal. Then for a nonsinusoidal wave, this instrument would have a fundamental error.

PROBLEM

4.41 An electronic voltmeter is designed so that the deflection of the moving element is proportional to the maximum value of any particular wave, but the scale of the instrument is calibrated to read effective volts for a *sinusoidal* wave. This voltmeter is used to measure the *rms* value of the triangular wave of Fig. 4.29. The voltmeter indication is 10 V. Find the true rms value of this triangular wave.

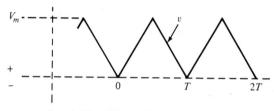

fig. 4.29   Triangular voltage wave.

Using Graphics to Estimate the Effective Value of a Particular Wave

As indicated by (4.75), the square of the effective value of a particular wave merely means the average value of the squared wave. Let us see how we could apply this concept to estimate quickly the effective values of waves. Let us return to the sinusoid (see Fig. 4.30). Without any consideration of the trigonometric identity used previously, we could have estimated the approximate shape of the i^2 as shown in Fig. 4.30. We could have easily spotted the points where this squared function is

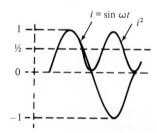

fig. 4.30   *i* and i^2 (*another example*).

zero and where it is a maximum. Then it would have been reasonable to estimate the average heights of the i^2 wave to be approximately one-half of its maximum value. This gives us a check on (4.86).

PROBLEMS

4.42 Consider the example of Fig. 4.31. Given the triangular voltage v, the v^2 wave is easily graphed approximately, as shown. One might estimate the average height of the v^2 wave to be about one-fourth of 64, or 16. This leads to an estimated rms value of 4. Compute the precise rms value of the voltage wave of Fig. 4.31 and compare this to the estimated value of 4.

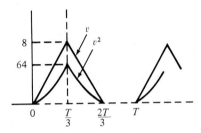

fig. 4.31 v and v² (another example) .

4.43 First estimate the rms value of the triangular wave of Fig. 4.29. Then compute its precise value.

Average Power and Effective Values

We must not lose track of the relationship between effective value of current and/or voltage in relationship to average power. You will remember from (4.74) that effective value of current was defined such that its square times the resistance through which the current flows gives average power in that resistance and, similarly, in Prob. 4.38, you demonstrated that the effective value of a voltage was defined such that V^2/R gave the average power in a resistor where the voltage v is impressed across that resistor. These definitions of effective current and effective voltage are summarized as follows:

$$P = I^2R \quad \text{and} \quad P = \frac{V^2}{R} \tag{4.89}$$

There are reasons to suspect that average power might be determined by the product of effective current and effective voltage. Instantaneous power is, of course, equal to the product of instantaneous current and instantaneous voltage. Where both current and voltage are steady, then average power is certainly the product of effective current and effective voltage. Of course where current or voltage is steady, the effective and the average values are identical. In this section, however, we are primarily concerned with the effective values of periodic currents and voltages, in general.

For the restricted situation of any periodic current and its periodic voltage in a resistance, effective voltage and effective current are related, as follows:

$$V = \sqrt{\frac{1}{T} \int_0^T v^2 \, dt} = \sqrt{\frac{1}{T} \int_0^T (iR)^2 \, dt} = R\sqrt{\frac{1}{T} \int_0^T i^2 \, dt} = RI \qquad (4.90)$$

This concludes that $V = IR$, but this is not the old familiar situation where V and I are steady values. These symbols here refer to effective values, where i is any periodic function. Then from (4.90) we get

$$P = I^2 R = I(IR) = IV \qquad (4.91)$$

Here we see that effective power in a resistor, for any periodic current or voltage, is equal to the effective value of the current times the effective value of the voltage. However, in a more general situation where L and/or C are present, the average value of power is not given by the product of effective voltage and effective current. For example, in Fig. 4.32c, we have an applied voltage at C having an effective value

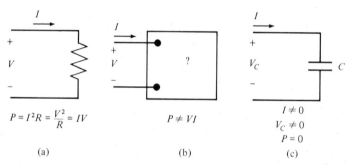

$$P = I^2 R = \frac{V^2}{R} = IV \qquad\qquad P \neq VI \qquad\qquad \begin{aligned} I &\neq 0 \\ V_C &\neq 0 \\ P &= 0 \end{aligned}$$

(a)   (b)   (c)

fig. 4.32   *Average power and effective current and voltage.*

V_C and the consequent effective current I. These two values are nonzero, but we will soon show again that the average power is equal to zero. This point was discussed earlier in the text where it was pointed out that the capacitor could store energy but not convert it to some other form, and therefore in a periodic situation it cannot consume or absorb average power. So, in general, where a periodic voltage having an average value of V is applied to an unknown system or "box" and there is a consequent, periodic current having an effective value I, the average power will not, except in special situations, be equal to VI. The situation of Fig. 4.24d is another illustration. Here both the i and v are periodic, and each will have its respective effective value. One certainly cannot expect that the average power will be equal to the product of the effective value of voltage and the effective value of current. For still another example, examine your solution to Prob. 4.31. Here you will see that average power P is not the product of effective current I and effective voltage V.

4.10   *AVERAGE POWER AND SINUSOIDS*

We found earlier that where any periodic current (sinusoidal included) flows through resistance, the average power P is the product $I \times V$, where I and V are effective values. For resistance $P = IV$. As was pointed out, this is a special case. Where L or C is involved, then $P \neq IV$. Now we wish to give special attention to average power P where i and v are sinusoidal but not in phase, a more general situation compared to the resistance case where i and v are in phase. Consider any situation where a sinusoidal voltage v is applied to some terminals, or to a box, and current i flows as a consequence, then

$$v = V_m \cos \omega t \qquad \text{and} \qquad i = I_m \cos(\omega t + \theta) \tag{4.92}$$

Notice that we have lost no generality in choosing v as a reference. Expressed as phasors,

$$\mathbf{V} = V \angle 0 \qquad \text{and} \qquad \mathbf{I} = I \angle \theta$$

where V and I are the effective values, respectively, or

$$V = \frac{V_m}{\sqrt{2}} \qquad \text{and} \qquad I = \frac{I_m}{\sqrt{2}}$$

It is important that you keep in mind the clear distinction in the meaning of $\mathbf{V}$ compared to V, and $\mathbf{I}$ compared to I. In effective values, the phasor (a complex number) is represented by $\mathbf{V}$ or $\mathbf{I}$, respectively. The magnitudes of these complex numbers are V and I, respectively.

Now it follows from (4.92) that

$$p = vi = V_m \cos \omega t \times I_m \cos(\omega t + \theta) \tag{4.93}$$

Then using a convenient trigonometric identity,

$$\cos a \cos b = \frac{1}{2} [\cos(a - b) + \cos(a + b)]$$

It follows that

$$p = \frac{V_m I_m}{2} [\cos \theta + \cos(2\omega t + \theta)] \tag{4.94}$$

Remember that

$$P = \frac{1}{T} \int_0^T p \, dt \tag{4.95}$$

We are already familiar with the fact that the integral of $\cos(2\omega t + \theta)$ from 0 to T is equal to zero; see (4.85). Therefore,

$$P = \frac{1}{T} \frac{V_m I_m}{2} \cos \theta \int_0^T dt = \frac{V_m I_m}{2} \cos \theta \tag{4.96}$$

Eliminating maximum values V_m and I_m in favor of effective values V and I,

$$P = \frac{(\sqrt{2}V) \times (\sqrt{2}I)}{2} \cos \theta = VI \cos \theta \text{ watts} \tag{4.97}$$

Notice the θ has the meaning of the phase difference between the phasor current **I** and the phasor voltage **V**. Again note that V and I are the effective values of the magnitudes of the voltage and current, respectively.

Example 4.2

Find the average power flow into the "box" of Fig. 4.33, where

$$i = 10 \cos(\omega t + 20°) \text{ A}$$

and

$$v = 30 \sin(\omega t + 158°) \text{ V}$$

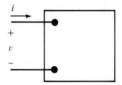

fig. 4.33   Sinusoidal v and i.

Expressed as phasors (effective values) in respect to the sine term $\mathbf{I} = 10/\sqrt{2} \angle 20°$ A, $\mathbf{V} = 30/\sqrt{2} \angle -90° + 158°$ V.

$$P = \frac{10 \times 30}{\sqrt{2} \times \sqrt{2}} \cos(-90° + 158° - 20°) = 150 \cos 48° = 100.4 \text{ W}$$

The phasor diagram gives another view of P as a function of **V** and **I**. Consider the values of this example (see Fig. 4.34). The phase difference between **V** and **I** is easily seen to be 48°. Therefore,

$$P = VI \cos \theta = 21.21 \times 7.07 \cos 48° = 100.3 \text{ W}$$

Notice that $I \cos \theta = a$ is the component of **I** in phase with **V**. Average power P is produced by the component of **I** in phase with **V** (or the component of **V** in phase with **I**). There is no average power associated with the combination of **V** and b,

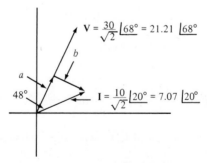

fig. 4.34   **V** and **I** of Example 4.2.

where b is the component of **I** normal to **V**. Then one can use the phasor diagram in estimating P. From the diagram of Fig. 4.34, it would be easy to estimate a as about 5. Then based on this estimate, P is approximately $5 \times 21 = 105$ W, as compared to the precise value of 100.3 W.

• • •

Phasors in Effective Values

Throughout the remaining part of this text, phasor currents and voltages will always be given in effective values unless specifically stated to the contrary. Maximum values will be the exception, not the rule. In engineering practice, phasor voltages and currents are specified in effective values unless specifically labeled as maximum.

For example, we refer to the residential ac voltage at the outlet as 120 V. This is an rms or effective value. The maximum value of this same voltage is $\sqrt{2} \times 120 = 169.7$ V.

PROBLEM

4.44   A voltage $\mathbf{V} = 8 \angle -20°$ V is applied to an impedance of $Z = 5{,}000 - j7{,}000 \ \Omega$.
   a.   To scale, draw a phasor diagram showing **V**, **I**, and then estimate the component of **I** in phase with **V**.
   b.   From the diagram, preestimate P.
   c.   Calculate P with precision.

Power Factor

Equation (4.97) can be written in a different form,

$$P = VI \cos \angle_{\mathbf{I}}^{\mathbf{V}} \tag{4.98}$$

where $\angle_{\mathbf{I}}^{\mathbf{V}}$ means the angle difference between **V** and **I**. Here the sign of this angle is immaterial because $\cos(-\theta) = \cos \theta$. Cos $\angle_{\mathbf{I}}^{\mathbf{V}}$ is called the *power factor* and $\angle_{\mathbf{I}}^{\mathbf{V}}$ is called the power factor angle. Notice that in a situation like that of Prob. 4.47, the power factor angle is the angle of the impedance **Z**. We will return to the significance of power factor in Chapter 6 when we study "complex power," a broader view.

PROBLEM

4.45   Find the power factor and power factor angle of each of the following situations:
   a.   Problem 4.44
   b.   Example 4.2.

Power in Impedance **Z**

A voltage **V** is applied to an impedance **Z** resulting in a current **I**. This is displayed in Fig. 4.35, where $\mathbf{Z} = (R - jX)$(capacitive); hence

$$\mathbf{V} = \mathbf{IZ} = \mathbf{I}R - jX\mathbf{I} \tag{4.99}$$

$$P = VI \cos \theta$$

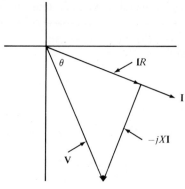

fig. 4.35 I and V for Z.

but

$$V \cos \theta = IR$$

so

$$P = I(IR) = I^2R \qquad (4.100)$$

Again we see that the product of (IR), the component of **V** in phase with **I**, and I yields the average power P. And because $-jX\mathbf{I}$ is normal to **I**, there is no average power associated with these two. This agrees with our previous experience that reactance (capacitive or inductive) cannot absorb average power. Or, all of the average power is absorbed in resistances. Even in a motor or a battery where electric power is absorbed or converted to another form (mechanical or chemical), we accordingly ascribe the property of resistance to the model of such a device to represent the power absorbed.

Example 4.3
The fact that the average power is absorbed only in resistance brings us to a powerful tool for checking consistency of problem solutions that involve the calculation of average power. Consider the situation of Prob. 4.44. Here

$$\mathbf{I} = \frac{\mathbf{V}}{\mathbf{Z}} = \frac{8 \angle -20°}{5,000 - j7,000} = \frac{8 \angle -20°}{8,602 \angle -54.46°} = 0.9300 \angle 34.46° \text{ mA} \qquad (4.101)$$

Then,

$$P = IV \cos \theta = 0.9300 \times 8 \cos(34.46° + 20°) = 4.325 \text{ mW}$$

Now let us make a consistency check as follows:

$$P = I^2R = (0.93 \times 10^{-3})^2 \times 5,000 = 4.325 \text{ mW}$$

● ● ●

PROBLEM
4.46 A sinusoidal voltage **V** is impressed on an element having admittance $\mathbf{Y} = g + jb$. Prove that $P = V^2g$.

Conservation of Average Power

The general principle of the conservation of average power is suggested in the previous paragraph. Consider some system or "box" forced by sinusoids as shown in Fig. 4.36. Notice the particular assignment of currents and voltages. Average powers are as follows:

$$P_1 = I_1 V_1 \cos\langle^{I_1}_{V_1} \qquad (4.102a)$$

$$P_2 = I_2 V_2 \cos\langle^{I_2}_{V_2} \qquad (4.102b)$$

$$P_3 = V_3 I_3 \cos\langle^{V_3}_{I_3} \qquad (4.102c)$$

$$P_4 = V_4 I_4 \cos\langle^{V_4}_{I_4} \qquad (4.102d)$$

It follows that P_1, P_2, and P_3 are average powers flowing into the "box"; P_4 is power *leaving*.

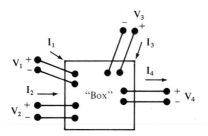

fig. 4.36 Conservation of average power.

The sum $(P_1 + P_2 + P_3 - P_4)$ is the net average electric power flow into the box. Inductance and capacitance can store instantaneous power, but not average power. So this net average electric power flowing into the box must be converted to some other form: electromagnetic radiation, thermal, mechanical, chemical. If the box contains only R, L, and C components, then

$$\underbrace{P_1 + P_2 + P_3}_{\substack{\text{Power} \\ \text{in}}} - \underset{\substack{\text{Power} \\ \text{out}}}{\downarrow}{P_4} = \underset{\substack{\text{Power} \\ \text{absorbed}}}{\downarrow}{\Sigma\, I^2 R} \qquad (4.103)$$

where $\Sigma\, I^2 R$ means the sum of all of the separate $I^2 R$ powers converted in the "box." If some of this converted power appears in a mechanical form as an example, then we could model this converted power as an equivalent $I^2 R$ power.

Then the principle of conservation of average power is:

> The sum of all electric average powers flowing into a closed volume is equal to the sum of all average electric powers leaving, plus internal average powers converted to some nonelectrical form, such as electromagnetic radiation, thermal, mechanical, chemical.

It is important to notice that *instantaneous powers* $v_1 i_1$, $v_2 i_2$, and so on are not so easily related because L and C elements do *not* absorb *average* power but *do* absorb *instantaneous* power.

Equations (4.102) for P_1, P_2, and so forth were written for sinusoidal currents and voltages; nevertheless, the principle of conservation of average power holds wherever the currents and voltages are periodic and thus *average power* has meaning.

4.11 COMPLEX POWER

A quantity known as *complex power* is represented by the symbol **S** and is given by

$$\mathbf{S} = \mathbf{VI^*} = P + jQ \qquad (4.104)$$

Notice that all three quantities, **S**, **V**, and **I**, in this equation are complex. The magnitude of **S** is called the *volt-amperes* or VI. The quantity **I*** is called the conjugate of **I**, or where $\mathbf{I} = I \angle \theta$, $\mathbf{I^*} = I \angle -\theta$. We will prove presently that the real part of **S** or (**VI***) is the average power P, as given by (4.98). The imaginary part Q of the complex power **S** is sometimes called vars (volt-amperes reactive). The reactive elements (L and C) only absorb vars, as we will soon see. Resistance elements only absorb average power P, which is the real part of the complex power **S**. Where $\mathbf{V} = V \angle \alpha$ and $\mathbf{I} = I \angle \beta$ then (4.104) gives **S** in a polar form.

$$\mathbf{S} = \mathbf{VI^*} = V \angle \alpha \times I \angle -\beta = VI \angle \alpha - \beta \qquad (4.105)$$

Changing **S** to its rectangular form gives us P and Q, as follows:

$$\mathbf{S} = VI \angle \alpha - \beta = VI \cos(\alpha - \beta) + jVI \sin(\alpha - \beta) \qquad (4.106)$$

and since

$$\mathbf{S} = P + jQ$$

then

$$P = VI \cos(\alpha - \beta) \quad \text{and} \quad Q = VI \sin(\alpha - \beta) \qquad (4.107)$$

We see that the angle $(\alpha - \beta)$ is the angle between **V** and **I**. It follows that the real part of **S** truly is the average power as we have known it previously.

PROBLEMS

4.47 Prove that in a series RL circuit, the Q, in vars, is given by $I^2 \omega L$, and the average power by $I^2 R$. Base your proof on (4.107).

4.48 Prove that capacitance consumes negative vars having magnitude $V^2(\omega C)$ or $I^2/(\omega C)$.

All of our electric motors and transformers, as well as some other devices, have inductance and therefore consume vars. The electric utility companies typically deliver both watts and vars to its customers. Of course the utility company must consume fuel to generate the watts, but not the vars. On the other hand, the vars do load the system and the components in the system, such as the generators, transformers, transmission lines, and so on. Thus the capacities, and therefore the cost, of power devices such as generators and transformers is determined by the volt-amperes.

Example 4.4
An industrial plant consumes 100 kW at 0.7071 power factor lagging (current lags voltage). Since the phase angle θ between $\mathbf{V}$ and $\mathbf{I} = \cos^{-1} 0.7071 = 45°$, $S = VI = P/\cos 45° = 1.414 \times 10^5$, and $Q = S \sin \theta = 10^5$ vars. See Fig. 4.37a. A load having negative vars, such as a bank of capacitors, may be added to increase the power factor and reduce the volt-amperes as illustrated in Fig. 4.37b. The cost of the

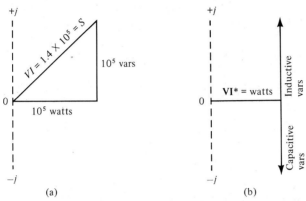

fig. 4.37   (a)   A power triangle showing the relationships among watts, vars, and **VI*** for the plant of Example 4.4. (b)   Illustration of the reduction in VI by adding capacitive vars to increase the power factor to unity.

capacitors must be weighed against the savings in the reduced electrical capacity of the utility system that delivers $\mathbf{S}$, the complex power. The greatest economic gain may be achieved by increasing the power factor to a value less than unity as may be perceived by solving Prob. 4.49.

● ● ●

The required voltage and capacitance ratings for the capacitor bank may be readily determined once the line voltage and the required vars are determined, since vars $= VI \sin \theta = I^2 X_C$.

PROBLEM
4.49   An industrial plant consumes 100 kW at 0.75 power factor lagging. The line voltage is 480 V.
   a.   Determine the VI and the vars for the plant.
   b.   Determine the capacitive vars and the capacitance required to increase the power factor to 0.87 lagging. What is the percentage reduction in VI?
   c.   Determine the required μF per reduced kVA in raising the power factor from 0.75 to 0.87.
   d.   Determine the required μF per kVA required per reduced kVA if the power factor were raised to 1.0 instead of 0.87.

4.12 AVERAGE VALUES OF CURRENT AND VOLTAGE

The average value of a current or voltage is useful in some situations. The average value of a current is no different from the average value of any other physical quantity or mathematical function; that is,

$$I_{av} = \frac{1}{T} \int_0^T i \, dt \qquad (4.108)$$

where i is periodic. The average value of the most typical wave is zero. Notice, because of the symmetry of each of the five waves of Fig. 4.24, the average value of each is equal to zero. Also, the average value of each of the waves of Figs. 4.25 and 4.26 is zero. However, the average value of the triangular wave of Fig. 4.29 is nonzero.

PROBLEMS

4.50 a. Find the average value of the wave of Fig. 4.29. View the symmetry of this wave and write the average value without performing an integration.

b. Apply the formal integration of (4.108) to this wave to find the average value.

c. This triangular wave of Fig. 4.29 is approximated as follows:

$$v_a = V_{av} + V_s \sin \omega t \qquad (4.109)$$

where V_{av} is the average value of the triangular wave. In this equation find the value of V_s such that the effective value of v_a is the same as the effective value of the real wave v. On one set of coordinates, superimpose plots of v and v_a for your value of V_s. These two waves should coincide reasonably closely.

4.51 Electronic rectifiers convert a sinusoidal voltage into a pulsating dc voltage, as indicated in Fig. 4.38. This process is called rectification.

a. Find the average value of this rectified sine wave.

b. Find its effective value. You should be able to relate the effective value of the rectified sine wave to the effective value of the unrectified sine wave without going through the formality of an integration.

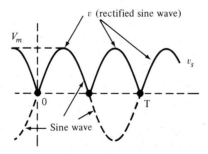

fig. 4.38   Rectified sine wave.

Your solution to Prob. 4.51 should have shown that the average value of a rectified voltage wave having a maximum value of V_m is $2V_m/\pi$.

The average value of a current or voltage wave is sometimes called the *dc value*.

PROBLEMS

4.52 A periodic current wave i having a dc value that is nonzero is applied to a parallel combination of RC, as shown in Fig. 4.39. The wave consists of a sinusoidal component having an angular velocity of 50,000 rad/s superimposed upon a dc value as shown. The value of the capacitance is 10^{-7} F and the value of the resistance is 20 kΩ. The current i has been flowing for an indefinite time such that the voltage v has reached the steady state. Under these conditions, the capacitor current i_C can have no dc component; otherwise, the voltage across the capacitor would increase indefinitely. All of the dc components of the current must be in i_R. The sinusoidal component of the current will divide between i_C and i_R. In the time domain, find the equation of v. You should find that v is largely dc, with a relatively small sinusoidal component.

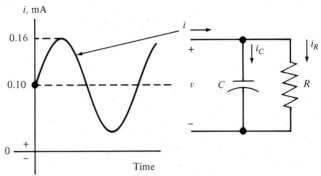

fig. 4.39   A wave having dc value is applied to parallel RC.

4.53 Figure 4.40 displays a model of a *parallel resonant circuit*, which plays an important role in radio, television, and in other signal-selecting situations. This circuit uses an inductor, which is modeled by the series combination of R_L and L. The capacitor model and other components of the circuit (not discussed here) lead to the value of R. The current source $\mathbf{I}$ *drives* the circuit. In this model, $R = 7$ kΩ, $R_L = 2$ Ω, $\omega L = 200$ Ω, and ωC is adjusted such that $\mathbf{I}_{LC}$ is in phase with $\mathbf{V}$. $I = 20.0$ mA.

  a.   Find the numerical values of V, I_L, I_C, I_{LC}. Use circuit reduction ideas. Note parallel branches.

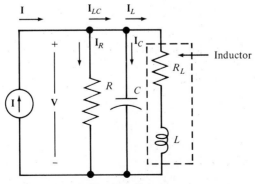

fig. 4.40   Parallel resonance.

b. From (a), calculate powers:

$$VI \cos\langle^{\mathbf{I}}_{\mathbf{V}}, \quad VI_R \cos\langle^{\mathbf{I}_R}_{\mathbf{V}}, \quad VI_C \cos\langle^{\mathbf{I}_C}_{\mathbf{V}}, \quad VI_L \cos\langle^{\mathbf{I}_L}_{\mathbf{V}}$$

Apply the principle of conservation of power to these as a consistency check.

c. Check consistency by calculating $I_R^2 R$ and $I_L^2 R_L$ to compare their sum with $VI \cos\langle^{\mathbf{I}}_{\mathbf{V}}$.

4.54 In the sinusoidal situation of Fig. 4.41, currents and voltages are sinusoidal.

a. Find $\mathbf{I}_R$ using a single loop equation. Knowing $\mathbf{I}_R$ and $\mathbf{I}$, find $\mathbf{I}_C$.

b. Find the power delivered by the voltage source $\mathbf{V}$, the power delivered by the current source $\mathbf{I}$, and the power absorbed by R. Are these three powers consistent?

c. Find $\mathbf{V}_C$. Now draw a phasor diagram, to scale, showing $\mathbf{V}, \mathbf{V}_R, \mathbf{V}_C, \mathbf{I}, \mathbf{I}_C,$ and $\mathbf{I}_R$. Now, in your diagram, show clearly (using a color) the component of $\mathbf{I}_R$ that is in phase with $\mathbf{V}$. This component multiplied by $\mathbf{V}$ is the power delivered by the source $\mathbf{V}$. From your diagram, approximate the power delivered by $\mathbf{V}$. In a similar fashion, make an approximate, graphical evaluation of the power delivered by the current source and the power absorbed by R. Are these estimates consistent with the calculated powers of part (b)?

d. Find $\mathbf{V}_C$ directly from given data by writing a single nodal equation.

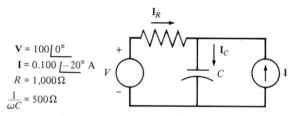

fig. 4.41   Average powers.

4.55 In the circuit of Prob. 4.54 (Fig. 4.41), circuit parameters are unchanged, but the voltages and currents are changed as follows: In the time domain, $v = 50 \sin(\omega t + 25°)$ V and $i = 0.050 \sin(\omega t + 5°)$ A.

a. Find the equation in the time domain for p_v, the power leaving the voltage source v. *Hint :* Note the similarities and differences between the v and i of this problem compared to those of Prob. 4.54. In solving Prob. 4.54, you did most of the arithmetic needed for the solution of Prob. 4.55.

b. Make an approximate plot of p_v, and by inspection of this plot, estimate the average value of p_v.

c. Calculate precisely the value of P_V, the average value of the power delivered by $\mathbf{V}$. Compare your estimated and precise values of P_V.

4.56 The current and voltage, respectively, at the terminals of a "box" (see Fig. 4.42) are

$$i = 1 \sin(\omega t + 32°) \text{ A}$$

and

$$v = 25 \cos(\omega t - 20°) \text{ V}$$

a Find the admittance Y at the terminals of this box.

b. Where $\omega = 5,000$ rad/s, find the numerical values of the two circuit elements (R, L, C) connected in parallel such that this parallel circuit has the same admittance as the "box."

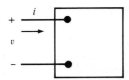

fig. 4.42 Configuration of Prob. 4.56.

   c. Find two circuit elements such that when the two are connected in series, the resultant model will have the same admittance as that of the box. As a check on this solution, first calculate the impedance of this series combination, and then calculate the admittance corresponding to that impedance.

4.57 In solving Prob. 4.56 you found, for the given "box," a model using parallel elements and then another model using series elements. On one set of coordinates, draw two phasor diagrams: one for each of these two models. Show on each of these diagrams the terminal voltage $\mathbf{V}$ and the terminal current $\mathbf{I}$ as given in Prob. 4.56. Of course, $\mathbf{V}$ and $\mathbf{I}$ are common to the two diagrams. For the model using parallel elements add the current phasors in each of the two parallel elements, and add for the series model the phasors for the voltages of the two separate elements. Of course your diagrams should show how the separate voltages of the series model add to $\mathbf{V}$, and how the separate currents of the parallel model add to $\mathbf{I}$.

4.58 The following laboratory measurements are made at the terminals of a device or "box": The sinusoidal voltage having frequency 7,000 Hz has a maximum value of 123 V, and the current leads the voltage by 67° and has a maximum value of 0.0100 A. Model this device with two circuit elements (R, L, C) connected in parallel. Find the necessary numerical value of each of these two elements in your model.

4.59 The impedance $\mathbf{Z}$ at the terminals of a network as shown in Fig. 4.43 is known to be $\mathbf{Z} = 122 \underline{/-32°}\ \Omega$ at a frequency of 10^5 rad/s.

   a. Model this network with a resistance R_s in series with a reactive element L_s or C_s. Find the numerical values of R_s and L_s or C_s.

   b. Model the same network with a resistance R_p in parallel with an L_p or C_p. Find the numerical values of the necessary R_p and L_p or C_p.

   c. These two models of the original network, one series and the other parallel, are equivalent at the one frequency of 10^5 rad/s. Now assume that we have the additional information that the terminal impedance of the network $\mathbf{Z}'$ is $89.88 \underline{/-51.34°}\ \Omega$ at $\omega = 2 \times 10^5$ rad/s. Is either one, both, or neither one a precise model for the network at both of these frequencies? Justify your conclusions with the necessary numerical calculations.

   d. Show the consistency of your solution by drawing four phasor diagrams: series, parallel, two frequencies. On each diagram (one model, one frequency) show $\mathbf{Z}$ and $\mathbf{Y}$ of that model.

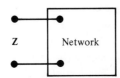

fig. 4.43 Network of Prob. 4.59.

Chapter 5

Total Response— RLC and Other Circuits

Introduction You will remember that the *total* response consists of the sum of the *natural* response and the *forced* response. Chapter 3 treated the total response of some first-order systems forced primarily by steady voltages or currents. In this chapter we will find the total response in higher-order systems forced by sinusoids and other functions. Chapter 4 gives us the background to write the forced response to a sinusoidal forcing function by inspection.

5.1 TOTAL RESPONSE, SINUSOIDAL FORCING FUNCTION

Consider the situation shown in Fig. 5.1. The current i is the sinusoid of (5.1). The switch S is open at $t = 0$. We desire the total response v for $t > 0$. The circuit elements are: $G_1 = 1/10{,}000 = 10^{-4}$ mho; $f = 10^6$ Hz; $C = 200$ pF; $G_2 = 1/500 = 2 \times 10^{-3}$ mho; and the forcing function is

$$i = 2\sqrt{2} \sin(\omega t - 35°) \text{ mA} \tag{5.1}$$

The switch is opened at $t = 0$, but prior to this time the circuit has come to a steady-state sinusoidal condition. Of course, immediately after the switch is opened, the response v will include an exponentially decaying term (the natural response) and a sinusoid (the forced response). Because of the capacitance C, the response v cannot "jump" at $t = 0$. We must know the initial value of v, $v(0^+) = v(0^-)$ in order to find the complete response v. This initial value of v is determined by the steady sinusoidal condition while the switch is still closed.

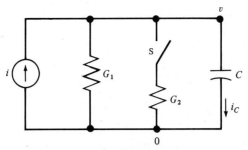

fig. 5.1   Find total response v.

Without the advantage of the phasor concepts of Chapter 4, the solution to this problem could be rather complex. One could argue that in order to find the initial value of v, one must write the differential equation for v that holds while the switch is still closed, then solve for the forced part of the response. Then this line of thought might require a second differential equation that applies while the switch is opened, leading to the complete solution for v. Our plan of attack will be much simpler than this approach that requires the solution to two differential equations. We will use the "solution by inspection" concept of Chapter 3 plus the phasor concept of Chapter 4.

We, of course, immediately recognize that the describing differential equation (both before and after the switch is closed) is first order, and also we recognize immediately the value of the time constant.

Let us now proceed to find the sinusoidal solution while the switch is still closed. The phasor equivalent of the forcing current is

$$\mathbf{I} = 2 \angle -125° \text{ mA} \qquad \text{effective} \tag{5.2}$$

To find $\mathbf{V}$, we might view the situation as a nodal problem, then

$$\mathbf{V} = \frac{I}{G_1 + G_2 + j\omega C} = \frac{2 \times 10^{-3} \angle -125°}{10^{-4} + (2 \times 10^{-3}) + (j1.257 \times 10^{-3})}$$

$$= \frac{2 \angle -125°}{2.1 + j1.257} = 0.8172 \angle -155.90° \text{ V} \tag{5.3}$$

Then transforming this phasor $\mathbf{V}$ back in the time domain,

$$v = 0.8172 \times \sqrt{2} \sin(\omega t - 65.90°) \qquad \text{for } t < 0 \tag{5.4}$$

Equation (5.4) gives us the initial value of v.

$$v(0^-) = v(0^+) = 0.8172 \times \sqrt{2} \sin(-65.90°) = -1.055 \text{ V} \tag{5.5}$$

The forced part of v after the switch has been opened is, of course, another sinusoid. Its phasor equivalent is

$$\mathbf{V} = \frac{I}{G_1 + j\omega C} = \frac{2 \times 10^{-3} \angle -125°}{10^{-4} + j1.257 \times 10^{-3}} = 1.586 \angle -210.5° \qquad \text{for } t > 0 \quad (5.6)$$

Then by letting $t = 0$, back in the time domain we have the value of the forced response for $t = 0^+$:

$$v_p(0^+) = 1.586 \times \sqrt{2} \sin(-120.5°) \text{ V} = 2.243 \sin(-120.5°) = -1.933$$

Then the complete solution after the switch has been opened is

$$v = v_p + Ke^{-t/\tau} = 2.243 \sin(\omega t - 120.5°) + Ke^{-t/\tau} \text{ V} \qquad (5.7)$$

where

$$\tau = R_1 C = \frac{C}{G_1} = \frac{200 \times 10^{-12}}{10^{-4}} = 2 \times 10^{-6} \text{ s} \qquad (5.8)$$

Now we can easily find the value of the arbitrary constant K, as follows:

$$K = v(0^+) - v_p(0^+) = -1.055 + 1.933 = 0.878 \qquad (5.9)$$

So the equation for v after the switch has been opened is

$$v = 2.243 \sin(6.283 \times 10^6 t - 120.5°) + 0.878 e^{-t/(2 \times 10^{-6})} \text{ V} \qquad (5.10)$$

A plot of v before and after the switch is closed is shown in Fig. 5.2. The "switch closed" part of the plot comes from (5.4), while the "switch open" part comes from (5.10). The particular solution (sinusoidal part in the "switch open") region is also plotted. The difference between v_p and v, of course, is the exponential term (natural response), so that after about three periods, v and v_p are nearly coincident.

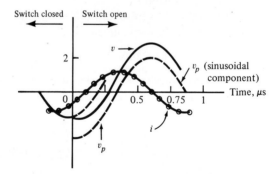

fig. 5.2   Complete response v in Fig. 5.1.

PROBLEMS

5.1   Note that there is no discontinuity (no jump) in v of Fig. 5.2. On the other hand, there is a sharp discontinuity in the *slope* of v at $t = 0$.
   a.   Starting with (5.4) and (5.10), calculate dv/dt at $t = 0^-$ and dv/dt at $t = 0^+$.
   b.   Do these two calculated derivatives seem to agree with the plot of Fig. 5.2?

5.2   For the situation centered around Fig. 5.1, start with (5.10) and other equations previously developed and find the equations for the capacitance current i_C. One of these equations will be valid for $t < 0$ and the other valid for $t > 0$. Plot i_C in the region

$0.25\ \mu s < t < 0.75\ \mu s$. Solve this problem twice, using two different approaches:
  a.   Apply $-i_C = i - vG$.
  b.   Apply $i_C = C\ dv/dt$.

5.3   Find the equations for i_C for the situation of Fig. 5.1 without using any part of the solution for v that has already been developed. One of these equations will be valid for $t > 0$ and the other for $t < 0$. Of course, your equations for this solution should agree with those in Prob. 5.2.

5.2   SERIES AND PARALLEL RLC CIRCUITS

A combination of a coil (inductor) in series with a capacitor and a combination of a coil in parallel with a capacitor are two common configurations in electric circuits. Figure 5.2a is a model of a voltage source v with its associated generator resistance R_s driving a series combination of a coil and capacitor. The coil or inductor has its own resistance R_c and inductance L_c. This model does not allow for loss in the capacitor through resistance in parallel with its capacitance.

Figure 5.3b is a model of a source i driving the parallel combination of an inductor and capacitor. Here, the generator has current source i and parallel conductance G. If the capacitor has significant loss, G could be adjusted accordingly.

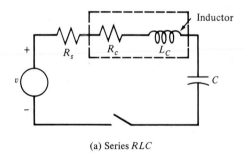

(a) Series *RLC*

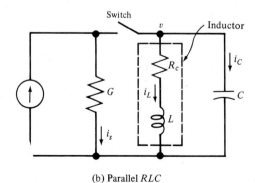

(b) Parallel *RLC*

fig. 5.3   RLC circuits.

Thinking in terms of a voltage source for the series combination as contrasted with a current source for the parallel combination is convenient, but not necessary. Remember that from Thévenin's and Norton's theorems, each generator has its equivalent in the other.

The text will now discuss at some length the parallel circuit, and you will analyze the series circuit in some of the problems that follow. The parallel compared to the series circuit is used more widely and is a little more difficult to analyze.

Let us find the complete response v where the switch of Fig. 5.3b is closed at $t = 0$ and the source i is a steady current having a value of I. This is *not* a first-order system, so we have no particular basis for writing the response v "by inspection." We will proceed much along the classical differential equation line of attack by first writing the pertinent differential equation for v, the desired response. We make considerable headway toward this end by relating v to i_L:

$$L \frac{di_L}{dt} + R_c i_L = v \tag{5.11}$$

Now we are motivated to eliminate i_L from (5.11). So, we write

$$i_L = i - Gv - C \frac{dv}{dt} \tag{5.12}$$

When we eliminate i_L between (5.11) and (5.12) and then rearrange terms, we get

$$\frac{d^2v}{dt^2} + \left(\frac{G}{C} + \frac{R_c}{L} \right) \frac{dv}{dt} + \left(\frac{R_c G + 1}{LC} \right) v = \frac{1}{C} \frac{di}{dt} + \frac{R_c}{LC} i \tag{5.13}$$

Then for our problem where $i = I$ (a constant), it follows that

$$\frac{d^2v}{dt^2} + \left(\frac{G}{C} + \frac{R_c}{L} \right) \frac{dv}{dt} + \left(\frac{R_c G + 1}{LC} \right) v = \frac{R_c I}{LC} \tag{5.14}$$

PROBLEM

5.4 The first term in (5.14) has dimensions of volts per second squared. As a consistency check on (5.14), show that each of the other terms in this equation has these same dimensions.

Equation (5.14) is obviously second order, and to proceed by classical differential equation methods we must find both the particular solution (forced part of the response) and the complementary function (natural part of the response). Because the right-hand side of (5.14) is constant, we immediately recognize the particular solution v_p:

$$v_p = \frac{R_c I}{R_c G + 1} \tag{5.15}$$

PROBLEM

5.5 Because v_p is also the steady-state solution, or that part of the solution that remains after the exponential terms have decayed to zero, one could look at the circuit itself without reference to (5.14) and write the value of v_p directly. Do this as a consistency check on both (5.14) and (5.15).

Equation (5.14) relates v in the circuit of Fig. 5.3b to time, the forcing function, and the circuit parameters. Let us generalize and treat the general second-order differential equation and its solution. Equation (5.16) has the same form as our particular equation (5.14).

$$\frac{d^2y}{dt^2} + a\frac{dy}{dt} + by = f \qquad (5.16)$$

where a and b are determined by the circuit parameters, f may be a function of time or constant, depending on the nature of the forcing function, and y is the desired response. In the particular situation of (5.14), f is the constant $(R_c I/LC)$. If f is a constant, then the particular solution is easily written as

$$y_p = \frac{f}{b} \qquad (5.17)$$

If f is a function of time, then it is not so easy to find y_p. This point will be discussed later.

The homogeneous part of (5.17) is

$$\frac{d^2y}{dt^2} + a\frac{dy}{dt} + by = 0 \qquad (5.18)$$

The complementary function (natural response) is the function that satisfies (5.18) and also contains two arbitrary constants. We can find this natural response quantity by substituting $y = Ae^{st}$ into (5.18) as follows:

$$As^2e^{st} + Aase^{st} + Abe^{st} = Ae^{st}(s^2 + as + b) = 0 \qquad (5.19)$$

To satisfy (5.19) for all values of time and for any arbitrary value of A, it is necessary that the parenthetical term be equal to zero. So we solve this quadratic in s. Of course, there are two different values of s that meet the requirements. These are

$$s_1, s_2 = -\frac{a}{2} \pm \sqrt{\left(\frac{a}{2}\right)^2 - b} \qquad (5.20)$$

So the complementary function is

$$y_h = A_1e^{s_1t} + A_2e^{s_2t} \qquad (5.21)$$

where A_1 and A_2 are arbitrary constants yet to be determined. The total solution to the original differential equation (5.17) is the sum of the particular solution y_p and the complementary function y_h; that is,

$$y = y_h + y_p = A_1e^{s_1t} + A_2e^{s_2t} + \frac{f}{b} \qquad (5.22)$$

where f is constant.

PROBLEM

5.6 Remember that the solution to a differential equation must:
   1. Contain the appropriate number of arbitrary constants.
   2. Satisfy the differential equation itself.
   Equation (5.22) contains the necessary number of arbitrary constants—two for this second-order equation. Now prove by means of a word argument that the value of v given by (5.22) does, in fact, satisfy (5.16).

Example 5.1

For the parallel circuit of Fig. 5.3b and where $R_c = 100\ \Omega$, $L = 0.10$ H, $C = 40$ nF, $G = 8 \times 10^{-5}$ mho, and $I = 0.5$ mA, find s_1 and s_2 of (5.22). Also evaluate v_p, the forced response.

$$a = \frac{G}{C} + \frac{R_c}{L} = \frac{8 \times 10^{-5}}{4 \times 10^{-8}} + \frac{100}{0.1} = 3,000 \tag{5.23}$$

$$b = \frac{R_c G + 1}{LC} = \frac{8 \times 10^{-3} + 1}{40 \times 10^{-10}} = 2.520 \times 10^8 \tag{5.24}$$

Then substituting these values of a and b in (5.20), we get

$$s_1, s_2 = -\frac{a}{2} \pm \sqrt{\left(\frac{a}{2}\right)^2 - b} = -1,500 \pm j\sqrt{2.52 \times 10^8 - 1,500^2} \tag{5.25}$$

$$s_1, s_2 = (-1,500 + j15,800),\ (-1,500 - j15,800) \tag{5.26}$$

Using (5.15), we find the forced response

$$v_p = \frac{R_c I}{R_c G + 1} = \frac{100}{(8 \times 10^{-3}) + 1} \times 5 \times 10^{-4} = 0.04960\ \text{V} \tag{5.27}$$

For the numerical values of s_1, s_2 from Example 5.1 as they apply to the parallel circuit of Fig. 5.3b, we can now write the total response. See (5.21) again.

$$v = v_C + v_p = A_1 e^{-1,500t + j15,800t} + A_2 e^{-1,500t - j15,800t} + 0.04960\ \text{V} \tag{5.28}$$

Then, from the identity $e^{a+b} = e^a \times e^b$,

$$v = e^{-1,500t}(A_1 e^{j15,800t} + A_2 e^{-j15,800t}) + 0.04960\ \text{V} \tag{5.29}$$

The arbitrary constants A_1 and A_2 of (5.29) can be determined from the appropriate initial conditions, but after these constants have been determined, (5.29) is not in a convenient form. Without further algebraic manipulation, (5.29) is not easily visualized or plotted.

• • •

In view of the fact that the response v must be a real quantity and yet the exponents are imaginary, (5.29) could be disturbing to the thoughtful person. But it happens that A_1 and A_2 are complex in such a manner that ultimately the imaginary parts of (5.29) cancel, leaving v as a real quantity.

Rather than pursuing (5.29) to find A_1 and A_2, then performing additional algebra to whip (5.29) into a convenient form, let us take a different route. In every second-order system where s_1 and s_2 are complex, as they are in Example 5.1, the complementary function y_c will have the form

$$y_c = e^{-\alpha t}(A_1 e^{j\omega t} + A_2 e^{-j\omega t}) \tag{5.30}$$

Now, if we use Euler's formula to convert each of the exponentials to the equivalent sine plus cosine terms and then collect terms, we obtain

$$y_c = e^{-\alpha t}[(A_1 + A_2)\cos \omega t + j(A_1 - A_2)\sin \omega t] \tag{5.31}$$

Because y_c must be real, both $(A_1 + A_2)$ and $j(A_1 - A_2)$ must be real; that is,

$$(A_1 + A_2) = B \qquad \text{and} \qquad j(A_1 - A_2) = D \tag{5.32}$$

where both B and D are real. Each of A_1 and A_2 alone is complex, but (5.32) must hold. Let $A_1 = a_1 + jb_1$ and $A_2 = a_2 + jb_2$. Then

$$A_1 + A_2 = a_1 + a_2 + j(b_1 + b_2) = B \qquad \text{real} \tag{5.33}$$

and

$$j(A_1 - A_2) = j(a_1 - a_2) + jj(b_1 - b_2) = D \qquad \text{real} \tag{5.34}$$

From (5.33) we see that $b_1 = -b_2$, and from (5.34) we see that $a_1 = a_2$. It follows that

$$\text{If } A_1 = a_1 + jb_1, \text{ then } A_2 = a_2 + jb_2 = a_1 - jb_1 \tag{5.35}$$

From (5.35) we see that A_1 and A_2 are the complex conjugates of each other. In your solution to Prob. 5.7, you will prove that

$$e^{-\alpha t}(A_1 e^{-j\omega t} + A_2 e^{-j\omega t}) = e^{-\alpha t}A \sin(\omega t + \theta) \tag{5.36}$$

where A_1 and A_2 are complex conjugates; A and θ are real constants.

PROBLEM
5.7 Prove (5.36). A and θ are determined by A_1 and A_2, but you need not find these relationships.

It follows from (5.29) and (5.36) that, in any second-order system where s_1 and s_2 are complex, any response in that system can be written as

$$y = Ae^{-\alpha t} \sin(\omega t + \theta) + y_p \tag{5.37}$$

where y and y_p are the total and forced responses, respectively, at some point in that system. When the generality of (5.37) is applied to our particular situation of (5.29), we have

$$v = Ae^{-1,500t} \sin(15,800t + \theta) + 0.04960 \text{ V} \tag{5.38}$$

From this we see that the natural part of the response is a damped sinusoid having magnitude A, time constant $1/1,500$, angular velocity $\omega = 15,800$, and phase θ. To find the values of the two arbitrary constants A and θ, we use two independent initial conditions. Return to Fig. 5.3 and note that if the circuit had reached steady state prior to time $t = 0$ when the switch was closed, then $v(0^-) = v(0^+) = 0$. So it follows from (5.38) that

$$v(0^+) = 0 = A \sin \theta + 0.04960$$

or

$$A \sin \theta = -0.04960 \tag{5.39}$$

For the other initial condition, note that because $v(0^+) = 0$, $i_s(0^+) = i_L(0^+) = 0$. It follows that $i_C(0^+) = I$. So

$$i_C(0^+) = C \left.\frac{dv}{dt}\right|_{t=0^+} = AC[-1500e^{-1,500t} \sin(15,800t + \theta)$$

$$+ 15,800e^{-1,500t} \cos(15,800t + \theta)]_{t=0^+}$$

$$i_C(0^+) = I = AC[-1,500 \sin \theta + 15,800 \cos \theta] \tag{5.40}$$

or

$$\frac{I}{C} = -1,500A \sin \theta + 15,800A \cos \theta \tag{5.41}$$

Using (5.39) to eliminate $A \sin \theta$ from (5.40) and then solving for $A \cos \theta$, we get

$$A \cos \theta = \frac{I/C - (1,500 \times 0.0496)}{15,800}$$

$$= \frac{(5 \times 10^{-4})/(4 \times 10^{-8}) - (1,500 \times 0.0496)}{15,800}$$

$$= 0.7864 \tag{5.42}$$

Then

$$\frac{A \sin \theta}{A \cos \theta} = \tan \theta = \frac{-0.04960}{0.7864} = -0.06307$$

or

$$\theta = -3.609° \tag{5.43}$$

Then, using (5.42) and (5.43),

$$A = \frac{0.7864}{\cos(-3.609°)} = 0.7880 \tag{5.44}$$

Using these values of the arbitrary constants A and θ, we have the solution that we seek.

$$v = 0.7880e^{-1,500t} \sin(15,800t - 3.609°) + 0.04960 \text{ V} \tag{5.45}$$

PROBLEMS

5.8 Look at the circuit of Fig. 5.3b while you close your eyes to the mathematical analysis that we have made and view the circuit directly to find the value of $v(\infty)$. Is this value consistent with the corresponding value coming from (5.45)?

5.9 Is (5.45) consistent with the initial values $v(0^+) = 0$ and $i_c(0^+) = 0.05$ mA?

In (5.45) we see that the coefficient of the damped sine term is more than 15 times larger than the final value 0.0496. That would have been difficult to anticipate without some pertinent experience.

PROBLEM

5.10 a. Determine the time constant associated with the exponential term in (5.45).
  b. Determine the period of the sine term.
  c. Calculate the number of periods of the sine term that is equal in time to the time constant.
  d. Using your solutions to parts (a), (b), and (c), prepare an approximate sketch of v through time equal to two time constants.

5.3 OPERATIONAL TECHNIQUES

Refer again to (5.13), the differential equation for v in the circuit of Fig. 5.3b. Equations (5.11) and (5.12) lead to (5.13). At the beginning, it was not necessarily obvious that these two equations would lead to the desired differential equation of (5.13). In complex situations especially, there is a better technique for finding the desired differential equation. You had experience with the operator technique in your study of differential equations. To illustrate, write (5.11) in a different form by letting some symbol such as s represent the differential operator (d/dt) so that (5.11) becomes

$$Lsi_L + R_c i_L = (sL + R_c)i_L = v \qquad (5.46)$$

we must remember that s standing alone does not have a physical interpretation as does L, R_c, i, and v. But, on the other hand, s operating on i_L has the physical interpretation of (di_L/dt). In (5.46) the term $(sL + R_c)$ has the dimensions of impedance and may be called the operational impedance of the coil. When this impedance operator operates on i_L appropriately, the product is the voltage as shown in (5.46). In the same way, (5.12) can be written in an operator form, as follows:

$$i_L = i - Gv - sCv \qquad (5.47)$$

Now we may eliminate i_L between (5.46) and (5.47) by the standard algebraic procedures.

PROBLEM

5.11 Eliminate i_L between (5.46) and (5.47); then change this equation in an operator form back in the time domain to the same format as (5.13).

By this operational technique, we may think of sL as the operational impedance of inductance, and sC as the operational admittance of capacitance. So we might see the circuit of Fig. 5.3b from an operational viewpoint, as shown in Fig. 5.4. Now

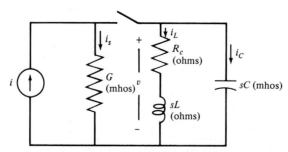

fig. 5.4 *Operational impedances R_C and sL, and operational admittances G_s and sC.*

that we have these impedances and admittances to manipulate, we can take advantage of all of the network reduction ideas and network theorems. For example, let us use the nodal-superposition idea to find v in the network of Fig. 5.4 as follows:

$$v\left(G + sC + \frac{1}{R_c + sL}\right) = i \tag{5.48}$$

PROBLEM

5.12 Perform algebraic manipulations on (5.48) as needed and write the result in the terms of t to obtain (5.13). Of course, $(s^2 v) = s(sv) = d^2 v/dt^2$.

So it happens that the use of the operator point of view can greatly simplify the process of developing the desired differential equation. But there is an additional payoff. Return to (5.18) to (5.21). To obtain (5.19), we let $y = Ae^{st}$, and this leads to the polynomial $(s^2 + as + b)$, which in turn leads to the roots s_1 and s_2, and then to the desired complementary function of (5.21). Rather than making this substitution, we only need to write the original differential equation in an operator form, then find the roots. And more than that, we need not write the differential equation in terms of time in the first place. See (5.48) as an example. In order to find the complementary function of (5.48), we merely need to find the roots of

$$G + sC + \frac{1}{R_c + sL} = 0 \tag{5.49}$$

Example 5.2

By the operator technique, find i_C in Fig. 5.3b.

Solution

Refer to Fig. 5.4 because it displays the operational impedances. Find i_C in terms of i and the operational impedances. Use the current divider ideas as in the following:

$$i_c = \frac{i_s\, C}{G + sC + \dfrac{1}{R_c + sL}} = \frac{i_s\, C(R_c + sL)}{LC\left[s^2 + s\left(\dfrac{G}{C} + \dfrac{R_c}{L}\right) + \dfrac{R_c G}{LC} + \dfrac{1}{LC}\right]} \qquad (5.50)$$

To maintain perspective, it is well to write (5.50) explicitly in terms of time. After simplifying we obtain:

$$\frac{d^2 i_C}{dt^2} + \left(\frac{G}{C} + \frac{R_c}{L}\right)\frac{di_C}{dt} + \left(\frac{R_c G}{LC} + \frac{1}{LC}\right)i_C = \frac{d^2 i}{dt^2} + \frac{R_c}{L}\frac{di}{dt} \qquad (5.51)$$

It is obvious from (5.51) that, when $i = I$ (a steady current),

$$\frac{d^2 i_C}{dt^2} + \left(\frac{G}{C} + \frac{R_c}{L}\right)\frac{di_C}{dt} + \left(\frac{R_c G}{LC} + \frac{1}{LC}\right)i_C = 0 \qquad (5.52)$$

Now, in order to write the complementary function part of the solution for i_C, we only need the roots of the bracketed term in the denominator of (5.50). We found these roots earlier in (5.26); thus

$$i_{Ch} = A_C e^{-1,500t}\, \sin(15,800t + \theta_C) \qquad (5.53)$$

where i_{Ch} is the complementary or homogeneous solution, but this is also the complete solution because the left-hand side of (5.52) is zero. So the total solution is

$$i_C = A_C e^{-1,500t}\, \sin(15,800t + \theta_C) \qquad (5.54)$$

Now we must use the initial conditions to find the arbitrary constants A_C and θ_C. At $t = 0^+$, $i_C = I$, the steady current. This is true because $v(0^+) = 0$, making i_s and i_L both equal to zero. So, using (5.54) at $t = 0^+$,

$$i_C(0^+) = I = A_C \sin \theta_C \qquad (5.55)$$

Now we seek the value of $[di_L/dt]_{t=0^+}$ as another initial condition. We note that

$$i_C = I - i_L - i_s \qquad (5.56)$$

and

$$\frac{di_C}{dt} = \frac{dI}{dt} - \frac{di_L}{dt} - \frac{di_s}{dt} \qquad (5.57)$$

Furthermore, $dI/dt = 0$ because I is steady, and from $v = 0 = L\, di_L/dt$ at $t = 0^+$, we see that $[di_L/dt]_{t=0^+} = 0$. Therefore, at $t = 0^+$,

$$\frac{di_C}{dt} = -\frac{di_s}{dt} = -\frac{dGv}{dt} = -G\frac{dv}{dt} = -\frac{GI}{C} \qquad (5.58)$$

We take the derivative of (5.54) and evaluate at $t = 0^+$.

$$\frac{di_C}{dt} = -\frac{GI}{C} = A_C[-1,500 \sin \theta_C + 15,800 \cos \theta_C] \qquad (5.59)$$

Using (5.55) to eliminate I in (5.59) and then solving for $A_C \cos \theta_C$,

$$A_C \cos \theta_C = \frac{(1{,}500 - G/C)A_C \sin \theta_C}{15{,}800} \tag{5.60}$$

Then

$$\tan \theta_C = \frac{A_C \sin \theta_C}{A_C \cos \theta_C} = \frac{15{,}800}{1{,}500 - G/C} = -31.60 \tag{5.61}$$

Hence

$$\theta_C = -88.19° \quad \text{or} \quad 91.81° \tag{5.62}$$

Now, from (5.55),

$$A_C = \frac{I}{\sin \theta_C} = \frac{5 \times 10^{-4}}{\sin 91.81°} = 5.002 \times 10^{-4} \tag{5.63}$$

So we write the solution

$$i_C = 5.002 \times 10^{-4} e^{-1{,}500t} \sin(15{,}800t + 91.81°) \tag{5.64}$$

For a consistency check on both (5.45) and (5.64), differentiate (5.45):

$$\frac{dv}{dt} = 0.7880 e^{-1{,}500t}[-1{,}500 \sin(15{,}800t - 3.609°)$$

$$+ 15{,}800 \cos(15{,}800t - 3.609°)] \tag{5.65}$$

Use the phasor concept to add the sine and cosine terms of (5.65). Then we can write the equation for i_C in terms of (dv/dt):

$$i_C = C \frac{dv}{dt} = 0.7880 \times 1.587 \times 10^4 \times 40 \times 10^{-9} e^{-1{,}500t} \sin(15{,}800t + 91.81°)$$

$$= 5.002 \times 10^{-4} e^{-1{,}500t} \sin(15{,}800t + 91.81°) \tag{5.66}$$

Because (5.64) and (5.66) are identical, we have good reason to have high confidence in both the equation for v and the equation for i_C.

● ● ●

PROBLEMS

5.13 Prove that (5.66) follows from (5.65).

5.14 To demonstrate the power and convenience of the operator technique, develop (5.52) without using the operator technique. To solve this problem you must find (5.52) by working in the time domain only; you may use only equations such as $i = i_s + i_L + i_C$, $v = R_c i_L + L \, di_L/dt$, and so forth.

You may be at a loss, as you proceed to solve this problem, as to what step should be taken next. However, by repeated trials you will be able to find a combination of equations and manipulations that will allow you to eliminate unwanted variables: i_s, i_L, and v. In contrast, the operator technique leading to (5.52) is very straightforward.

PROBLEM

5.15 Find the equations that describe i_s both before and after the switch is closed in the parallel coil-capacitor situation of Fig. 5.4.

a. For an easy and quick solution to this problem, use the equation for v as in (5.45). Use the results of this effort as the answer to the following situation, which is more challenging.

b. Assume that none of the previous analysis is available to you, and you must find i_s for $t > 0$. Begin with the network and find the differential equation for i_s in terms of i and the circuit parameters. If you do not lean heavily on the equations already developed, this problem gives you the opportunity to practice intensively on the concepts already treated at length.

c. Prepare an approximate plot of i_s for t both greater and less than zero.

5.4 INITIAL CONDITIONS

For second- and higher-order systems, it is sometimes challenging to find the necessary initial conditions that lead to the determination of a particular complete response.

To find the values of these initial conditions, it is necessary to analyze the state of the circuit at the initial instant, namely, at $t = 0^+$. There is really "no royal road" to finding initial conditions other than the careful and selective application of the following familiar principles:

1. The current i_L in inductance cannot jump; that is, $i_L(0^+) = i_L(0^-)$.
2. The voltage across an inductance may jump; therefore $[di_L/dt]_{t=0+}$ may differ from $[di_L/dt]_{t=0+}$ by any finite amount.
3. The voltage across capacitance v_C cannot jump; that is, $v_C(0^+) = v_C(0^-)$.
4. The current in capacitance may jump; therefore $[dv_C/dt]_{t=0+}$ may differ from $[dv_C/dt]_{t=0-}$ by any finite amount.
5. Kirchhoff's laws (current and voltage) must hold at $t = 0^+$ as well as at any other instant of time.
6. One may take the derivative of a voltage or current equation. For example, if $i_1 = G_2 v_2 + i_3$, then $di_1/dt = G_2\, dv_2/dt + di_3/dt$. Of course, higher derivatives may be taken also.
7. *Caution:* The value of a current, or voltage, at $t = 0^+$ or any other instant of time tells nothing of the derivative of that current at the same instant of time. For example, if $i_1(0^+) = 0$, then it follows that $[di_1/dt]_{t=0+}$ may have any finite value. A careless analyst might argue incorrectly that if $i_1(0^+) = 0$, then $[di_1/dt]_{t=0+} = d(0)/dt = 0$. One must take the derivative as a function of *time*, then evaluate that derivative at $t = 0^+$.

Example 5.3

Assume that in the network of Fig. 5.5, we desire the response quantity i. The switch is closed at $t = 0$ and the applied voltage v is a sinusoid as shown. At $t = 0^-$,

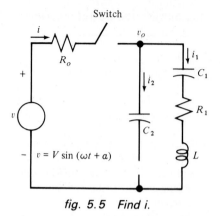

fig. 5.5   Find i.

$i = i_1 = i_2 = 0$ and the two capacitance voltages are each equal to zero. Our problem is to find the necessary initial conditions that would ultimately lead to the equation for i. In other words, we must find $i(0^+)$, $[di/dt]_{t=0+}$, and $[d^2i/dt^2]_{t=0+}$.

The analyst does not have a preconceived approach to finding these three initial conditions in a very simple fashion but rather faces a sort of puzzle or miniresearch

TABLE 5.1   Quantities Involved in Example 5.3

Item	Value at $t = 0^-$	Value at $t = 0^+$	Step
v	$V \sin \alpha$*	$V \sin \alpha$*	
i	0*	$\dfrac{V \sin \alpha}{R_o}$	
v_{C_1}	0*	0	1
$v_{C_2} = v_o$	0 *	0	1
i_1	0*	0	3
i_2	0	$v(0^+)/R_o$	4
$\dfrac{dv_{C_2}}{dt}$		$\dfrac{V \sin \alpha}{C_2 R_o}$	5
$\dfrac{di}{dt}$		$\dfrac{\omega V \cos \alpha}{R_o} - \dfrac{V \sin \alpha}{C_2 R_o^2}$	6
$\dfrac{di_1}{dt}$		0	7
$\dfrac{d^2i}{dt^2}$	$\dfrac{V \sin \alpha}{R_o}\left(\dfrac{1}{C_2 R_o^2} - \omega^2\right) - \dfrac{\omega V \cos \alpha}{R_o^2 C_2}$		8

* These values were given in the statement of the problem.

project. It is usually helpful to enter into a tabulation all of the initial currents and voltages as they are determined, as shown in Table 5.1. The steps are as follows:

1. The capacitance voltages cannot jump; therefore $v_{C_1}(0^+) = v_{C_1}(0^-) = 0$ and $v_{C_2}(0^+) = v_{C_2}(0^-) = 0$. The values of these two capacitance voltages, and other voltages and currents, at $t = 0^+$ are not needed directly, but almost surely they will be needed in order to find $i(0^+)$, $[di/dt]_{t=0+}$, and $[d^2i/dt^2]_{t=0+}$. So for future reference we tabulate initial values as they become known.

2. Because we now know the values of $v(0^+)$ and $v_o(0^+)$, there is a strong suggestion that we apply Kirchhoff's voltage law to the (v, R_o, C_2) loop. It follows that $v(0^+) = R_o i(0^+) + v_o(0^+)$ or $i(0^+) = v(0^+)/R_o = (V \sin \alpha)/R_o$. Now we enter this value for $i(0^+)$ in our tabulation.

3. Almost surely it will be helpful to know the value of $i_1(0^+)$. $i_1(0^+)$ must be equal to $i_1(0^-) = 0$ because of the inductance L.

4. It follows from Kirchhoff's current law and from step 3 that $i_2(0^+) = i(0^+)$.

5. We can find $[dv_{C_2}/dt]_{t=0+}$ from

$$C_2 \frac{dv_{C_2}}{dt} = i_2$$

or

$$\left. \frac{dv_{C_2}}{dt} \right|_{t=0+} = \frac{i_2(0^+)}{C_2} = \frac{i(0^+)}{C_2} = \frac{V \sin \alpha}{C_2 R_o}$$

6. Because we now know the value of $[dv_{C_2}/dt]_{t=0+}$, it appears useful to take the derivative of the voltage terms in the (v, R_o, C_2) loop. We have

$$\frac{dv}{dt} = R_o \frac{di}{dt} + \frac{dv_{C_2}}{dt}$$

or

$$\left. \frac{di}{dt} \right|_{t=0+} = \frac{\omega V \cos \alpha}{R_o} - \frac{V \sin \alpha}{C_2 R_o^2}$$

7. Because the value of $[di_1/dt]_{t=0+}$ probably will be useful, we seek its value:

$$v_o = v_{C_2} = R_1 i_1 + L \frac{di_1}{dt} + \frac{1}{C_1} \int_0^{0+} i_1 \, dt$$

But $i_1(0^+) = 0$, $v_{C_2}(0^+) = 0$, and $\int_0^{0+} i_1 \, dt = 0$. Therefore, $[di_1/dt]_{t=0+} = 0$.

8. To find $[d^2i/dt^2]_{t=0+}$, we take the second derivative of the voltages around the (v, R_o, C_2) loop, as follows.

$$\frac{d^2v}{dt^2} = R_o \frac{d^2i}{dt^2} + \frac{d^2v_{C_2}}{dt^2} = R_o \frac{d^2i}{dt^2} + \frac{1}{C_2} \frac{di_2}{dt}$$

or

$$R_o \frac{d^2i}{dt^2} = \frac{d^2v}{dt^2} - \frac{1}{C_2} \frac{di_2}{dt}$$

Now, because

$$[di_1/dt]_{t=0^+} = 0, \quad [di_2/dt]_{t=0^+} = [di/dt]_{t=0^+}$$

Then

$$\left.\frac{d^2i}{dt^2}\right|_{t=0^+} = \left.\frac{1}{R_o}\frac{d^2v}{dt^2}\right|_{t=0^+} - \left.\frac{1}{R_o C_2}\frac{di}{dt}\right|_{t=0^+} = \frac{V\sin\alpha}{R_o}\left(\frac{1}{C_2 R_o^2} - \omega^2\right) - \frac{\omega V \cos\alpha}{R_o^2 C_2}$$

• • •

That was a rather tedious task to find the necessary three initial conditions. By the method of Laplace transforms (Chapter 8), the determination of the initial conditions is a matter of routine. Soon you will see how the Laplace approach greatly simplifies the task of finding initial conditions.

The following problem is simpler than the one of the previous example. Nevertheless, you will find significant challenge in this problem.

PROBLEM

5.16 a. For the second-order system of Fig. 5.6, find the initial values of $v_o(0^+)$ and $[dv_o/dt]_{t=0^+}$, where $i = At + B$ and where the switch is closed at $t = 0$. For $t < 0$, all currents and voltages are zero. It is suggested that in order to solve this problem, you complete a tabulation of initial values as was done in Example 5.3.

b. Apply, in the same manner, Kirchhoff's voltage and current laws using your initial conditions to demonstrate the consistency of your solution.

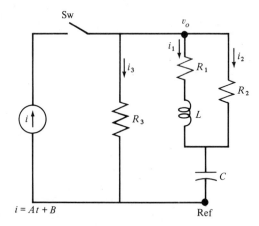

fig. 5.6   A second-order system.

5.5   UNDERDAMPED AND OVERDAMPED SITUATIONS

The equation for i_C (5.64) and the equation for v (5.45) are equations for an *underdamped* or oscillatory situation. Here we see an exponentially decreasing or damped sinusoid. In this same circuit, the parameters can be adjusted so that the damped sinusoid does not occur. This is called the *overdamped* case or situation. In the second-order, underdamped situation the roots s_1 and s_2 are complex conjugates as in the previous analysis. In the overdamped situation these roots are real and unequal.

From (5.20) we see that s_1 and s_2 are complex when $(a/2)^2 < b$. Notice in (5.24) that b for our particular case is almost independent of G and R_c, but a is highly dependent upon G and R_c. Let us adjust three of the parameters as follows: $G = 1 \times 10^{-3}$ mho, $R_c = 400$ Ω, and $L = 0.200$ H. Let C remain at 40 nF. Then it follows from (5.23) that

$$a = \frac{G}{C} + \frac{R_c}{L} = \frac{1 \times 10^{-3}}{40 \times 10^{-9}} + \frac{400}{0.2} = 2.700 \times 10^4 \tag{5.67}$$

and, from (5.24),

$$b = \frac{R_c G + 1}{LC} = \frac{(400 \times 10^{-3}) + 1}{0.2 \times 40 \times 10^{-9}} = 1.750 \times 10^8 \tag{5.68}$$

Then, from (5.25),

$$s_1, s_2 = -\frac{a}{2} \pm \sqrt{\left(\frac{a}{2}\right)^2 - b} = \left(-1.350 \pm \sqrt{1.350^2 - 1.750}\right)10^4$$

$$s_1, s_2 = (-1.350 \pm 0.2693)10^4 = -1.081 \times 10^4, -1.619 \times 10^4 \tag{5.69}$$

Now we can write the solution as in (5.70).

$$v = A_1 e^{-1.081 \times 10^4 t} + A_2 e^{-1.619 \times 10^4 t} + v_p \tag{5.70}$$

where v_p is the forced response or particular solution; A_1 and A_2 are the two arbitrary constants. Equation (5.70) describes an overdamped situation in a second-order system because there are two exponentially decaying, nonsinusoidal terms in the natural part of the response.

PROBLEMS
5.17 a. In (5.70) find the numerical values of the arbitrary constants A_1 and A_2 and also the forced response v_p so that you have a completely numerical solution. Does your solution make sense in view of the fact that v must begin at zero at $t = 0$ and finally rise in some manner to the steady value v_p where the steady current divides between the coil and G?
    b. Prepare an approximate plot of v.
    c. Is your solution consistent with the two initial conditions?

5.18 Check your solution to Prob. 5.17 against the following facts:
    1. $v > 0$ for all $t > 0$.
    2. $v(0^+) = 0$.
    3. $v(\infty) = v_p$.

Mathematically, one may speak of a *critically damped* situation in addition to the overdamped and underdamped situations. The second-order system is said to be critically damped when $(a/2)^2 = b$ (see (5.20)), and therefore there is only one distinct root. You will remember from your differential equations that the complementary functions take a special format for the critically damped situation. The critically damped situation is of no particular practical importance because, in the real world, $(a/2)^2$ cannot be *precisely* equal to b over a finite time interval.

PROBLEMS

5.19   Figure 5.7 is a model of an automobile ignition system. The "ignition coil" really consists of two separate adjacent coils L_1 and L_s that experience approximately the same magnetic flux. The useful or "spark" voltage is v_s. The switch S corresponds to the "breaker points." The resistance and inductance of the "primary coil" are R and L_1, respectively.

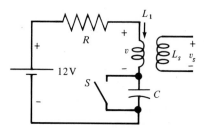

fig. 5.7   *Model of an automobile ignition system.*

The following numerical values are not necessarily typical of those in your car, but these values will serve as a base for the numerical problem that follows.

The switch S is closed long enough for the current in L_1 to reach, for practical purposes, its steady value. Soon after S is opened, voltage v rises to a peak of several hundred volts. The secondary coil L_s has 40 times as many turns as the primary coil and thus has $v_s = 40 \times v$, as you will learn in Chapter 9.

Assuming that $L_1 = 0.02$ H, $R = 10\ \Omega$, and $C = 0.1\ \mu$F, find the maximum value to which v_s will rise after switch S is opened.

Suggestions:

1. Find the differential equation that relates v to the parameters and the 12-V source. To do this, you may want to implement the voltage divider concept by means of the operator technique. To make sure that you understand the meaning of your operational expression for v, write its equivalent in the traditional differential equation form.

2. Before solving your differential equation, sketch your anticipated feel for v. Where does it start and where does it end if it is oscillatory?

3. Your whole solution hangs on the correct initial conditions. Make sure that they are correct. Keep in mind the fact that although a current or voltage can be zero at $t = 0^+$, its derivative can be any finite value at that instant.

4. Your instructor could make a very effective laboratory exercise built around this problem. Someone in the class might already know, or be willing to find, typical parameter values in some actual automobile ignition system. Or alternately, the parameters of a real ignition system might be measured in the laboratory. The oscillogram of v_s, or v because it is lower, might be observed in the laboratory. This laboratory observation might be done on a real ignition system or alternately

on some "mockup." *Caution :* Beware of dangerously high voltages. Consult with your instructor if in doubt about the safety of yourself and the measuring equipment.

5.20 In Prob. 5.19, it was assumed that S, the breaker points, were closed long enough for the coil current to reach its steady value. This is not necessarily true, especially at high engine speeds. Assume that our ignition system is applied to an eight-cylinder, four-cycle engine running at 4,000 r/min (revolutions per minute). For four-cycle operation, remember that each piston fires once in two revolutions of the engine. Further assume that the breaker points are closed 75 percent of the time.

    a. Determine whether the breaker points are closed long enough at 4,000 r/min for the coil current to truly reach the final steady value.

    b. If not, calculate the maximum value of v_s at this speed.

    c. Determine whether steady-state current is essentially reached at 1,000 r/min.

5.21 Assume that the capacitor in the automobile ignition circuit of Prob. 5.19 is removed from across the breaker points.

    a. Qualitatively discuss the operation of the ignition system without the capacitor.

    b. Could a resistor be used to replace the capacitor? If so, what resistance value is needed to produce the same peak ignition voltage as provided by the capacitive circuit of Prob. 5.19?

    c. Does the capacitor have any advantages over the resistor in producing high ignition voltages? If so, what are the advantages?

5.22 In the circuit of Fig. 5.8, the switch S has been open long enough for the circuit to reach a steady condition. Note that the forcing function is a sinusoid. The switch is closed at $t = 0$. Find numerical equations that describe v_o for both $t > 0$ and $t < 0$.

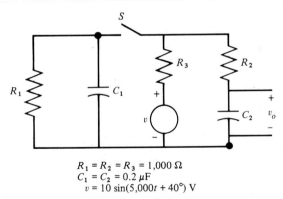

$$R_1 = R_2 = R_3 = 1,000 \ \Omega$$
$$C_1 = C_2 = 0.2 \ \mu\text{F}$$
$$v = 10 \sin(5,000t + 40°) \ \text{V}$$

fig. 5.8   Second - order, sinusoidal forcing function.

Prepare a plot that clearly shows what happens to v_o for the significant time before and after $t = 0$.

5.23 Where a sinusoidal voltage is applied suddenly to an inductor having both R and L, the transient or natural part of the response current is a function of the point on the voltage cycle where the switch is closed. Suppose, as shown in Fig. 5.9, the switch is closed at $t = 0$, but the phase angle α of the applied voltage is adjusted so that the switch is closed at any desired point on the sinusoidal voltage wave.

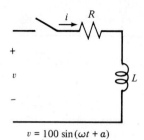

$$v = 100 \sin(\omega t + a)$$

fig. 5.9   Inductor circuit.

a. Where $R = 20\ \Omega$ and $L = 1$ H, find the value of α that makes the transient part of i equal to zero.

b. Find the value of α that makes the natural transient part i the maximum possible.

c. Plot the total current i and the applied voltage wave for each of these two conditions. Plot all three waves on one set of coordinates. Your plot should show clearly the point on the voltage wave where the current begins. Also plot the transient part of the current where it is maximum, as in part (b).

5.24 The current i is a known function of time.

a. Write the differential equation for v_2 as a function of i, C_2, C_1, R_2, R_1, and L. Before proceeding to write the differential equation, ask yourself whether this equation should be first, second, or third order.

b. Remove R_2 from the network of Fig. 5.10. Should the differential equation for v_2 of this new circuit be first, second, or third order?

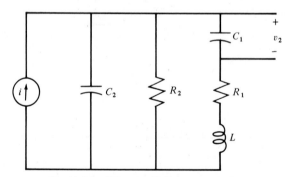

fig. 5.10   The circuit for Prob. 5.24.

Chapter 6

Networks

Introduction In this chapter we will extend the sinusoidal analysis of Chapter 4 to more complex and meaningful situations. We will also combine the principles of Chapters 2, 3, and 5 to acquire a more general view of circuits, especially in the steady state, but also in terms of the total response. Earlier, we restricted the application of superposition, Thévenin's theorem, network reduction ideas, and other fundamental concepts to resistive networks; in this chapter we will view these notions in a much more general sense.

6.1 ANOTHER VIEW OF VOLTAGE AND CURRENT SOURCES

You will remember from Chapter 2 that a voltage source (or current source) is a concept used to model devices and not a device itself. A true voltage source as a concept delivers a voltage that is independent of its current or the impedance to which it is connected; however, no practical device can deliver a voltage at its terminals that is independent of the current that it delivers. In a similar sense, the current delivered by any practical generator must be dependent in some manner on the impedance seen by the device.

In Chapter 2 we restricted the application of voltage and current sources to resistive networks. Now with the additional background of Chapters 4 and 5, we can broaden the application of these concepts of sources. For example, in solving Example 6.1, we must find a model that uses a voltage source to represent an ac generator with internal impedance that might include either an inductive or capacitive element as well as a resistive element. To solve this problem, one must combine the ideas of the voltage source as treated in Chapter 2, with the ideas of impedance in Chapter 4.

Example 6.1

A sinusoidal ac generator is tested in the laboratory. The results of these measurements are summarized as shown in Fig. 6.1. Find a voltage source or generator model (Thévenin equivalent) of the generator using the data measured in the laboratory (see Fig. 6.1). Because real-life measurements include errors of some magnitude, we must not expect to find an *exact* model. It is given that the internal impedance of the generator includes both resistive and capacitive (no inductive) effects. Notice that the magnitude only of the voltage **V** is given (in effective volts). Of course in this

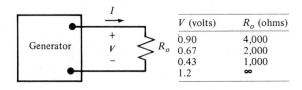

V (volts)	R_o (ohms)
0.90	4,000
0.67	2,000
0.43	1,000
1.2	∞

fig. 6.1   Generator circuit and laboratory test data.

sinusoidal situation, current, voltage, and impedances are complex quantities. The frequency was held constant at 5,000 Hz.

Solution
$V_s = V$ when $R_o = ∞$; therefore $V_s = 1.20$ V. See Fig. 6.2. We can calculate I for each R_o. For $R_o = 4,000$ Ω,

$$I = \frac{V}{R_o} = \frac{0.90}{4,000} = 0.2250 \text{ mA}$$

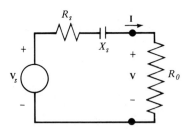

fig. 6.2   Generator model.

The square of the magnitude of the total impedance for $R_o = 4,000$ is

$$Z^2_{4,000} = (R_s + 4,000)^2 + X_s^2 = \left(\frac{1.2}{0.2250 \times 10^{-3}}\right)^2 = 5,333^2 \text{ Ω}$$

Similarly,

$$Z^2_{2,000} = (R_s + 2,000)^2 + X_s^2 = \left(\frac{1.2}{(0.67)/2,000}\right)^2 = 3,582^2 \text{ Ω}$$

Then eliminating X_s^2 between these two equations, we have

$$R_s = \frac{5,333^2 - 3,582^2 - 4,000^2 + 2,000^2}{(8,000 - 4,000)} = 902.5 \text{ Ω}$$

Then solving for $|X_s|$, we get

$$|X_s| = [5,333^2 - 4,902^2]^{1/2} = 2,100 \text{ Ω}$$

So the Thévenin equivalent of the generator is as shown in Fig. 6.2, where

$$R_s = 902.5 \ \Omega \quad \text{and} \quad X_s = -2,100 \ \Omega \quad \text{(capacitive)}$$

$$\bullet \quad \bullet \quad \bullet$$

PROBLEMS

6.1 In Example 6.1 we reached a solution by using the $R_o = 4,000\text{-}\Omega$ and $R_o = 2,000\text{-}\Omega$ lines of data. Find the Thévenin equivalent of the generator by using the $R_o = 4,000\text{-}\Omega$ and $R_o = 1,000\text{-}\Omega$ lines of data. Because all measured data (including that given in the problem) contain error, you should expect the solution to Prob. 6.1 to agree reasonably close, but not precisely, with the solution to Example 6.1.

6.2 Find two current source models (Norton) of the generator of Example 6.1: one model with correct R and C elements in parallel, and another with different R and C elements in series.

6.3 Make the necessary calculations and then plot both the magnitude and the phase of **V** of the circuit of Example 6.1 against R_o. Plot both of these functions on one set of coordinates. Assume that the phase of the internal source is zero.

Example 6.1 and Prob. 6.2 could give an erroneous impression. There might be the implication here that the series RC and parallel RC models, developed in these solutions, perform equally for all frequencies. Not so; these models are not equivalent at frequencies other than the frequency (5,000 Hz) at which the "laboratory data" were observed.

PROBLEM

6.4 For both the RC series and the RC parallel models of Prob. 6.2, calculate the generator internal impedance at 2,000 Hz.

In solving Example 6.1 and Prob. 6.2, we found three models of the given generator. All are valid at the one single frequency of 5,000 Hz. Additional information or data are required to find one or more models valid over a band of frequencies. We will return to this kind of problem later in the chapter.

> The concepts of the current source, the voltage source, Norton's theorem, and Thévenin's theorem (previously applied to resistive networks) apply to all linear networks where they may contain any combination of capacitance and inductance elements as well as resistance elements.

6.2 MORE GENERAL APPLICATIONS OF SUPERPOSITION

The concept of superposition is very powerful in linear networks and other linear systems. In Chapter 2, superposition was applied to resistive networks only. Here we will see how superposition applies to any linear RLC network. A linear network is one in which each of the values of the R, L, and C elements in the network is independent of its current or voltage. Resistance values of semiconductors are typically nonlinear to some degree. Depending on the degree of nonlinearity, circuits using these devices may be treated as though linear, or perhaps as nonlinear. The L and C elements, depending on the situation, may be treated as linear or nonlinear. For the most part, this text is constrained to linear networks; however, occasionally we will analyze a nonlinear situation.

Consider the system of Fig. 6.3a, where there are three forcing functions, v_1, v_2, and i_3 (any time domain), and we desire some response such as i_n or v_m. The concept of superposition tells us that we could proceed to find the response at i_n (or v_m) to v_1 alone. Call this response i_{n_1}. Then find the corresponding response i_{n_2} due to v_2 acting alone, and i_{n_3} due to i_3 alone. Then the total response due to all three forcing functions acting simultaneously is

$$i_n = i_{n_1} + i_{n_2} + i_{n_3} \tag{6.1}$$

When we say v_1 is acting alone, this means that v_2 is reduced to zero or v_2 is "shorted" (short-circuited) and i_3 is zero or the i_3 circuit is "open." See Fig. 6.3b.

The current i_{n_1} is that component of i_n when all sources except v_1 have been *killed*. A *killed* or *dead* source is one that is reduced to zero. We kill a voltage source by placing a short at its terminals; we kill a current source by *opening* the lead that carries the normal current.

> The principle of superposition in linear electric networks is: A current or voltage at any point is the sum of all separate components where each component is the consequence of one live current or voltage source with all other sources killed.

Superposition has broad application in other linear systems as well as electric networks. It is also important to note that in discussing superposition, no constraint was placed on the forcing functions. In the situation of Fig. 6.3, the following is a reasonable situation to which superposition is applicable: $i_3 = 2.1$ mA, $v_1 = 7e^{-0.01t}$ sin $10^4 t$ V, and $v_2 = 25 \sin(0.1t + 20°)$ V.

Superposition will be used in many situations that follow throughout this text. You will find it to be a powerful weapon. No formal proof will be made in this text,

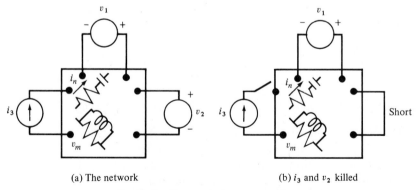

(a) The network   (b) i_3 and v_2 killed

fig. 6.3   Illustrating superposition.

but you will see that it is valid in many particular situations that will follow. You will solve many problems where you will use superposition for one method of solution, and then by some other method to achieve the same solution.

Example 6.2

In Fig. 6.4, there is a current source i_1 and a voltage source v_2. Where $v_C(0+) = 0$, find the total response v including natural and forced components. First we shall find v by using the principle of superposition, then we shall find v by using the

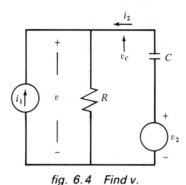

fig. 6.4 *Find v.*

method of branch currents and comparing the two solutions. For both of these methods, we will use the operator technique discussed in Section 5.3. You may wish to quickly review this section.

a. By superposition, find v. Due to i_1 alone $v_2 = 0$, in the s domain

$$v' = i_1\left(R\|\left(\frac{1}{Cs}\right)\right) = \frac{i_1 R/Cs}{R + 1/Cs} = \frac{i_1 R}{1 + RCs} \tag{6.2}$$

where v' is the component of v due to i_1 alone. Due to v_2 alone ($i_1 = 0$), using the voltage divider equation

$$v'' = v_2\left(\frac{R}{R + 1/Cs}\right) = \frac{v_2 RCs}{1 + RCs} \tag{6.3}$$

where v'' is the component of v due to v_2 alone. Then from the principle of superposition,

$$v = v' + v'' = \frac{R}{1 + RCs}(i_1 + Csv_2) \tag{6.4}$$

b. Solving for v by branch currents,

$$v = (i_1 + i_2)R \tag{6.5}$$

$$i_2 = \frac{v_2 - v}{1/Cs} = Cs(v_2 - v) \tag{6.6}$$

Eliminating i_2 between (6.5) and (6.6) and rearranging, we obtain

$$v = \frac{R}{1 + RCs}(i_1 + Csv_2) \tag{6.7}$$

and it is evident that (6.7) and (6.4) agree.

• • •

PROBLEM

6.5 To show that you understand the true meaning of (6.4), show that this equation has the meaning of the following differential equation:

$$RC\frac{dv}{dt} + v = Ri_1 + RC\frac{dv_2}{dt} \tag{6.8}$$

Our goal was to find v. To complete the solution, we would solve the differential equation implied by (6.4) or the explicit differential equation of (6.8).

Commonly in electronic and communication circuits, a current wave (or voltage wave) has an average component or steady component plus a sinusoidal component as in Fig. 6.5, where $i = 0.1 + 0.06 \sin 10^5 t$ mA. In some of the situations, we wish to produce a voltage from this current that, in the steady-state condition, is nearly steady, with very little "ripple." The circuit of Fig. 6.5 can be used to accomplish the goal. By an appropriate choice of R and C, v can be largely steady, with only a small sinusoidal component. The application of superposition greatly simplifies the analysis of this situation. Think of a single-current current source as having the two components of Fig. 6.5. Now find v', the component of v due to i_1 alone. Then find v'', the component of v due to i_2 alone. It follows that $v = v' + v''$. Voltage v' is constant, and v'' is sinusoidal. By appropriate selection of R and C, v'' can be made sufficiently small.

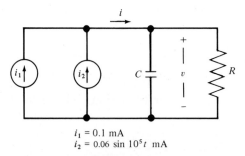

$i_1 = 0.1$ mA
$i_2 = 0.06 \sin 10^5 t$ mA

fig. 6.5   Application of superposition.

PROBLEMS

6.6 Given a current source $i = 0.1 + 0.06 \sin 10^5 t$ mA, select R and C of Fig. 6.5 such that v has maximum and minimum values of 12.05 and 11.95 V, respectively, in the steady-state condition. Apply the concept of superposition. Write the equation of v.

6.7 Suppose that you must solve Prob. 6.6 without using superposition; outline a procedure for finding the solution. It should occur to you that the method of superposition is elegant and relatively simple.

Superposition is not valid in every conceivable situation and must be applied with caution. Read again page 229. Problem 6.8 illustrates an incorrect application of superposition.

PROBLEM

6.8 Two current sources are connected as shown in Fig. 6.6. Show that superposition will yield the correct value of v, but the power p in R is not $(p_1 + p_2)$, where p_1 and p_2 are the separate powers due to i_1 alone and i_2 alone, respectively. Notice that without reference to superposition, we know that $v = (i_1 + i_2)R$ and $p = (i_1 + i_2)^2 R$.

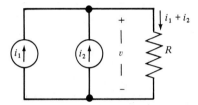

fig. 6.6   Two current sources.

6.3   MORE COMPLEX NETWORKS—METHOD OF BRANCH CURRENTS

So far in this text, we have applied Kirchhoff's voltage and current laws to rather simple resistance networks. Now we are ready to consider more complicated situations that involve L and C and, in the process, to develop additional techniques. Consider the situation of Fig. 6.7, where v_1 and v_2 are known. Each of the currents i_1, i_2, and i_3 flows in a distinct *branch* of the circuit and is therefore called a *branch current*, as was discussed in Chapter 2. The loop current was also introduced in Chapter 2 and will be considered in the light of more complex circuits in the next section.

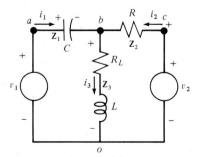

fig. 6.7   Three branch currents.

Let us review the method of branch currents in finding the three currents of Fig. 6.7. Assume that v_1 and v_2 are sinusoids represented by $\mathbf{V}_1$ and $\mathbf{V}_2$, respectively, and we desire the forced or steady-state responses represented by $\mathbf{I}_1$, $\mathbf{I}_2$, and $\mathbf{I}_3$. We write three independent equations involving these three currents. First, we apply Kirchhoff's law to the closed path or loop *aboa* to obtain one of these three.

$$\mathbf{V}_1 = \left(-\frac{j}{\omega C} \right)\mathbf{I}_1 + 0 \times \mathbf{I}_2 + (R_L + j\omega L)\mathbf{I}_3 \qquad (6.9)$$

Because $\mathbf{I}_2$ does not contribute to the voltage in this loop, one might ask why write the $0 \times \mathbf{I}_2$ term into this equation. For purposes of ultimate solution of the three equations by determinants or matrices, there is merit in explicitly displaying all of the current coefficients (including the zeros). Also, although it is not obvious at this point, there is convenience in writing the coefficient of the current preceding rather than following the current that it multiplies. Ultimately (for matrix solutions) you will see some advantage in convenience and in writing the term $(R_L + j\omega L)\mathbf{I}_3$ rather than $\mathbf{I}_3(R_L + j\omega L)$.

To continue in writing the necessary three equations, we apply Kirchhoff's law to loop *cboc*,

$$\mathbf{V}_2 = 0\mathbf{I}_1 + R\mathbf{I}_2 + (R_L + j\omega L)\mathbf{I}_3 \qquad (6.10)$$

The needed third equation comes from the application of Kirchhoff's current law at node *b*.

$$0 = \mathbf{I}_1 + \mathbf{I}_2 - \mathbf{I}_3 \qquad (6.11)$$

The solution of (6.9), (6.10), and (6.11) leads to the values of $\mathbf{I}_1$, $\mathbf{I}_2$, $\mathbf{I}_3$ in terms of $\mathbf{V}_1$, $\mathbf{V}_2$, ωC, R_L, ωL, and R. However, before showing these solutions, let us simplify the nomenclature by letting $\mathbf{Z}_1 = -j/\omega C$, $\mathbf{Z}_2 = R$, and $\mathbf{Z}_3 = R_L + j\omega L$. Then these equations reduce to

$$\mathbf{V}_1 = \mathbf{Z}_1\mathbf{I}_1 + 0\mathbf{I}_2 + \mathbf{Z}_3\mathbf{I}_3 \qquad (6.12)$$

$$\mathbf{V}_2 = 0\mathbf{I}_1 + \mathbf{Z}_2\mathbf{I}_2 + \mathbf{Z}_3\mathbf{I}_3 \qquad (6.13)$$

$$0 = \mathbf{I}_1 + \mathbf{I}_2 - \mathbf{I}_3 \qquad (6.14)$$

It follows that, by determinants,

$$\mathbf{I}_1 = \frac{\begin{vmatrix} \mathbf{V}_1 & 0 & \mathbf{Z}_3 \\ \mathbf{V}_2 & \mathbf{Z}_2 & \mathbf{Z}_3 \\ 0 & 1 & -1 \end{vmatrix}}{|\mathbf{D}|} \qquad \mathbf{I}_2 = \frac{\begin{vmatrix} \mathbf{Z}_1 & \mathbf{V}_1 & \mathbf{Z}_3 \\ 0 & \mathbf{V}_2 & \mathbf{Z}_3 \\ 1 & 0 & -1 \end{vmatrix}}{|\mathbf{D}|} \qquad \mathbf{I}_3 = \frac{\begin{vmatrix} \mathbf{Z}_1 & 0 & \mathbf{V}_1 \\ 0 & \mathbf{Z}_2 & \mathbf{V}_2 \\ 1 & 1 & 0 \end{vmatrix}}{|\mathbf{D}|} \qquad (6.15)$$

$$|\mathbf{D}| = \begin{vmatrix} \mathbf{Z}_1 & 0 & \mathbf{Z}_3 \\ 0 & \mathbf{Z}_2 & \mathbf{Z}_3 \\ 1 & 1 & -1 \end{vmatrix} \qquad (6.16)$$

By expanding the determinants for I_1, we get

$$I_1 = \frac{V_1(Z_2 + Z_3) - V_2 Z_3}{Z_1 Z_2 + Z_1 Z_3 + Z_2 Z_3} \tag{6.17}$$

Let us now make some consistency checks on (6.17):

1. The terms $V_1(Z_1 + Z_2)$ and $V_2 Z_2$ each have dimensions of volts × ohms. Each of the product terms that is added to other product terms in the denominator has dimensions of ohms². The right-hand side has dimensions of (volts × ohms)/ohms² = volts/ohms = amperes.
2. The $V_1(Z_2 + Z_3)$ must carry a positive sign, whereas the $V_2 Z_3$ must carry a negative sign. This is evident when one views the assigned directions of V_1, V_2, and I_1.
3. If we allowed Z_3 to approach zero, looking at the resultant simplified circuit, we see that $I_1 = V_1/Z_1$. Then when we allow $Z_3 \to 0$ in (6.17), we see that

$$I_1 \to \frac{V_1 Z_2}{Z_1 Z_2} = \frac{V_1}{Z_1} \qquad \text{as } Z_3 \to 0 \tag{6.18}$$

So we see that (6.17) passes at least three consistency checks.

PROBLEMS

6.9 To apply still another consistency check to (6.17), allow $Z_2 \to 0$ and write the limiting value of I_1 by analyzing the simplified circuit resulting from letting $Z_2 \to 0$. Then determine from (6.17) the limiting value of I_1 as $Z_2 \to 0$. Of course, these two limiting values of I_1 should be identical.

6.10 By the method of branch currents, solve for I_3 of the circuit of Fig. 6.7. Apply at least three consistency checks to your solution.

Example 6.3

Solve for I_1 of Fig. 6.7 by the application of superposition. Let I_1' and I_1'' be the current due to V_1 acting alone and V_2 acting alone, respectively.

$$I_1' = \frac{V_1}{Z_1 + (Z_2 \parallel Z_3)} = \frac{V_1}{Z_1 + \dfrac{Z_2 Z_3}{Z_2 + Z_3}} = \frac{V_1(Z_2 + Z_3)}{Z_1 Z_2 + Z_1 Z_3 + Z_2 Z_3} \tag{6.19}$$

$$I_1'' = \underset{\substack{\uparrow \\ \text{Total} \\ \text{current}}}{\frac{-V_2}{Z_2 + (Z_1 \parallel Z_3)}} \times \underset{\substack{\uparrow \\ \text{Current} \\ \text{division}}}{\frac{Z_3}{Z_1 + Z_3}} = \frac{-V_2 Z_3}{Z_1 Z_2 + Z_2 Z_3 + Z_1 Z_3} \tag{6.20}$$

Then,

$$I_1 = I_1' + I_1'' = \frac{V_1(Z_2 + Z_3) - V_2 Z_3}{Z_1 Z_2 + Z_2 Z_3 + Z_1 Z_3} \tag{6.21}$$

Of course, (6.21) is identical to (6.17).

• • •

6.11 Solve for I_3 of Fig. 6.7 by the application of superposition. This solution, of course, must be identical to the solution to Prob. 6.10.

6.4 METHODS OF LOOP CURRENTS AND NODAL EQUATIONS

The *branch current* method applied to the circuit of Fig. 6.7 leads to a family of three simultaneous equations. One of these equations was an expression of Kirchhoff's current law and might be used to eliminate one current such as I_3 immediately before writing solutions in a determinant form, as was shown in Chapter 2. Doing this by eliminating I_3 from (6.12) and (6.13) gives us

$$V_1 = (Z_1 + Z_3)I_1 + Z_3 I_2 \tag{6.22}$$

$$V_2 = Z_3 I_1 + (Z_2 + Z_3)I_2 \tag{6.23}$$

Now if we wished to proceed by determinants, we would encounter only determinants of second rather than third order. Then after I_1 and I_2 have been found, we can easily find I_3 from $I_1 + I_2$.

6.12 From (6.22) and (6.23) find I_1, I_2, and then I_3. Of course these solutions should agree identically with corresponding values found by previous methods.

You will remember from Chapter 2 (resistive networks) that we can implement the line of thought that leads to (6.22) and (6.23) in a highly organized technique by assigning *loop currents* as shown in Fig. 6.8. Here the loop currents I_1 and I_2 are the previous branch currents I_1 and I_2, respectively, but we think of each of these currents passing through branch 3 to close on themselves. After the two loop currents have been found, the branch current I_3 is found from $I_1 + I_2$. The loop cur-

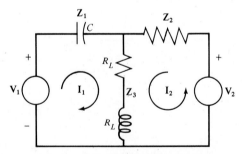

fig. 6.8   Loop currents.

rents yield a neatly organized approach to the writing of the simultaneous equations, as was previously discussed. One sums the voltages around the closed path for each loop current. Notice in (6.22) how the coefficient of I_1 is the total impedance traversed by I_1 in the closed loop of I_1 and that the coefficient of I_2 is that impedance that is traversed by I_2.

The direction a current is assigned in a given loop is arbitrary.

PROBLEM

6.13 In Fig. 6.8, assign I_2' the reversed direction in respect to I_2 so that I_2' flows from $+$ to $-$ internally in V_2. Now find I_2' and show that $I_2' = -I_2$. The two equations are

$$V_1 = (Z_1 + Z_3)I_1 - Z_3 I_2' \qquad (6.24)$$

$$-V_2 = -Z_3 I_1 + (Z_2 + Z_3)I_2' \qquad (6.25)$$

Solve (6.24) and (6.25) for I_1 and I_2'. Solve (6.22) and (6.23) for I_1 and I_2. Of course the two solutions for I_1 must be identical, and I_2 must be equal to $-I_2'$.

You will remember from Chapter 2 that we have some choice in selecting the closed paths for the loop currents. For example, suppose we had *no* interest in the branch current in Z_2, but we wished the branch currents in Z_1 and Z_3. Then loops might be assigned or defined as shown in Fig. 6.9. Notice that I_1 is the branch current in Z_1 as before, I_2 is the branch current in Z_3, and $I_2 - I_1$ is the branch current in Z_2. Elsewhere in the discussion relative to Figs. 6.7, 6.8, and 6.9 (different approaches to the same networks) a particular current or voltage takes on one meaning for one figure and a different meaning in the next figure. For example, V_a of Fig. 6.9 is identical to V_1 of Fig. 6.8.

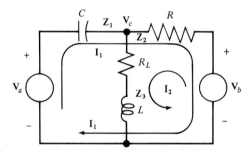

fig. 6.9   *A different assignment of loop currents.*

PROBLEM

6.14 Solve for I_1 and I_2 of Fig. 6.9 and show that these loop currents yield the same branch currents as do the loop currents of Fig. 6.8.

Figure 6.10 is the model of a particular power system problem. Here there are two generators supplying power through long transmission lines (line 1 and line 2) to one large load. The load might consist of a composite of a metropolitan-industrial area. Each of the two transmission lines might be 200 km long. Each generator might have a capacity of 100 MW. A transmission line can be modeled in terms of (for line 2) R_2, L_2, C_2, and half of C_3. There is a transformer bank for

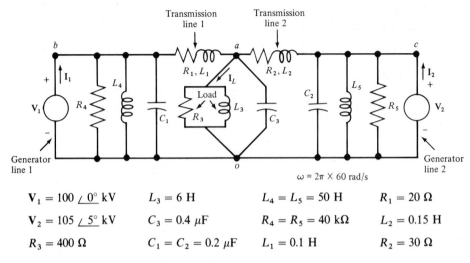

$$V_1 = 100 \underline{/\ 0°}\ \text{kV} \qquad L_3 = 6\ \text{H} \qquad L_4 = L_5 = 50\ \text{H} \qquad R_1 = 20\ \Omega$$

$$V_2 = 105 \underline{/\ 5°}\ \text{kV} \qquad C_3 = 0.4\ \mu\text{F} \qquad R_4 = R_5 = 40\ \text{k}\Omega \qquad L_2 = 0.15\ \text{H}$$

$$R_3 = 400\ \Omega \qquad C_1 = C_2 = 0.2\ \mu\text{F} \qquad L_1 = 0.1\ \text{H} \qquad R_2 = 30\ \Omega$$

fig. 6.10   Model of a power system.

changing voltage levels between each generator and its respective transmission line. The model of the transformer at the right end of line 2 influences the parameters R_2, L_2, L_5, and R_5. For some purposes, the model of the generator includes significant series impedances; however, for this situation, the terminal voltages of these two generators V_{bo} and V_{co} are held at particular values by automatic, voltage regulating systems.

PROBLEM

6.15   For the data given in respect to the model of Fig. 6.10 solve to find the numerical value of the phasor load current I_L, load voltage V_{ao}, and the two respective generator currents I_1 and I_2. In a later problem relative to this same system, you will use these calculated values to compute generated and load powers and other significant quantities in this system.

Suggestions : If you approach this problem from a brute force point of view, you might assign ten different loop currents in the respective ten *windows*. However, here is another example where certain parameters (R_4, L_4, C_1, C_2, L_5, and R_5) do not in any way influence load current and load voltage. For further simplification, you may like to work with admittances (rather than impedances) of all the vertical branches. Then the equivalent of R_3, L_3, and C_3 could be reduced to one single admittance. Likewise, R_4, L_4, and C_1 could be reduced to another single admittance. Conditions at the load could be computed from two simultaneous loop equations. The two line currents could be found from these same two equations. Subsequently, you could find the two generator currents without becoming involved in further simultaneous equations.

If you solved Prob. 6.15 through two simultaneous loop equations, as was suggested, then it should occur to you that a single nodal equation could be used to find the nodal voltage V_{ao}. So in this situation, the nodal approach eliminates the simultaneous equation aspect of the problem.

PROBLEM

6.16 If you used two simultaneous loop equations to solve Prob. 6.15, solve now to find the numerical values as stated in Prob. 6.15 by writing a single nodal equation (use the nodal-superposition technique) that yields the value of V_{ao}. On the other hand, if you used the nodal approach in solving Prob. 6.15, solve this problem now by using two simultaneous loop equations. Obviously, the two approaches should yield the same results.

In the nodal approach to this problem, try to write the equation that is *explicit* in V_{ao} without any preliminary writing.

6.5 MATRIX FORM OF LOOP AND NODE EQUATIONS

Circuit equations written in a matrix form are not only amenable to determinant or matrix solutions, but also, with a little practice, these equations in this form can be written easily, accurately, and in a highly organized fashion. The loop equations in matrix form are written as shown in (6.26) to (6.28). See similar equations, (2.55) to (2.57), for resistive networks.

$$\mathbf{V}_1 = \mathbf{Z}_{11}\mathbf{I}_1 + \mathbf{Z}_{12}\mathbf{I}_2 + \mathbf{Z}_{13}\mathbf{I}_3 + \cdots + \mathbf{Z}_{1n}\mathbf{I}_n \qquad (6.26)$$
$$\mathbf{V}_2 = \mathbf{Z}_{21}\mathbf{I}_1 + \mathbf{Z}_{22}\mathbf{I}_2 + \mathbf{Z}_{23}\mathbf{I}_3 + \cdots + \mathbf{I}_{2n}\mathbf{I}_n \qquad (6.27)$$
$$\vdots \qquad \vdots \qquad \vdots \qquad \vdots \qquad \vdots$$
$$\mathbf{V}_n = \mathbf{Z}_{n1}\mathbf{I}_1 + \mathbf{Z}_{n2}\mathbf{I}_2 + \mathbf{Z}_{n3}\mathbf{I}_3 + \cdots + \mathbf{Z}_{nn}\mathbf{I}_n \qquad (6.28)$$

In these equations, $\mathbf{V}_1, \mathbf{V}_2, \mathbf{V}_3, \ldots, \mathbf{V}_n$ are the known impressed voltages in the respective closed paths 1, 2, 3, ..., n. The currents $\mathbf{I}_1, \mathbf{I}_2, \mathbf{I}_3, \ldots, \mathbf{I}_n$ are the respective unknown loop currents whose values are sought. In order to find the currents in any loop equation situation, we need only identify the $\mathbf{V}$'s and the $\mathbf{Z}$'s of this family of equations. To see clearly the meaning of the $\mathbf{Z}$'s of these equations, consider the principle of superposition. Then we see that $\mathbf{Z}_{11}\mathbf{I}_1$ is the voltage in closed path number 1 produced by loop current number 1 acting alone. Further, $\mathbf{Z}_{12}\mathbf{I}_2$ is the voltage in closed path number 1 produced by loop current number 2 acting alone. The meaning of each $\mathbf{ZI}$ term in these equations is easily identified; it follows that $\mathbf{Z}$ is found in a straightforward manner.

Equations (6.26) to (6.28) can be interpreted in the time domain where needed. In the time domain, the $\mathbf{Z}$'s are operational impedances such as $R + sL$ for a series RL device. On the other hand, if the forcing functions were sinusoidal and we desired the steady-state values of the loop currents, then the $\mathbf{Z}$'s would be expressed as complex values such as $R + j\omega L$ for the same series RL device.

Example 6.4

For the network of Fig. 6.9 in sinusoidal steady state where the loop currents are assigned as shown, the $\mathbf{V}$'s and $\mathbf{Z}$'s are as follows: $\mathbf{V}_1 = \mathbf{V}_a - \mathbf{V}_b$, $\mathbf{V}_2 = \mathbf{V}_b$, $\mathbf{Z}_{11} = R - j/\omega C$, $\mathbf{Z}_{12} = -R$, $\mathbf{Z}_{21} = -R$, $\mathbf{Z}_{22} = R + R_L + j\omega L$.

● ● ●

It is important to note that to find the loop currents, we need not necessarily write the loop equations as shown in (6.26) to (6.28). It is only necessary to find all of the **Z**'s and **V**'s of these equations. To organize our thinking, we arrange all of the **Z**'s in an array as shown in (6.29). This bracketed array is called a matrix. The symbol [**Z**] is called the matrix **Z**.

$$[\mathbf{Z}] = \begin{bmatrix} \mathbf{Z}_{11} & \mathbf{Z}_{12} & \mathbf{Z}_{13} & \cdots & \mathbf{Z}_{1n} \\ \mathbf{Z}_{21} & \mathbf{Z}_{22} & \mathbf{Z}_{23} & \cdots & \mathbf{Z}_{2n} \\ \vdots & \vdots & \vdots & & \vdots \\ \mathbf{Z}_{n1} & \mathbf{Z}_{n2} & \mathbf{Z}_{n3} & \cdots & \mathbf{Z}_{nn} \end{bmatrix} \tag{6.29}$$

Notice the order of the subscripts among the elements of the [**Z**], the **Z** matrix. The first or left subscript attached to each element identifies the row that the element occupies; the second subscript identifies the column. Of course the elements of the nth row are the **Z** elements associated with the equation for the nth closed path. Compare the first row of [**Z**] with (6.26) to (6.28).

For the sinusoidal, steady-state situation of Fig. 6.9, the matrices [**Z**], [**V**], and [**I**] are

$$[\mathbf{Z}] = \begin{bmatrix} (R - j/\omega C) & -R \\ \\ -R & (R + R_L + j\omega L) \end{bmatrix} \quad [\mathbf{V}] = \begin{bmatrix} (\mathbf{V}_a - \mathbf{V}_b) \\ \\ \mathbf{V}_B \end{bmatrix} \quad [\mathbf{I}] = \begin{bmatrix} \mathbf{I}_1 \\ \\ \mathbf{I}_2 \end{bmatrix} \tag{6.30}$$

Given [**Z**] and [**V**] as in (6.31), one may by matrix theory solve for [**I**]. Those not familiar with matrix theory will find the following brief comments obscure but perhaps interesting. We may write loop equations in matrix form as

$$[\mathbf{V}] = [\mathbf{Z}] \times [\mathbf{I}] \tag{6.31}$$

When the special rules of matrix multiplication (not treated in this text) are applied, (6.26) to (6.28) follow from (6.31). Remember that each of [**V**], [**Z**], and [**I**] is a matrix or array of values. The elements in matrix [**V**] as an example are the separate known voltages, as we have seen before. From matrix theory (not treated in this text), we can solve for [**I**] by

$$[\mathbf{I}] = [\mathbf{Z}]^{-1} \times [\mathbf{V}] \tag{6.32}$$

where $[\mathbf{Z}]^{-1}$ is called the inverse of [**Z**]. The matrix inverse and the matrix multiplication is easily computed with the aid of the computer. Of course, the elements in [**I**] are the desired loop currents. Matrix solution is amenable to computer solution of complex situations where there are perhaps 6 to 50 elements in the **I** matrix (6 to 50 simultaneous equations). In the examples treated in this text, solution by determinants or by elimination will be convenient.

PROBLEMS

6.17 Using the values V_1, V_2, Z_{11}, Z_{12}, ... of Example 6.4, expand the appropriate determinants to find an explicit expression for the phasor current in the L branch of Fig. 6.9. In a later problem you will check this solution against a nodal solution for the same current.

6.18 a. For the situation of Fig. 6.9, calculate the numerical value of the phasor V_c where $\omega L = 1,000$ Ω, $1/\omega C = 500$ Ω, $R_L = 30$ Ω, $R = 500$ Ω, $V_a = 10 \angle 20°$ V, $V_b = 10 \angle -30°$ V.

b. Draw a phasor diagram approximately to scale showing the three branch currents, V_c, and the two source voltages. Make checks on your diagram such as: branch current I_1 must lead $(V_a - V_c)$ by 90°.

Now let us face a different situation in respect to the circuit of Fig. 6.9. Assume that we must know the total solution for the branch current in L where the two voltages V_a and V_b are any functions of time. To do this, we first write the $[Z]$ where the elements are the appropriate operational impedances. Then we expand the appropriate determinants whose elements are drawn from the $[Z]$ and $[V]$. This leads us to an explicit equation for the operational value of I_2. This operational equation for I_2 is the equivalent of the time domain differential equation that relates i_2 to the two time-domain voltages v_a and v_b. Now we may solve this differential equation to find the desired, explicit equation for i_2. It is important to note that in order to solve the differential equation, it is not necessary to actually write the differential equation; the equivalent operation equation contains the needed information. However, until you become sufficiently confident of the proper interpretation of the operational equation, you should write the equivalent differential equation.

PROBLEM

6.19 In the circuit of Fig. 6.9, v_a is a step voltage of 10 V suddenly applied at $t = 0$; $v_b = 20e^{-0.001t}$ V suddenly applied at $t = 0$. Find the explicit equation for the total response current in L using the numerical values of the parameters in Prob. 6.18. For $t < 0$, the current in L is 0 and the voltage across C is 0 at $t = 0$.

Suggestions

a. Find the differential equation for i_L by using operator techniques. Be reasonably sure that this equation is correct before proceeding.

b. Finding the natural response should be straightforward.

c. Finding the forced response may be more difficult. Applying the principle of superposition may simplify your effort to finding the forced response. In fact, one can write the component of forced response in i_L due to V_a (10 volts steady) by inspecting the circuit and without referring to the different equation. Remember that the total forced response must satisfy the differential equation. Herein lies an important consistency check.

General Matrix Nodal Equations

In any linear network, the nodal equations can be written as follows:

$$I_1 = Y_{11}V_1 + Y_{12}V_2 + Y_{13}V_3 + \cdots + Y_{1n}V_n \tag{6.33}$$
$$I_2 = Y_{21}V_1 + Y_{22}V_2 + Y_{23}V_3 + \cdots + Y_{2n}V_n \tag{6.34}$$

$$I_n = Y_{n1}V_1 + Y_{n2}V_2 + Y_{n3}V_3 + \cdots + Y_{nn}V_n \tag{6.35}$$

Known
currents
entering
nodes

Unknown currents
leaving nodes

These are much like the nodal network equations of (2.68) to (2.70) for resistive circuits. The current $Y_{11}V_1$ is the current leaving node 1 due to node voltage V_1 acting alone. The term $Y_{23}V_3$ is the component of current leaving node 2 due to node voltage V_3 acting alone. The term I_4 is the known current entering node 4. Other terms in these equations are easily identified. Again we need not write the nodal equations as shown in (6.33) to (6.35). We *do* need to identify the elements of [Y] and [I] in order to solve for the nodal voltages, the elements of the [V]. These matrix elements are defined as

$$[Y] = \begin{bmatrix} Y_{11} & Y_{12} & \cdots & Y_{1n} \\ Y_{21} & Y_{22} & \cdots & Y_{2n} \\ Y_{31} & Y_{32} & \cdots & Y_{3n} \\ \vdots & \vdots & & \vdots \\ Y_{n1} & Y_{n2} & \cdots & Y_{nn} \end{bmatrix} \qquad [V] = \begin{bmatrix} V_1 \\ V_2 \\ V_3 \\ \vdots \\ V_n \end{bmatrix} \qquad [I] = \begin{bmatrix} I_1 \\ I_2 \\ I_3 \\ \vdots \\ I_n \end{bmatrix}$$

Like the loop equations, we may write the nodal equations in matrix form as

$$[I] = [Y] \times [V] \tag{6.36}$$

From matrix theory, or by using a computer, we can solve for the node-voltage matrix [V] from the relationship

$$[V] = [Y]^{-1} \times [I] \tag{6.37}$$

PROBLEM

6.20   a.   Apply the nodal approach to the circuit of Fig. 6.7, and show that for the steady-state sinusoidal case,

$$Y_{11} = G + \frac{1}{R_L + j\omega L} + j\omega C$$

$$I_1 = V_{ao}\, j\omega C + V_{co}\, G$$

and

$$V_{bo} = \frac{I_1}{Y_{11}}$$

where $G = 1/R$.

b.   Using this value of V_{bo}, solve for the phasor current in L and check this against your solution to Prob. 6.10.

Example 6.6

The circuit of Fig. 6.11 might be found in electronics. Given the phasors $I = 1 \angle 0°$ mA and $V = 5 \angle -160°$ mV, we desire to find the nodal voltage V_1, where $G_1 = 0.02$ mho, $\omega C_2 = 0.0002$ mho, $\omega C_1 = 0.10$ mho, and $G_2 = 0.005\ \Omega$. One might think of this as a one-node or a two-node problem.

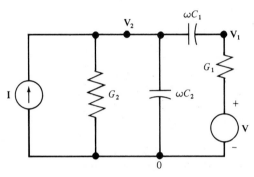

fig. 6.11   Find V_1.

Find V_1 in the circuit of Fig. 6.11 by the matrix nodal approach, where both V_1 and V_2 are unknown nodal voltages. [Y] and [I] are as follows:

$$[\mathbf{Y}] = \begin{bmatrix} G_1 + j\omega C_1 & -j\omega C_1 \\ -j\omega C_1 & G_2 + j\omega(C_1 + C_2) \end{bmatrix} \qquad [\mathbf{I}] = \begin{bmatrix} VG_1 \\ I \end{bmatrix}$$

By expansion of the appropriate determinants extracted from these matrices, we get

$$V_1 = \frac{VG_1(G_2 + j\omega(C_1 + C_2)) + j\omega C_1 I}{D} \qquad (6.38)$$

where

$$D = (G_1 + j\omega C_1)(G_2 + j\omega(C_1 + C_2)) + \omega^2 C_1^2$$
$$= G_1 G_2 - \omega^2 C_1 C_2 + j\omega(C_1 G_2 + C_1 G_1 + C_2 G_1)$$

● ● ●

PROBLEMS

6.21   For a consistency check on (6.38), show that each term on the right-hand side of this equation has dimensions of voltage.

6.22   Find the nodal voltage V_1 in the situation of Example 6.6 through a single-node analysis where V_1 is the unknown nodal voltage. Admittance Y_{11} is found by "killing" I and V only. Also, both I and V contribute to the known current into node 1. Show that your solution agrees with (6.38).

6.23 a. Calculate the numerical value of V_1 in the situation of Prob. 6.22.

b. From this value, calculate the numerical value of V_2.

c. From these nodal voltages, calculate the three remaining branch currents.

d. Display all of the phasor currents and voltages on a single phasor diagram to show the consistency of your solution. Obviously, the branch current in G_2 must be in phase with V_2, while the current in C_2 leads V_2 by 90°. Your diagram should show the consistency of other phase relationships and that appropriate voltages and currents add according to Kirchhoff's laws.

6.6 MAXIMUM POWER TRANSFER

There are many situations in signal processing (communication, electronics) where there is a source with its internal voltage V_1 and internal impedance Z_1 (sinusoidal steady state), and we desire to select the load impedance Z_2 for maximum power in the load impedance. See Fig. 6.12. This implies that V_1 and Z_1 have already been optimized and we only have the freedom to adjust Z_2, the load impedance. One example might be an audio amplifier driving a loudspeaker having impedance Z_2 that will be selected for maximum power in Z_2 for given source voltage and impedance. In another example, Z_2 might be the input impedance to a second stage of a two-stage amplifier.

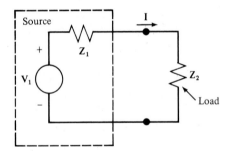

fig. 6.12 Source and load.

Z_1 and Z_2 have both real and imaginary parts as

$$Z_1 = R_1 + jX_1 \quad \text{and} \quad Z_2 = R_2 + jX_2$$

Of course P_2, the power in the load, is given by

$$P_2 = I^2 R_2 = \frac{V_1^2 R_2}{(R_1 + R_2)^2 + (X_1 + X_2)^2} \tag{6.39}$$

Equation (6.39) shows that to achieve maximum P_2 for any given value of R_2, if possible one would adjust X_2 so that $(X_1 + X_2) = 0$ or $X_2 = -X_1$. It is not quite so obvious how to select R_2 for maximum P_2 because R_2 is in both the denominator and numerator of (6.39).

PROBLEMS

6.24 a. Show from (6.39) that $P_2 \to 0$ as $R_2 \to 0$, and $P_2 \to 0$ as $R_2 \to \infty$.

      b. Turn your attention away from (6.39) and view the circuit of Fig. 6.12. Justify the same conclusion.

6.25 Given that $X_1 + X_2 = 30$, $R_1 = 20$, and $V_1 = 1$; calculate about ten points on the curve $P_2 = f(R_2)$ and plot. Assume that it is not practical to adjust X_1, X_2, or R_1 for maximum power.

In solving Prob. 6.25 you should have found a rather flat-topped curve, as suggested in Fig. 6.13. It is not obvious at this point, but $P_{2\,\text{max}}$ occurs at $R_2 = 36.06$ in this situation. Let us now proceed to find the value of R_2 that makes P_2 have its maximum when X_1, X_2, and R_1 are fixed. To do this, let $dP_2/dR_2 = 0$ where V_1, R_1, and $X_1 + X_2$ are constants; that is,

$$\frac{dP_2}{dR_2} = V_1^2 \left(\frac{1}{(R_1 + R_2)^2 + (X_1 + X_2)^2} - \frac{2(R_1 + R_1)R_2}{[(R_1 + R_2)^2 + (X_1 + X_2)^2]^2} \right) = 0$$

$$(6.40)$$

It follows that

$$(R_1 + R_2)^2 + (X_1 + X_2)^2 = 2(R_1 + R_2)R_2 \qquad (6.41)$$

Solving (6.41) for R_2, we find the value of R_2 for maximum power in $\mathbf{Z}_2$, or in R_2:

$$R_2^2 = R_1^2 + (X_1 + X_2)^2 \qquad \text{or} \qquad R_2 = \sqrt{R_1^2 + (X_1 + X_2)^2} \qquad (6.42)$$

Equation (6.42) tells us that the maximum power occurs in $\mathbf{Z}_2$ (or R_2) when R_2 is equal to the magnitude of the impedance of the closed circuit after R_2 has been extracted. Or we may say that for maximum power transfer when only R_2 can be adjusted, we must adjust R_2 until it matches the magnitude of the remaining part of the closed-path impedance. And, of course, (6.39) makes it clear that for any value of R_2, decreasing the value of $X_1 + X_2$ increases P_2, the power in R_2.

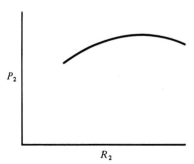

fig. 6.13   Power in R_2 as a function of R_2; $X_1 + X_2 = 30$, $R_1 = 20$.

PROBLEM

6.26 Given a source where $V_1 = 50$ V (effective), $\mathbf{Z}_2 = R_2 + j5$ Ω, and $\mathbf{Z}_1 = 5 + j10$ Ω.

      a. The resistance R_2, but not the reactance X_2, of the load can be adjusted for maximum power in the load. Select R_2 for maximum P_2. Calculate P_2. Calculate the efficiency of power transfer $\mathscr{E}$ where $\mathscr{E} = P_2/P_i$ and P_i is the power delivered by the voltage source V_1.

b. Now assume that X_2 can be adjusted for another $P_{2\,\text{max}}$ while R_2 is held constant at the value you selected in (a). For this value of X_2, find the new $P_{2\,\text{max}}$ and the new value of $\mathscr{E}$.

c. For the value of X_2 of (b), again adjust R_2 for still another $P_{2\,\text{max}}$. Calculate this new $P_{2\,\text{max}}$ and new value of $\mathscr{E}$.

The efficiency of power transfer as demonstrated in the solution to Prob. 6.26 might seem very low, but in many communication situations, the efficiency is not nearly as important as the magnitude of P_2 itself.

From this previous treatment of the situation of Fig. 6.12, we see that for maximum power transfer (maximum power P_2 in R_2 or Z_2):

1. If X_2 can be adjusted, make $(X_1 + X_2)$ as small as possible.
2. Let $R_2 = |R_1 + j(X_1 + X_2)|$.
3. If both R_2 and X_2 can be adjusted independently, let $R_2 = R_1$ and $X_2 = -X_1$ or $R_2 + jX_2 = R_1 - jX_1$.

There are many other situations similar to 1, 2, and 3 above where there is some constraint on R_2 and/or X_2. Consider the following case.

Refer again to Fig. 6.12 where $Z_1 = R_1 + jX_1$ and $Z_2 = R_2 + jX_2 = Z_2 \underline{/\alpha_2}$. There are important situations where the angle α_2 must be held constant, but Z_2 (magnitude) is variable. Then $R_2 = Z_2 \cos \alpha_2$ and $X_2 = Z_2 \sin \alpha_2$. Thus

$$P_2 = \frac{V_1^2 R_2}{(R_1 + R_2)^2 + (X_1 + X_2)^2}$$

$$= \frac{V_1^2 Z_2 \cos \alpha_2}{(R_1 + Z_2 \cos \alpha_2)^2 + (X_1 + Z_2 \sin \alpha_2)^2} \tag{6.43}$$

where R_1, X_1, and α_2 are constant, and

$$\frac{dP_2}{dZ_2} = V^2 \cos \alpha_2 \left[\frac{1}{(R_1 + Z_2 \cos \alpha_2)^2 + (X_1 + Z_2 \sin \alpha_2)^2} \right.$$

$$\left. - \frac{2Z_2 [\cos \alpha_2 (R_1 + Z_2 \cos \alpha_2) + \sin \alpha_2 (X_1 + Z_2 \sin \alpha_2)]}{[(R_1 + Z_2 \cos \alpha_2)^2 + (X_1 + Z_2 \sin \alpha_2)^2]^2} \right] = 0 \tag{6.44}$$

or

$$(R_1 + Z_2 \cos \alpha_2)^2 + (X_1 + Z_2 \sin \alpha_2)^2 = 2Z_2 [\cos \alpha_2 (R_1 + Z_2 \cos \alpha_2)$$

$$+ \sin \alpha_2 (X_1 + Z_2 \sin \alpha_2)]$$

or

$$R_1^2 + X_1^2 + Z_2^2 + 2Z_2 (R_1 \cos \alpha_2 + X_1 \sin \alpha_2)$$

$$= 2Z_2 (R_1 \cos \alpha_2 + X_1 \sin \alpha_2) + 2Z_2^2$$

Then

$$R_1^2 + X_1^2 = Z_2^2 \quad \text{or} \quad Z_1^2 = Z_2^2 \quad \text{or} \quad Z_1 = Z_2 \qquad (6.45)$$

From (6.45), we see that if the angle of the load impedance Z_2 is held constant, there is maximum power in Z_2 when the Z_2 is adjusted to Z_1. This is an important concept in using a transformer to achieve maximum power transfer. In this situation, Z_2 can be adjusted but α_2 cannot.

PROBLEMS

6.27 For the circuit of Fig. 6.12, $V_1 = 10$ V effective, $R_1 = 5\ \Omega$, and $X_1 = 10\ \Omega$. For the cases listed in Table 6.1, X_2, R_2, and (R_2/X_2) are constrained as indicated. Fill in the blanks of Table 6.1 with the values of R_2, X_2, or R_2/X_2 that yield the maximum value of P_2, and then record the maximum P_2.

TABLE 6.1   Values for Prob. 6.27

Case No.	X_2	R_2	R_2/X_2	$P_{2_{max}}$
1	4			
2				
3	−4			
4			0.5	
5			−0.5	
6		5		
7		10		

6.28 A generator with internal voltage V_1 and impedance $Z_1 = 100 + j200\ \Omega$ delivers power P_L to $R_L = 300\ \Omega$ as in Fig. 6.14. A capacitor having reactance X_C is added in parallel with R_L with the hopes of increasing P_L. Only X_C can be adjusted for maximum P_L. Impedance Z_{ao} corresponds to the Z_2 of Fig. 6.12, where $Z_{ao} = Z_2 = R_2 + jX_2$, and both R_2 and X_2 are functions of X_C in such a way that R_2 cannot be adjusted independently of X_2.

a.   Find the equations that give each of R_2 and X_2 as an explicit function of R_L and X_C.

From your solution to (a) we see that rules for selecting either R_2 and/or X_2 do not apply directly to this situation. Equation (6.39) is applicable here where R_2 and X_2 are appropriate functions of X_C. The subsequent evaluation of dP_2/dX_C could lead to moderate complexity. Furthermore, it may not be clear to you whether the addition of a capacitance of any magnitude will increase P_L. So an alternate route is proposed. This proposed analysis may not seem elegant, but it is valid and gives important insight. In the coming pages, we will find an additional concept that yields a straightforward approach to our problem. Suppose we examine the question: Does the addition of capacitance current decrease V_1 for a given V_{ao}? If so, then the addition of the capacitance would increase V_{ao} for a given V_1, thus increasing I_R and also P_L accordingly. For convenience, think of V_{ao} as reference independent of X_C. Then find X_C that yields the minimum V_1. This value of X_C is

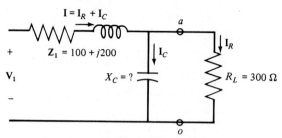

fig. 6.14   Select X_C for maximum P_L .

the value we seek for maximum P_L. Prepare a phasor diagram as a guide for our thinking. For purposes of selecting optimum X_C, $\mathbf{V}_{ao}$ may have any value. Here we let $\mathbf{V}_{ao} = 300 \angle 0°$ V. Then $\mathbf{I}_R = 1 \angle 0°$ and $\mathbf{I}_C = j I_C$. This phasor diagram of Fig. 6.15 shows that there is a value of I_C that will make $\mathbf{V}_1 < \mathbf{V}'_1$ where $\mathbf{V}'_1$ is the value of $\mathbf{V}_1$ for $\mathbf{I}_C = 0$. Notice that $\mathbf{V}_1$ must terminate on the *a-b* line as I_C is increased. We can write

$$\mathbf{V}_1 = \mathbf{V}_{ao} + \mathbf{I}_R \mathbf{Z}_1 + \mathbf{I}_C \mathbf{Z}_1 = 300 + 100 + j200 - 200 I_C + j100 I_C \qquad (6.46)$$

This reduces to

$$\mathbf{V}_1 = 400 - 200 I_C + j(200 + 100 I_C) \qquad (6.47)$$

b. Differentiate the appropriate expression to find the optimum I_C, then X_C, and then P_L.
c. Check your solution by solving for P_L for three different values of X_C, one of which includes your optimum value.
d. Check your solution to (b) by a graphical technique applied to the phasor diagram drawn to scale. See Fig. 6.15. For minimum V_1, note that $\mathbf{V}_1$ is normal to $(\mathbf{I}_C \mathbf{Z}_1)$.

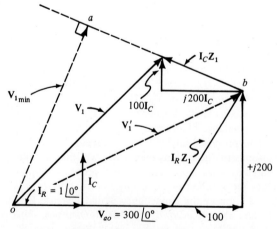

fig. 6.15   Phasor diagram for circuit of Fig. 6.35.

6.7   MAXIMUM POWER FROM CURRENT SOURCE

The previous discussion is related to the power transfer from a voltage source V_1 and its internal impedance Z_1 to a load impedance Z_2. It is natural to ask about the transfer from a current source I_1 with its admittance Y_1 to a load admittance Y_2. See Fig. 6.16. We should expect some very close parallels between these two situations. One might compare the Thévenin equivalent of this current source to the voltage source and thus find the various conditions for maximum power transfer from a current source. But we will choose a different route.

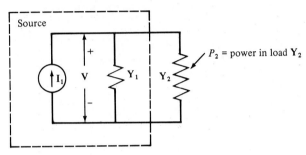

fig. 6.16   *Current source* (I_1, Y_1) *and load* Y_2.

TABLE 6.2   *Summary of Some Conditions for Maximum Power Transfer*

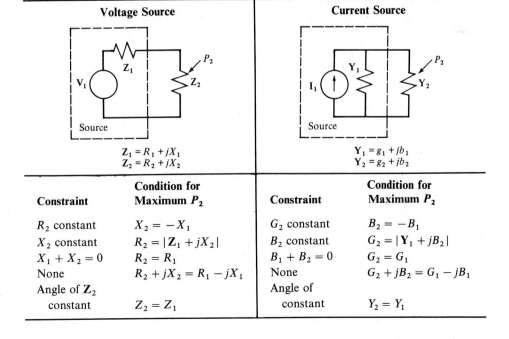

	Voltage Source		**Current Source**
	$Z_1 = R_1 + jX_1$		$Y_1 = g_1 + jb_1$
	$Z_2 = R_2 + jX_2$		$Y_2 = g_2 + jb_2$

Constraint	Condition for Maximum P_2	Constraint	Condition for Maximum P_2
R_2 constant	$X_2 = -X_1$	G_2 constant	$B_2 = -B_1$
X_2 constant	$R_2 = \lvert Z_1 + jX_2 \rvert$	B_2 constant	$G_2 = \lvert Y_1 + jB_2 \rvert$
$X_1 + X_2 = 0$	$R_2 = R_1$	$B_1 + B_2 = 0$	$G_2 = G_1$
None	$R_2 + jX_2 = R_1 - jX_1$	None	$G_2 + jB_2 = G_1 - jB_1$
Angle of Z_2 constant	$Z_2 = Z_1$	Angle of constant	$Y_2 = Y_1$

Where $\mathbf{Y}_1 = G_1 + jB_1$ and $\mathbf{Y}_2 = G_2 + B_2$,

$$P_2 = V^2 G_2 \qquad \text{and} \qquad \mathbf{V} = \frac{I}{\mathbf{Y}_1 + \mathbf{Y}_2} \tag{6.48}$$

It follows that

$$P_2 = \frac{I^2 g_2}{(G_1 + G_2)^2 + (B_1 + B_2)^2} \tag{6.49}$$

Now when (6.39) and (6.49) are compared, we see that we can immediately cite the condition for maximum power transfer for a current source from the corresponding condition for a voltage source. These conditions for both sources are summarized in Table 6.2.

PROBLEM

6.29 With the added background of the concept of power transfer from a current source, you are in a position to view Prob. 6.28 in a different light. Convert the voltage source as given in Prob. 6.28 to its equivalent current source and, in light of Table 6.2, proceed to find X_C for optimum power in R_L. This result should, of course, agree with your previous solution.

6.8 SENSITIVITY TO CHANGE OF FREQUENCY OR CHANGE OF PARAMETERS

In radio and television receivers and transmitters and many, many other situations, *tuning* circuits are used to select or reject some signals in preference to others. The radio or television station selection procedure is an example. Telephone and other signals are channeled, rejected, and selected by means of circuits that are highly sensitive to change of frequency or change of one or more parameters: inductance, capacitance, resistance. It is not too difficult to design a circuit whose input impedance might change by a factor of 10^4 with a 5 percent change of frequency. In fact, in practically every situation where the engineer is concerned with sinusoidal performance, he needs to know how performance varies in respect to a frequency change. As a different kind of example, the frequency of the current in the rotating member of a common induction motor varies from about 2 Hz to 60 Hz. Its torque and speed performance are highly dependent upon the effect of this range of frequencies.

Graphical or Locus Displays

The engineer, analyzing and designing circuits that are forced by sinusoids, has a great need for a graphical view of complex impedance, phasor current, or some other complex quantities. Consider the circuit of Fig. 6.17, where $\mathbf{V}$ is a sinusoid.

$$v = V_m \sin \omega t \qquad \mathbf{V} = V_m \angle \theta = V_m \, e^{j\theta} \tag{6.50}$$

where, as you will remember, the symbolic form $\angle \theta$ has identically the same mean-

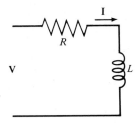

fig. 6.17   Coil having R and L.

ing as the true mathematical form $e^{j\theta}$. The coil current, of course, is

$$I = \frac{V}{R + j\omega L} \tag{6.51}$$

Now we would like to visualize how **I** varies in respect to frequency ω. Of course, **I** is complex having real and imaginary parts—or, alternatively, magnitude and angle. We need to view both components of this complex quantity as shown in Fig. 6.18. Here we are using **V** as a reference. In the limit as frequency approaches zero, the reactance ωL approaches zero, so that **I** approaches V/R, as shown in Fig. 6.18. On the other hand, as ω increases, the magnitude of **I** will decrease while the angle will increase in magnitude in the fourth quadrant. For the other limit as ω approaches ∞, the magnitude of **I** will approach zero and its phase angle will approach $-90°$.

The line traced by the end point of **I** is called the *locus* of **I**. It is not obvious at this point, but this locus is a semicircle, as shown. It is true that, with your present background, you cannot necessarily predict the precise locus; on the other hand, however, you can predict the general shape of the locus.

It is important that you have a clear understanding of the coordinate system of Fig. 6.18. Here the real axis is the real part of **I**, and the vertical axis is the imaginary part of that phasor.

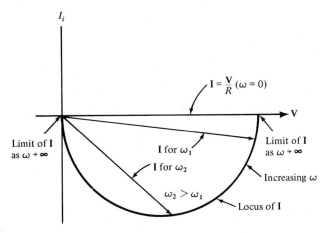

*fig. 6.18   Root locus of **I** in Fig. 6.17.*

Soon, we will prove that the locus of **I** as shown in Fig. 6.18 is truly a semicircle. A simpler proof for another situation is involved in Prob. 6.30.

PROBLEMS

6.30 For the circuit of Fig. 6.17, prove that the locus of the complex input impedance **Z** as a function of ω is a vertical straight line segment lying in the first quadrant only. Graphically, display this locus showing the limiting points from the locus as ω approaches zero and as ω approaches ∞. Label the axes of this coordinate system to clarify your own picture.

6.31 For the coil of Fig. 6.17, the frequency is fixed but R is variable. Find the equation of the locus of the complex input impedance **Z** as a function of R. Label the axes of the system. Show limiting points as R approaches zero and as R approaches ∞.

Proof of the Circular Locus of Fig. 6.18

For the network of Fig. 6.17, the locus of **I** is truly a semicircle as shown in Fig. 6.18. This will be demonstrated. Multiplying numerator and denominator of the right-hand side of (6.51) by the conjugate of the denominator, we get

$$\mathbf{I} = \frac{\mathbf{V}(R - j\omega L)}{R^2 + \omega^2 L^2} = \frac{\mathbf{V}R}{R^2 + \omega^2 L^2} - \frac{j\omega L \mathbf{V}}{R^2 + \omega^2 L^2} \tag{6.52}$$

Real and imaginary parts of **I** are as follows where **V** is the reference:

$$I_r = \frac{VR}{R^2 + \omega^2 L^2} \qquad I_i = \frac{-V\omega L}{R^2 + \omega^2 L^2} \tag{6.53}$$

where I_r and I_i are the real and imaginary parts of **I**, respectively. These two equations of (6.53) are the parametric equations of the locus that we seek, where ω is the parameter. Since we desire to eliminate this parameter, we form the sum of the squares

$$I_r^2 + I_i^2 = \frac{V^2(R^2 + \omega^2 L^2)}{(R^2 + \omega^2 L^2)^2} = \frac{V^2}{R^2 + \omega^2 L^2} \tag{6.54}$$

The right-hand side of (6.54) can be rewritten using (6.53).

$$I_r^2 + I_i^2 = \frac{V}{R} I_r \tag{6.55}$$

Rearranging terms,

$$I_r^2 - \frac{V}{R} I_r + I_i^2 = 0 \tag{6.56}$$

Adding to both sides of the equation and completing the square, we get

$$\left(I_r - \frac{V}{2R}\right)^2 + I_i^2 = \frac{V^2}{4R^2} \tag{6.57}$$

Equation (6.57) is the equation of a circle with the center at

$$I_r = \frac{V}{2R} \qquad I_i = 0 \qquad \text{radius} = \frac{V}{2R} \tag{6.58}$$

PROBLEM

6.32   Plot a few points of the circle of (6.57) to demonstrate that this is consistent with Fig. 6.18.

Equation (6.57) is the equation of a full circle, whereas the locus of Fig. 6.18 is a semicircle. In the real world, ω is positive only, and that constraint was not inserted into the development that leads to (6.57). Or, in other words, if we allowed for negative values of ω, then the locus of **I** would be a complete circle.

PROBLEM

6.33   A resistor $R = 1{,}000\ \Omega$ is in parallel with a capacitor $C = 200$ pF. Where frequency is a variable, find the locus of the admittance of this circuit. Of course, frequency is not allowed to be negative. Show limiting values as ω approaches zero and as ω approaches $+\infty$, respectively. Calculate the frequency when the angle of the admittance is 42°.

Capacitor and Coil in Series

A coil having resistance R and inductance L is in series with capacitance C, as shown in Fig. 6.19. Of course the impedance of this circuit is given as follows:

$$\mathbf{Z} = R + j\!\left(\omega L - \frac{1}{\omega C}\right) = R + jX \tag{6.59}$$

where

$$X = \omega L - \frac{1}{\omega C} \tag{6.60}$$

Here the reactance X may be either positive or negative, depending on the value of ω, the frequency.

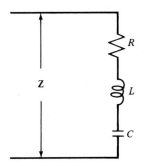

fig. 6.19   Series resonance.

PROBLEMS

6.34   Find the locus of the input impedance **Z** in the network of Fig. 6.19 where frequency varies from 0 to $+\infty$. Notice that in this case, X ranges through negative as well as positive values. Check the consistency of your locus against (6.59).

6.35   Find the locus of the admittance **Y** of the circuit of Fig. 6.19 where ω varies from zero to $+\infty$. Notice that there is a close correspondence to this problem compared to the development of the locus of Fig. 6.18. The reactance X of (6.59) is a more complicated function of ω than is the reactance of the series circuit of Fig. 6.17, as expressed in (6.51). But this does not complicate the analysis to find the locus. There is an important difference here, however, in the comparison of the locus of Fig. 6.18. In the former situation, ωL could assume only positive values. Here, X can assume negative as well as positive values, as indicated by (6.60).

In solving Prob. 6.34 you should have found the locus of the impedance of the series circuit of Fig. 6.19 to be as indicated in Fig. 6.20. In an *RLC* series circuit such as Fig. 6.19, the circuit is said to be in resonance when the complex value of the input impedance **Z** is equal to the resistance R; that is, at resonance,

$$\mathbf{Z} = R \qquad \omega_0 L = \frac{1}{\omega_0 C} \qquad \omega_0 = \frac{1}{LC} \tag{6.61}$$

where ω_0 is the resonance frequency in radians per second.

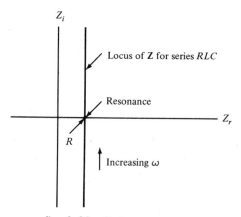

fig. 6.20   Series resonance.

PROBLEM

6.36   For the series circuit of Fig. 6.19, where $R = 100\,\Omega$, $L = 0.5\,\text{mH}$, and $C = 100\,\text{pF}$, calculate and show some pertinent numerical values on the locus of the input impedance of this circuit. For example, calculate the value of ω where the input impedance is the minimum, $100\,\Omega$. Then perhaps calculate the two values of an ω where the magnitude of the impedance is perhaps $1{,}000\,\Omega$. Then, to get a different view of this situation, plot on one set of coordinates both the magnitude and the phase of **Z** as a function of frequency.

Impedance as a Function of Frequency

As we have seen, the locus of the **Z** (complex admittance), **Y**, **V**, or **I** as a function of frequency gives important insight into circuit performance. The plot of Z (magnitude only) and/or the plot of the phase of **Z** gives still another view of circuit

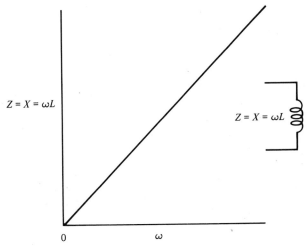

$Z = X = \omega L$

$Z = X = \omega L$

0   ω

fig. 6.21   Z of inductance as a function of frequency ω.

performance. For example, the plot of Z (magnitude only) or X (reactance) of inductance alone is a straight line, as shown in Fig. 6.21.

The plot of the impedance magnitude as a function of ω for an inductor having resistance and inductance is shown in Fig. 6.22.

Where reactance is the imaginary part (including sign) of impedance Z, the reactances of inductance and capacitance, respectively, are as shown in Fig. 6.23. Of course, inductive reactance is $X_L = \omega L$ while capacitive reactance is $X_C = -1/\omega C$.

PROBLEM

6.37   An inductor having 20 Ω of resistance and 0.1000 H of inductance is in series with a 0.0100-μF capacitor with negligible resistance.

a.   Prepare for this circuit numerical plots of the total reactance and magnitude of the total impedance as functions of ω.

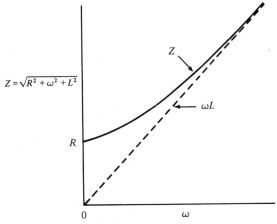

$Z = \sqrt{R^2 + \omega^2 + L^2}$

Z

ωL

R

0   ω

fig. 6.22   Z of an inductor, where $\mathbf{Z} = R + j\omega L$.

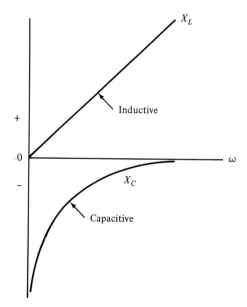

fig. 6.23 *Inductive and capacitive reactances as a function of* ω.

b. As a consistency check on your solution, note that both the circuit reactance and impedance must approach the inductive reactance at high frequencies and approach the magnitude of the capacitive reactance at low frequencies as shown in Fig. 6.23. Note also that $Z > 0$ for all values of ω, and X (circuit reactance) assumes positive, negative, and zero values.

Suppressing Frequencies

Many of our signals (voltages or currents) have components of more than one frequency (dc has frequency zero). Situations arise where we wish to suppress a component of one frequency in respect to other components of different frequencies. For example, an electronic rectifier may convert the standard sinusoidal voltage of 60 Hz to an output having a desired dc voltage plus, among others, an undesirable component voltage having a frequency of 120 Hz. Ideally, the rectifier output would contain no ac components. The rectifier has an internal voltage v_1 containing both the dc component and the 120-Hz component. See Fig. 6.24. The resistance R_1

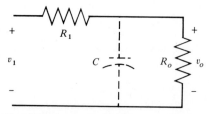

fig. 6.24 *Add C to reduce 120-Hz component in* v_o.

represents the internal resistance of the rectifier, while R_0 is the load resistance. We might add the capacitance C to suppress the ac component of 120 Hz in v_o, the useful output voltage.

PROBLEM

6.38 In the situation of Fig. 6.24, $R_1 = 2$ kΩ and $R_o = 20$ kΩ.
  a. Find a numerical value of C such that the 120-Hz component in $v_o = 0.1$ of the 120-Hz component in v_1. For a simple but approximate solution, note the X_C must be approximately $0.1R_1$ where $R_o = 10R_1$.
  b. For your approximate value of C, calculate the precise ratio of the 120-Hz component in v_o compared to that in v_i.
  c. Calculate the ratio of the dc component in v_o to that in v_i.

Series Resonance

We have seen how the total reactance of a series RLC circuit such as that in Fig. 6.19 is given by (6.60), and when this reactance is zero, the circuit is said to be in resonance. Problems 6.39 and 6.40 demonstrate how the series resonant circuit can be used to an advantage.

PROBLEMS

6.39 Assume that we have a coil having resistance R and inductance L, as given in Prob. 6.36. Both the resistance and the inductance are intermixed internally in the coil such that the voltage across the resistance alone and the voltage across the inductance alone cannot be measured directly. This resistance is a function of frequency and therefore cannot be measured with a multimeter or some other dc device. Where your instruments will measure magnitude only of alternating current and voltage, design an experiment to show how to measure the resistance of the coil at 1 MHz. You are not allowed to use a bridge measurement; you may use a signal generator of variable frequency; you have an unlimited number of capacitors available. The inductance of the coil is given in Prob. 6.36, but the resistance is assumed to be unknown. Describe your experimental setup and your procedure for making this measurement.

6.40 The signal source of Fig. 6.25 has three sinusoidal terms of different frequencies as indicated. This source also has an internal resistance of 5 kΩ. We desire to use a series resonant circuit to suppress the second-harmonic ($\omega = 2 \times 10^6$) term in the output signal v_o. You have available a coil with inductance and resistance, as indicated in the

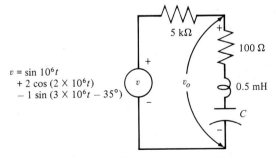

$v = \sin 10^6 t$
$+ 2 \cos (2 \times 10^6 t)$
$- 1 \sin (3 \times 10^6 t - 35°)$

5 kΩ

100 Ω

0.5 mH

C

fig. 6.25 Signal source v_i ; $R_i = 5$ kΩ.

figure. You may select a capacitor to suit your needs. Design the circuit such that the ratio of the second-harmonic term to the fundamental ($\omega = 10^6$) in v_o is a minimum. Now calculate the performance of your circuit by calculating the magnitude of v_o at the three different frequencies of v. Compare the ratio of v_o/v at the frequency $\omega = 2 \times 10^6$ to the ratio v_o/v at the other two frequencies.

If you could select an internal resistance different from the 5 kΩ, would you change that resistance to give a more effective suppression of the second-harmonic term in the input signal?

The series resonant circuit was used with a high-resistance source in Fig. 6.25 to suppress a particular frequency or narrow band of frequencies. This same circuit may also be used to select a narrow band of frequencies if the output voltage v_o is the voltage across either the inductor or the capacitor, as shown in Fig. 6.26, and the driving source resistance is low. This type of circuit is often used to select a desired radio signal from among the many being broadcast.

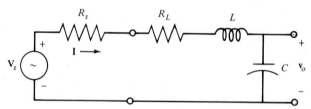

fig. 6.26   A series resonant circuit arranged to select a narrow band of frequencies from a broad frequency spectrum.

For the circuit of Fig. 6.26, the voltage divider concept leads us to

$$\frac{V_o}{V_s} = \frac{-jX_C}{R_s + R_L + j(X_L - X_C)} \qquad (6.62)$$

The magnitude of the output voltage V_o as a function of frequency f current is sketched in Fig. 6.27. The voltage V_o is a maximum at the resonance frequency f_o where the capacitive reactance is equal to, and cancels, the inductive reactance, so the total impedance seen by the voltage source V_s is simply $R_s + R_L = R$, the total

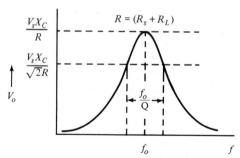

fig. 6.27   Output voltage V_o.

resistance of the series circuit. This may be seen from either Fig. 6.26 or (6.62). Therefore, at resonance, the voltage ratio V_o/V_s is, from (6.62),

$$\frac{V_o}{V_s} = \frac{X_C}{R} \tag{6.63}$$

This ratio of either reactance X_C or X_L (since they are equal at resonance) to the series resistance R may be very large, and therefore a large voltage gain (V_o/V_s) may be realized for frequencies at or near resonance. Because this ratio determines the *quality* of the circuit as a frequency-selective device, it is given a special symbol Q, where

$$Q = \frac{X_C}{R} = \frac{X_L}{R} = \frac{\omega_0 L}{R} = \frac{1}{\omega_0 CR} \tag{6.64}$$

The Q given by (6.64) is known as the *circuit Q*. For a given reactance X, it is limited by the circuit resistance. The ratio X_L/R_L of a coil at the desired resonance frequency is a figure of merit for the coil and is known as the *coil Q*. Let us use Q_L for the coil Q symbol where

$$Q_L = \frac{\omega_0 L}{R_L} \tag{6.65}$$

The circuit bandwidth is usually defined as the frequency range over which the output power delivered to some resistance at V_o does not drop to less than half of the midband or maximum power, assuming the source voltage V_s to be independent of frequency. Since power is proportional to either the square of the current or the square of the voltage, the power delivered by the circuit to V_o drops to half its maximum value when either the current or the voltage at V_o drops to half its maximum value, providing the load resistance remains constant. In the circuit of Fig. 6.26, the actual load resistance would be across the terminals at V_o, but it is not shown because it is normally very high so as not to reduce the Q of the circuit materially. It will become clearer as you continue to study tuned circuits that a high resistance in parallel with a reactive element, capacitance C in this instance, is approximately equivalent to a small resistance in series with that element. The power delivered to this equivalent series resistance is proportional to I^2, and I is inversely proportional to the total impedance Z. The impedance at the half-power point will be $\sqrt{2}Z_{min}$ or $\sqrt{2}R$ where $R = R_S + R_L + R_o'$, and where R_o' is the series equivalent of the load resistance.

$$Z_{min} = R \qquad Z_{1/2} = \sqrt{2}R \tag{6.66}$$

At the half-power frequencies, the reactance must be equal to R to make $Z_{min} = \sqrt{2}R$. So at ω_1 (the lower half-power frequency),

$$\frac{1}{\omega_1 C} - \omega_1 L = R \tag{6.67}$$

and at the upper half-power frequency ω_2,

$$\omega_2 L - \frac{1}{\omega_2 C} = R \tag{6.68}$$

Multiplying both sides of (6.67) by ω_1/L,

$$\frac{1}{LC} - \omega_1^2 = \frac{R\omega_1}{L} \tag{6.69}$$

Then, because $\omega_0^2 = 1/LC$,

$$\omega_0^2 - \omega_1^2 = \frac{R\omega_1}{L} \tag{6.70}$$

A similar manipulation of (6.68) yields

$$\omega_2^2 - \omega_0^2 = \frac{R\omega_2}{L} \tag{6.71}$$

Adding (6.70) and (6.71),

$$\omega_2^2 - \omega_1^2 = \frac{R}{L}(\omega_1 + \omega_2) \tag{6.72}$$

Factoring the left-hand side,

$$(\omega_2 - \omega_1) \times (\omega_2 + \omega_1) = \frac{R}{L}(\omega_1 + \omega_2) \tag{6.73}$$

Dividing both sides of (6.73) by $(\omega_1 + \omega_2)$, we obtain the *bandwidth B* in radians per second:

$$B_1 = (\omega_2 - \omega_1) = \frac{R}{L} = \frac{\omega_0 R}{\omega_0 L} = \frac{\omega_0}{Q} \tag{6.74}$$

Dividing both sides of (6.74) by 2π, the bandwidth in hertz is obtained:

$$f_2 - f_1 = B = \frac{f_0}{Q} \quad \text{hertz} \tag{6.75}$$

Ordinarily we use the resonant circuit to reject or pass a signal of a single frequency or narrow band of frequencies; therefore, minimum bandwidth is desired. In other words, maximum circuit Q is advantageous.

PROBLEM

6.41 Let us assume that an antenna designed to receive a given radio station can be modeled as a voltage source having $R_s = 36\ \Omega$. The station we desire to receive is operating at 1 MHz and induces a 0.10-mV signal into the antenna. Let us further assume that we have chosen an inductor having $L = 5 \times 10^{-4}$ H and $Q_L = 100$ at $f = 1$ MHz to select this station with a circuit similar to Fig. 6.26.

    a. Determine the required tuning capacitance, the output voltage V_o at the resonance frequency, and the bandwidth, assuming there is no resistive loading across the V_o terminals.

    b. Determine the effect on the output voltage and bandwidth if a 150-kΩ resistive load is placed across the output terminals (in parallel with C in Fig. 6.26).

6.9 PARALLEL RESONANCE

Just as a series combination of a coil and a capacitor is highly useful in rejecting or in selecting a signal of a particular frequency, a parallel combination of a coil and a capacitor in other situations is also highly useful in either selecting or rejecting a given frequency or band of frequencies. The circuit of Fig. 6.28 is a typical frequency selection circuit. Here we have a voltage source V_s with its internal resistance R_s, and a load resistance R_2. The frequency-sensitive network combination consisting of a coil in parallel with a capacitor is used as shown between the source and the load. The coil has significant resistance R_L, as well as its inductance L. If the capacitor has significant power loss, which might be true unless high-quality capacitors are used, then the capacitor might be modeled with a parallel combination of a capacitance C and a resistor R_C (see Fig. 6.28). We will soon discover that the load voltage $\mathbf{V}_2$ reaches a rather sharp maximum at a particular frequency or narrow band of frequencies. Thus this circuit is useful in rejecting all frequencies except this single frequency or narrow band of frequencies. Again, radio and television selection or tuning is characterized by this situation, where we wish to reject all *broadcasting stations* except the one that is desired.

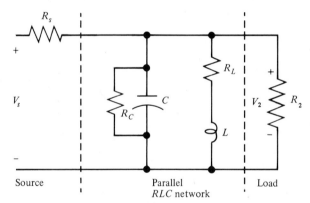

fig. 6.28 A typical parallel RLC circuit used to select a desired frequency or band of frequencies.

The analysis of the circuit of Fig. 6.28 is simplified significantly if we convert the voltage source to its equivalent current source, as indicated in Fig. 6.29. Now we might be tempted to write the equation for the *a-b* impedance of this network of Fig. 6.29. It happens to be much simpler to treat this circuit in terms of the admittance at the *a-b* terminals. Call this admittance $\mathbf{Y}$.

$$\mathbf{Y} = G_s + G_C + G_2 + \mathbf{Y}_{LC} \tag{6.76}$$

where each G in this equation is the reciprocal of its equivalent R, and $\mathbf{Y}_{LC}$ is the admittance of the parallel combination of the coil and capacitor, where the coil has

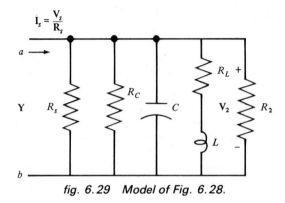

fig. 6.29   Model of Fig. 6.28.

resistance R_L. The admittance of the coil is

$$Y_L = \frac{1}{R_L + j\omega L} \tag{6.77}$$

This reduces to

$$Y_L = \frac{R_L}{R_L^2 + \omega^2 L^2} - \frac{j\omega L}{R_L^2 + \omega^2 L^2} \tag{6.78}$$

In many situations, R_L^2 is negligibly small compared to $(\omega L)^2$. For example, if the coil $Q_L \geq 10$, $(\omega L)^2 \geq 100\ R_L^2$ and the error in neglecting R_L^2 is less than 1 percent. Then for $Q_L \geq 10$, the high-Q case, (6.78) reduces to

$$Y_L = \frac{R_L}{(\omega L)^2} - \frac{j}{\omega L} \tag{6.79}$$

Then the circuits of Figs. 6.29 and 6.30 are equivalent, where

$$G' = G_s + G_2 + G_C$$

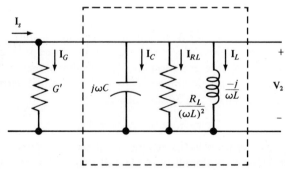

fig. 6.30   Admittance model of Fig. 6.29 when $Q_L \geq 10$.

The admittance model of Fig. 6.30 may be further simplified by combining the coil loss conductance $R_L/(\omega L)^2$ with G' as shown in Fig. 6.31, where

$$G = G' + \frac{R_L}{(\omega L)^2}$$

The total admittance becomes

$$\mathbf{Y} = \left(G_s + G_C + G_2 + \frac{R_L}{(\omega L)^2} \right) + j\left(\omega C - \frac{1}{\omega L} \right) \qquad (6.80)$$

It is important to note that the element having conductance $R_L/(\omega L)^2$ is abstract in the sense that its conductance or resistance is a function of frequency. The circuit equivalence is valid, but we must remember this is a special element having resistance proportional to ω^2. For many purposes, we are interested in the performance of the parallel coil and capacitor in a narrow band of frequencies (near resonance) only. So in these cases we can calculate $R_L/(\omega L)^2$ at the resonance frequency, then assume this value is constant over the narrow range of frequencies of interest.

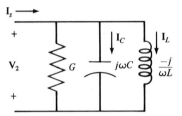

fig. 6.31   Simplified admittance model of Fig. 6.30.

If we are concerned with designing the tuned circuit (coil and capacitor) to fit a particular application, then G_s is fixed by the voltage source. At this point we assume that G_C is negligible. In any event, it typically has a minor effect on G. Ultimately we wish to relate C, L, and R_L with the performance of this tuned circuit. In order to get a feeling for the effect of each of the elements of the parallel tuned circuit, consider the following typical values: $G_s = 10^{-3}$ mho, $G_2 = 10^{-4}$ mho, $\omega = \omega_0 = 10^6$ rad/s, $\omega_0 L = 200$, $L = 2 \times 10^{-4}$ H, $R_L = 10\ \Omega$, $C = 5 \times 10^{-9}$ F. Then, at the resonance frequency ω_0, the conductance $(R_L/(\omega L)^2) = 0.00025$. Also, $1/\omega_0 L = 1/200 = 0.005$ mho. The susceptance of C is $\omega_0 C = 0.005$ mho. So the total susceptance, of course, is zero at ω_0. Therefore, the total conductance is 0.00135 mho, which is also the total admittance. If we let $\mathbf{I}_s = 1.35\ \angle\ 0°$ mA, then

$$\mathbf{V}_2 = \frac{\mathbf{I}_s}{G} = \frac{0.00135}{0.00135} = 1.00\ \angle\ 0°\ \text{V} \qquad \text{at } \omega_0 \text{ (resonance)} \qquad (6.81)$$

A parallel circuit such as this is said to be in resonance (at least by one definition) when the susceptance is zero or when the admittance is real only or the impedance is real only. Sometimes a different definition of resonance is used. This later definition says the circuit is in resonance when the magnitude of the admittance is a minimum. For most practical cases, the two definitions lead to practically, but not exactly, the same resonance frequency. The definition of zero susceptance will be used here because of its simplicity in concept.

To better visualize the resonance condition, consider the phasor diagram of Fig. 6.32, where $\mathbf{V}_2 = 1 \angle 0°$ V. At resonance, the susceptance is zero,

$$\mathbf{I}_G = \mathbf{I}_S = \mathbf{V}_2\,G = 1\angle 0° \times 0.00135\angle 0° = 1.35\angle 0° \text{ mA} \tag{6.82}$$

$$\mathbf{I}_C = \mathbf{V}_2\,\mathbf{Y}_C = 1\angle 0° \times 0.005\angle 90° = 5.00\angle 90° \text{ mA} \tag{6.83}$$

$$\mathbf{I}_L = \mathbf{V}_2\,\mathbf{Y}_L = 1\angle 0° \times 0.005\angle -90° = 5.00\angle -90° \text{ mA} \tag{6.84}$$

$$\mathbf{I}_{RL} = \mathbf{V}_2\,\frac{R_L^2}{\omega L} = 1\angle 0° \times 0.00025\angle 0° = 0.25\angle 0° \text{ mA} \tag{6.85}$$

and the coil current is

$$\mathbf{I}_{\text{coil}} = \mathbf{I}_L + \mathbf{I}_{RL} = 0.25 - j5.00 = 5.006\angle -87.14° \text{ mA} \tag{6.86}$$

Notice that $\mathbf{I}_{RL}$ is a current in the pure L of the model of Fig. 6.26, but is not a real, physical current that could be measured directly. Likewise, I_{RL} is fictitious. On the other hand, $\mathbf{I}_{\text{coil}}$ is the real current in the coil. This current has the two fictitious components $\mathbf{I}_{RL}$ and $\mathbf{I}_L$.

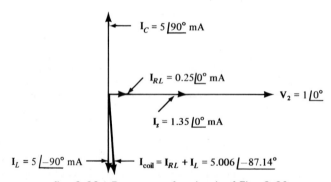

fig. 6.32   Resonance for circuit of Fig. 6.30.

It is also important to note that the total current I_s is much smaller than either I_L or I_C. In this example, I_s is only 27 percent of I_C or I_L. A single source $\mathbf{V}_2$ delivers a total current I_s to the three branches of the circuit of Fig. 6.31 such that each of two of the branch currents is much larger than the total. Of course, $\mathbf{I}_L$ and $\mathbf{I}_C$ at resonance cancel each other, and $\mathbf{I}_{\text{coil}}$ and $\mathbf{I}_C$ nearly cancel each other.

Because the susceptance, and thus admittance, of the circuit will be larger at a frequency either greater or less than ω_0, the useful voltage $\mathbf{V}_2$ for a given I_s is greatest at ω_0, where Y is minimum. Equation (6.80) leads to the locus of $\mathbf{Y}$ as shown in Fig. 6.33. Notice that the conductance G includes the term $R_L/(\omega L)^2$ and thus total conductance G decreases (slightly for this case) with increasing frequency.

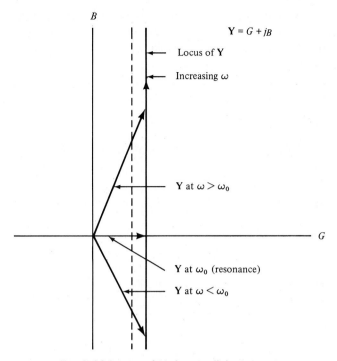

fig. 6.33 Locus of **Y** *for parallel resonance.*

Were it not for this term in the conductance G, the locus of **Y** would be a vertical straight line. Notice that for $\omega > \omega_0$, the susceptance $B > 0$ and for $\omega > \omega_0$, $B < 0$.

PROBLEM

6.42 a. To determine how much the locus of Fig. 6.33 departs from the vertical, calculate at both $\omega = 1.2\omega_0$ and at $\omega = 0.8\omega_0$ points on the locus.

  b. For a reasonable approximation, assume that this locus is a straight line and draw the locus through these three points: ω_0, $1.2\omega_0$, $0.8\omega_0$.

It is of some interest to note that the angle of **Y** at resonance is zero; at a frequency greater than resonance frequency, the angle is positive; at a frequency less than resonance, this angle is negative. This should not be surprising because at resonance the current in C is equal in magnitude to the current in L. As a consequence, the circuit at resonance appears to be a high resistance. At higher frequencies, the capacitance current will dominate, making **Y** have a positive angle. And, of course, the circuit will appear inductive at frequencies below the resonance frequency. Of course V_2 is proportional to Z (since $\mathbf{Z} = 1/\mathbf{Y}$), which is displayed in Fig. 6.34.

In a different form, Fig. 6.34 displays the same information as does the admittance locus of Fig. 6.33.

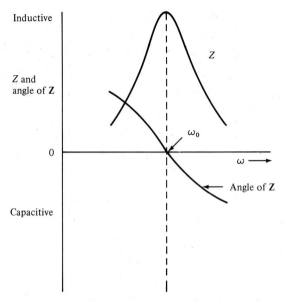

fig. 6.34   Parallel resonance.

One should suspect that the performance of our parallel resonance circuit is very much a function of the L and C combination used. We should be reminded that the resonant frequency is given by

$$\omega_0 = \frac{1}{\sqrt{LC}}$$ (6.87)

PROBLEM

6.43 To make sure that you understand the proof or basis of (6.87), develop the proof here.

It is obvious from (6.70) that there is an indefinite number of combinations of L and C that will yield a desired resonance frequency. Then it is natural to ask: How does one select the optimum value of each of these? Your solution to Probs. 6.44 and 6.45 will give insight here. Rather than the previous coil, assume that we have one whose inductance $L = 50\ \mu H$ and $R_L = 0.625\ \Omega$ instead of $L = 200\ \mu H$ and $R_L = 10\ \Omega$ as in the preceding example. In this new situation we have reduced both L and R_L such that $R_L/(\omega L)^2$ is unchanged. The coil designer cannot necessarily achieve, with ease, this reduction of coil resistance. We will come back to this point.

PROBLEM

6.44 Using a coil having an inductance of 50 μH and a resistance $R_L = 0.625\ \Omega$, find the necessary capacitance to make the parallel circuit resonant at an angular velocity of 10^6 rad/s as in the previous case. Insert this coil and capacitor in the network of Fig.

6.30 in place of the previously used coil and capacitor. Then make calculations and plot as functions of ω both magnitude and phase of V_2 for both coil-capacitor combinations. Plot these four functions on one set of coordinates so that performance can be compared easily.

Where we are concerned with passing the one resonance frequency, your solution to Prob. 6.44 should clearly show that the case of the reduced inductance has much better rejection of the unwanted frequencies. Decreasing L and simultaneously increasing C increases the respective admittances Y_L and Y_C. So, as the frequency changes from ω_0, the sum $Y_L + Y_C$ changes more dramatically.

To summarize these two previous situations and examine still another set of parameters, see Table 6.3.

TABLE 6.3 Three Different Sets of Parameters — Parallel Resonance

	First Coil	Second Coil	Increased R_s and V_s
L	0.2×10^{-3}	0.05×10^{-3}	0.05×10^{-3}
R_L	10	0.625	0.625
$R_L/(\omega L)^2$	0.25×10^{-3}	0.25×10^{-3}	0.25×10^{-3}
G_s	1×10^{-3}	1×10^{-3}	0.25×10^{-3}
R_s	1,000	1,000	4,000
G_2	0.1×10^{-3}	0.1×10^{-3}	0.1×10^{-3}
G	1.35×10^{-3}	1.35×10^{-3}	0.6×10^{-3}
Y_L, Y_C*	5×10^{-3}	20×10^{-3}	20×10^{-3}
I_s	1.35×10^{-3}	1.35×10^{-3}	0.6×10^{-3}
V_2^*	1	1	1
V_s	1.35	1.35	2.4
ω_0	10^6	10^6	10^6

* At $\omega = \omega_0$ (resonance).

The generator internal impedance R_s also has an important bearing on the performance of this resonant circuit, as will become evident in solving Prob. 6.45.

PROBLEM

6.45 Increase R_s to 4,000 Ω from 1,000 Ω and increase V_s by the amount necessary to keep V_2 (at resonance) at 1 V. Then calculate the necessary value of V_s. Leave G_2 at 10^{-4}, Y_C (resonance) at 20×10^{-3}, RL at 0.625, and ω_0 at 10^6. See the third column of Table 6.3. On the coordinate system used for Prob. 6.44, plot V_o as a function of ω in the region of ω_0.

Your plots of Probs. 6.44 and 6.45 show:

1. How the lower inductance and resistance of the second coil improved the circuit's ability to reject frequencies off resonance.
2. How the increased resistance of the source (right-hand column of Table 6.3) also improved the rejection ability of the circuit.

A brief note about the ac resistance of a coil is significant. Perhaps you have asked yourself the question, why can't the designer make the coil resistance as low as desired by increasing the cross-sectional area of the conductor? This line of argument is sound for direct current and for alternating currents of sufficiently low frequency. But when the frequency is sufficiently high, the current does not flow with uniform density throughout the cross section of a conductor. This phenomenon—called the *skin effect*—is typically treated in the context of electromagnetic fields, but will not be developed here. Suffice it to say that where frequency is sufficiently high, increasing the cross-sectional area of a conductor does not necessarily decrease resistance significantly.

It is interesting to note that the parallel conductance that represents the coil loss may be written as

$$G_L = \frac{R_L}{(\omega_0 L)^2} = \frac{1}{\left(\dfrac{\omega_0 L}{R_L}\right)\omega_0 L} = \frac{1}{Q_L \omega_0 L} \tag{6.88}$$

Thus the coil Q_L may be expressed as

$$Q_L = \frac{1}{\omega_0 L G_L} \tag{6.89}$$

Then, if we combine all of the conductances into one total conductance G, as was done in Fig. 6.31, this G represents the total power loss in the circuit, and the circuit Q may be obtained by the relationship

$$Q = \frac{1}{\omega_0 LG} = \frac{\omega_0 C}{G} \tag{6.90}$$

Since the parallel RLC circuit is the dual of the series RLC circuit, its bandwidth may be derived in a manner similar to that of the series RLC circuit. The only significant difference is that V replaces I, G replaces R, C replaces L, and L replaces C in the parallel derivation. The result is that $B = f_0/Q$ hertz in the parallel circuit as well as in the series circuit.

PROBLEMS
6.46   Prove that $B = f_0/Q$ hertz in a parallel tuned circuit.
6.47   Determine the bandwidth for each of the cases summarized in Table 6.3.

6.10 MULTIPLE RESONANCE—REACTANCE CURVES

If a circuit has a sufficient number of L and C components, it may exhibit *multiple resonance*, or more than one resonance frequency. Consider the lossless circuit of Fig. 6.35 as an example. The L_1 and C_1 parallel combination may be in parallel resonance at frequency ω_1; then the circuit will exhibit series resonance at a lower frequency where the inductive reactance of the L_1C_1 combination is equal in magnitude to the capacitive reactance of C_2. Thus the input impedance of this network (no resistance involved) will be infinite at one frequency and zero at another frequency. In this section we study interesting situations of this kind. To assure precision of communication, let us make sure that we understand the meaning of the terms resistance, reactance, conductance, and susceptance. An impedance is given by

$$\mathbf{Z} = R + jX = 10 - j20 \ \Omega \qquad (6.91)$$

The *resistance* associated with this circuit is 10 Ω, and the *reactance* is -20 Ω. Then the corresponding admittance is

$$\mathbf{Y} = G + jB = \frac{1}{10 - j20} = 0.020 + j0.040 \ \text{mho} \qquad (6.92)$$

where 0.020 is the conductance and 0.040 is the susceptance. For this particular case, while the reactance is negative, the susceptance is positive. This element or circuit is capacitive because the reactance is negative or because the susceptance is positive.

It is also important to note that the conductance and susceptance of (6.92) are *not* the reciprocals of the resistance and reactance, respectively, of (6.91).

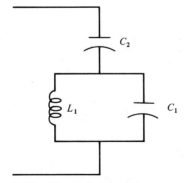

fig. 6.35   Circuit for double resonance.

The circuit of Fig. 6.35 is lossless, since no resistance elements are involved. Of course our coils especially have resistance or loss, but in addition, all capacitors have some loss, usually much less than the loss in a coil. In many simple to moderately complex networks, we can learn much about resonance conditions where resistances are ignored. The frequency at which a resonance phenomenon

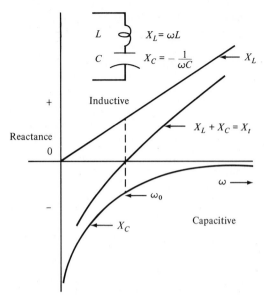

fig. 6.36   Reactance curves for series resonance.

takes place is usually nearly independent of the resistance or resistances in the circuit. On the other hand, the impedance or admittance at or near resonance of the network is highly dependent on the resistance. So for a few pages that follow, we will consider resistanceless circuits as we search for the frequencies at which series and parallel resonant phenomenon occur. This leads to a great simplification of thought.

Consider the reactance curves of a series LC circuit as displayed in Fig. 6.36. The reactance of the inductor alone, X_L, is a straight line through the origin as illustrated. Of course this reactance is positive. On the other hand, the reactance of the capacitor is negative and decreases in magnitude with frequency as shown. The total reactance of the circuit passes through zero at a frequency where the inductive reactance is equal in magnitude to the capacitive reactance. And of course this frequency, where the total reactance is zero, is called the series resonance frequency.

It is convenient to treat the parallel LC in terms of admittance rather than impedance. The admittance of the parallel circuit of Fig. 6.37 is

$$\mathbf{Y} = j\left(\omega C - \frac{1}{\omega L}\right) \tag{6.93}$$

With no resistance elements in the circuit, the admittance has no real part (no conductance). The total susceptance B_t is

$$B_t = \omega C - \frac{1}{\omega L} \tag{6.94}$$

The capacitive susceptance B_C is a straight line passing through the origin and having positive values only, as shown in Fig. 6.37, whereas inductive susceptance is

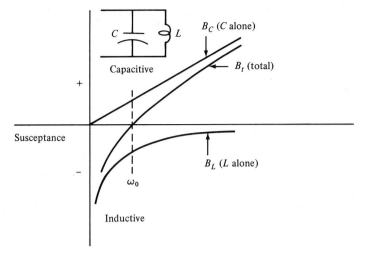

fig. 6.37 Susceptance curves for parallel resonance.

negative and decreases in magnitude as frequency increases. At the frequency where the total susceptance is zero, we have parallel resonance. At this point, admittance is zero and impedance is infinity. As we proceed through the following situation, it is well to keep in mind that at frequencies below the resonance frequency of a parallel circuit, the current in the inductance L dominates (I_L is larger than I_C), and therefore the circuit appears to be inductive. In other words, the susceptance is negative. And, of course, at frequencies above the resonance frequency, the capacitive current dominates and so the susceptance is capacitive or positive. Returning to the series circuit, this circuit appears capacitive at frequencies below the resonance frequency because the capacitor voltage is greater than the inductor voltage.

We found it convenient to treat the parallel circuit in terms of its admittance or susceptance, but where the parallel circuit is not standing alone but in series with some other circuit, as in Fig. 6.35, then we have some need for treating the reactance of the parallel circuit. Also, we must be able to think in terms of susceptance of the series circuit. Consider a circuit whose impedance is given by

$$\mathbf{Z} = R + jX \tag{6.95}$$

The admittance of this circuit is

$$\mathbf{Y} = \frac{R}{R^2 + X^2} - \frac{jX}{R^2 + X^2} \tag{6.96}$$

We note that a circuit that has positive reactance will have negative susceptance.

When the parallel LC circuit is in series with some other reactive elements, it may be more convenient to treat this parallel circuit in terms of impedance and reactance rather than admittance or susceptance. For the parallel circuit of Fig. 6.38, the admittance is

$$\mathbf{Y} = j\left(\omega C - \frac{1}{\omega L}\right) \tag{6.97}$$

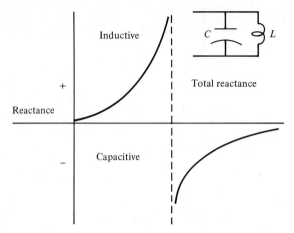

fig. 6.38   Reactance of parallel circuit.

and the impedance is

$$\mathbf{Z} = \frac{1}{j\left(\omega C - \dfrac{1}{\omega L}\right)} = \frac{j}{\dfrac{1}{\omega L} - \omega C} = \frac{j\omega L}{1 - \omega^2 LC} \tag{6.98}$$

so the reactance is

$$X = \frac{\omega L}{1 - \omega^2 LC} \tag{6.99}$$

The plot of the reactance of a parallel LC circuit is as shown in Fig. 6.38. A reactance plot such as this is easily reconstructed without resorting to equations by merely thinking of such considerations as:

1. As $\omega \to \omega_0$ (resonance), X_t (total) $\to \infty$.
2. For $\omega < \omega_0$, $I_L > I_C$. So the effect of L dominates and the circuit is inductive.
3. For $\omega > \omega_0$, $I_C > I_L$. So C dominates and the circuit is capacitive.

So with some brief thought, one can sketch these reactance and susceptance curves such as Figs. 6.37 and 6.38 without resorting to equations.

PROBLEM
6.48 Prepare a plot showing the susceptance of the series LC circuit for a range of ω passing through ω_0. Is your plot consistent with the fact that this circuit is capacitive for $\omega < \omega_0$ and inductive for $\omega > \omega_0$?

Now with this background, let us discover the double-resonance phenomenon in the circuit of Fig. 6.35. To do this we could develop either susceptance or reactance curves for the total circuit. Since we are probably a little more skilled in thinking in terms of reactance than we are in terms of susceptance, let us develop the reactance curves for this circuit. First, it is important to develop an *overview* before diving into the detail. Obviously the L_1C_1 parallel circuit will be resonant at the appropriate frequency, $\omega_1 = 1/\sqrt{L_1C_1}$.

PROBLEM

6.49 Focusing on the resonance frequency of the parallel L_1C_1 part of the circuit of Fig. 6.35, prepare the reactance curve of this parallel combination without again referring to Fig. 6.38, which shows these curves. This exercise will give you some needed experience and build your self-confidence.

Now C_2 is in series with the parallel combination L_1C_1, so we will need to add the reactance of C_2 to the reactance of the L_1C_1 combination.

PROBLEM

6.50 Add to the coordinate system of Prob. 6.49 the reactance curve for C_2. Now by adding the two reactance curves point by point, develop the curve for the total reactance of the complete circuit.

Figure 6.39 displays the authors' solution to Prob. 6.50. Compare your solution to this.

Notice that our reactance curves tell us that this circuit is in series resonance at ω_2 and parallel resonance at ω_1. Or that the circuit has zero impedance at frequency ω_2 and infinite impedance at frequency ω_1. Shortly, we will see how we could use this circuit to reject frequency ω_2 and pass frequency ω_1, or conversely, pass ω_2 and reject ω_1, depending on how we apply the $L_1C_1C_2$ configuration with respect to other components in the total circuit.

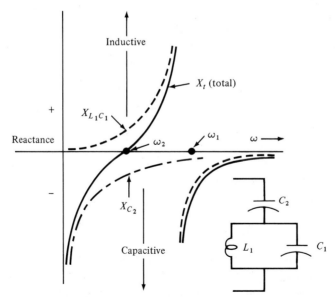

fig. 6.39 Reactance curves for C_2 in series with parallel $L_1 C_1$.

Now at this point let us prepare ourselves to be more effective and reliable in analyzing other multiple-resonance systems. To do this, let us try to improve our ability to check consistency. Again consider our solution of Fig. 6.39. Notice that our solution says that the total circuit is capacitive as ω approaches zero and also capacitive as ω approaches infinity. Now, without referring to equations, merely concentrate on viewing the circuit. We see that as ω approaches zero, the voltage across L_1 approaches zero and the voltage across C_2 approaches the applied voltage. At this extreme frequency, the circuit acts as though neither L_1 nor C_1 is present and the external terminals see only the presence of C_2. Now, as frequency becomes sufficiently large, C_1 *shorts* L_1 and the resultant circuit consists of C_1 in series with C_2. So again the total circuit is capacitive. The reactance curves of Fig. 6.35 show this current to be capacitive as $\omega \rightarrow 0$ and as $\omega \rightarrow \infty$.

As another consistency check, notice that the $L_1 C_1$ circuit is inductive at frequencies below its resonance frequency ω_1. For series resonance, the parallel circuit must have inductive reactance because, of course, C_2 always has capacitive reactance. This agrees with the solution shown in Fig. 6.39 because ω_2 is less than ω_1.

It is also important to note the asymptotic approaches of the various segments of the reactance curves.

PROBLEMS

6.51 a. To further develop our skills in rapidly sketching reactance curves for various multiresonant circuits, prove that the reactance of an inductor X_L as a function of ω is a straight line through the origin. See the illustration of Fig. 6.36.

  b. Prove that the reactance X_C of a capacitor as a function of frequency is a hyperbola. See the illustration of Fig. 6.36.

6.52 Where $C = 10^{-7}$ F and $L = 10^{-5}$ H, plot, on one set of coordinates, X_C and X_L, each as a function of frequency ω. Plot a few points, especially including the region where the two reactances are approximately equal in magnitude.

6.53 For the circuit of Fig. 6.39, select values of L_1, C_1, and C_2 such that f_1 (parallel resonance) falls at 10^5 Hz and f_2 (series resonance) is at 0.8×10^5 Hz. After you have thought about this problem briefly, you will conclude that there is an indefinite number of solutions to this problem. This is because the parallel resonance frequency is fixed by the product $L_1 \times C_1$. There is an indefinite number of individual values of L_1 that would lead to this unique product. Depending on the application of the circuit, there would be other considerations that would guide the designer to a unique combination of L_1 and C_1. For purposes of this problem, select $L_1 = 5 \times 10^{-5}$ H, and then proceed to unique values of C_1 and C_2.

6.54 Using the numerical values of C_1 and L_1 of Prob. 6.52 in the parallel configuration, add another reactive element in some manner such that series resonance occurs at 1.2×10^5 Hz. As a means to an end in this problem, draw reactance curves for your circuit configuration. Finally, calculate the numerical value of the circuit element or elements that you have added.

6.55 In the circuit of Fig. 6.40, $L_1 C_1 = 0.8 L_2 C_2$. Draw reactance curves for this configuration to illustrate the location of all of the series and parallel resonant frequencies. Identify them clearly on your sketches. Run consistency checks on your final solution. Does the total circuit appear to be inductive or capacitive at very low frequencies? Does the total circuit appear to be capacitive at frequencies well above both of the parallel resonant frequencies?

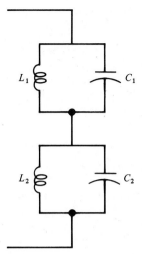

fig. 6.40 Circuit with two parallel resonances.

Effect of Resistance

Throughout all of this discussion on multiple resonance, we have ignored the effect of resistance in inductors and capacitors and elsewhere in the circuit. This led to great simplifications in finding various resonance frequencies. However, the impedance of these circuits near the resonant points is highly sensitive to any resistance that a coil or capacitor may have. For many situations one can neglect resistance effects in capacitors for essentially all purposes, but ordinarily one must consider the effect of coil resistance upon circuit impedance or admittance especially close to the resonance frequencies. A solution to Prob. 6.56 will make the point clear.

PROBLEM
6.56 Return to Prob. 6.53 and calculate the impedance of the circuit at both the parallel resonance point and the series resonance point where the resistance of the coil is 6 Ω.

Referring to your solution to Probs. 6.56 and 6.53, we see that where coil resistance is zero, the impedance of this circuit was infinite at the parallel resonance frequency. Now adding the resistance to the coil made this impedance finite, but still rather large. Likewise, the resistance of the coil made the impedance of the circuit finite rather than zero at the series resonance point.

PROBLEM
6.57 We desire to select parameters in the circuit of Fig. 6.28 to achieve maximum circuit Q or minimum bandwidth. We must live with the following constraints: $\omega_0 = 10^6$ rad/s, Q of the coil alone is 100, L may range from 10 to 1,000 μH, and each of R_s and R_2 may range from 100 to 500,000 Ω. R_C may range from 10^6 to 10^8 Ω. Select a set of optimum parameters. If a precise solution requires a differentiation, you may wish to settle for a reasonably approximate solution.

Perhaps this discussion on Q and bandwidth implies that Q or bandwidth of the circuit is the criterion of merit. As you will see in the next problem, you can make modifications in the circuit that increases the Q but at the same time decreases the magnitude of the output signal for all frequencies and voltage across the load resistance R_2. It is not obvious from this study of circuits that for many communication situations, the loss of useful output voltage is not nearly as significant as the gain in selectivity or decrease in bandwidth.

PROBLEM

6.58 a. For the parallel resonant circuit of Fig. 6.28, $R_s = 5,000 \ \Omega$, $R_2 = 100,000 \ \Omega$, $L = 50 \ \mu H$, the Q of the coil is 80, R_C is infinite, and C is such that the coil and capacitor are in resonance at 10^6 rad/s. Where the effective value of V_s is 1 V, sketch approximately the curve of V_2 as a function of frequency. Keep the effort to a reasonable minimum by calculating V_2 at resonance and at one of the half-power points.

   b. Increase R_s by adding a resistor externally to the source so that the new value of R_s is 50,000 Ω. On the same coordinate system as the above, sketch the new curve of V_2. For the higher R_s, the curve should exhibit sharper selectivity (smaller bandwidth), but at the same time should be at a lower level.

   c. Calculate the bandwidth and circuit Q for each case.

One might ask, what is the effect of a complex impedance at the load and/or in the source upon the performance of the parallel circuit of Fig. 6.28? Problem 6.59 gives us the opportunity to investigate such a situation.

PROBLEMS

6.59 The circuit of Fig. 6.28 has the parameters given in Prob. 6.58 except that the impedance in series with the voltage source is $10,000 - j10,000 \ \Omega$ at $\omega = 10^6$ rad/s. The capacitor is selected again such that the coil and capacitor are in resonance at 10^6 rad/s. Find the frequency at which V_2 is a maximum. *Caution:* The practical answer is not necessarily 10^6 rad/s. Furthermore, the reactance curve concept is not necessarily adequate unless it is applied with caution. Do not make too much of a deal of this problem; a differentiation to find the desired maximum may be quite laborious.

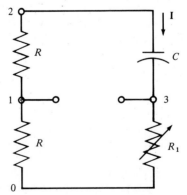

fig. 6.41 *A phase-shifting network.*

6.60 A circuit is needed to reject a 5-kHz band of frequencies centered at 500 kHz, where the multifrequency driving source has $R_s = 50 \, \Omega$ and the load resistance is 50 Ω. Design the rejection circuit using a parallel resonant circuit. This type of circuit is commonly known as a *trap*.

6.61 In this network of Fig. 6.41, the phase of voltage V_{31} can be changed or shifted in respect to the applied voltage of V_{20} by varying R_1; and as the phase is changed, the magnitude of V_{31} remains constant. If R_1 were varied from zero to infinity, the phase of V_{31} would vary from 0° through 90° to 180°. Draw a phasor diagram to help you show that these statements are true. *Suggestions :* Use voltage V_{20} as a reference, as shown in Fig. 6.42. Then show node 1 clearly on this reference at the midpoint. For all values of R_1, I will lead V_{20}. Next find node 3 on this diagram by showing V_{30} and then V_{23}. On your phasor diagram, show the locus of node 3 as R_1 is varied.

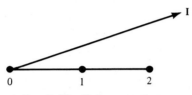

fig. 6.42   Reference phasor.

6.62 For practical use of the voltage V_{31} in the phase shifter of Fig. 6.41, some finite impedance must be connected between nodes 3 and 1. Under this circumstance, the magnitude of V_{31} will be dependent upon R_1. Assume that you have two 5-kΩ fixed resistors and one potentiometer that can be varied from 10 to 100 kΩ. We desire to shift the voltage V_{31} from 25° to 140°, where 400 kΩ of resistance must appear between terminals 3 and 1. Where the frequency is 5,000 Hz, select the capacitance of C.

6.63 For the situation of Prob. 6.62, calculate the magnitude of V_{31} at two widely separated values of R_1 to demonstrate that this magnitude is a function of R_1 due to the presence of the 400 kΩ.

6.64 An industrial load of 100 MVA has a lagging power factor of 0.75. An ac generator alone or alternately a combination of a generator and capacitor bank must be selected to supply this load at a minimum investment of dollars. The cost of generators and capacitors in the range of 5 to 100 MVA is assumed to be 4 cents and 1.2 cents per volt-ampere of capacity, respectively. Determine the minimum cost of purchasing the generator alone or, alternatively, the generator plus capacitor bank. Specify generator VA rating and capacitor VA rating for the minimum total cost.

Chapter 7

Wave Analysis

Introduction In previous chapters we discussed methods for finding the forced response to constant, exponential, or sinusoidal forcing functions. In this chapter we now extend our treatment to include another class of forcing functions—those that are periodic, but not necessarily sinusoidal. An example of such a forcing function is shown in Fig. 7.1, which is a periodic waveform of the kind that might occur in audio systems. A method of finding the response to such a forcing function is based on the representation of the function by a series of sinusoidal terms, called a *Fourier* series, because of a theorem published by Jean-Baptiste Joseph Fourier in the early 1800s.

Fourier's theorem, which is discussed in detail in the next section, states that a periodic function such as $f(t)$ in Fig. 7.1 can be written as the sum of an infinite series like

$$f(t) = a_0 + a_1 \cos \omega_0 t + a_2 \cos 2\omega_0 t + \cdots + a_n \cos n\omega_0 t + \cdots$$
$$+ b_1 \sin \omega_0 t + b_2 \sin 2\omega_0 t + \cdots + b_n \sin n\omega_0 t + \cdots$$

These sinusoidal terms are called *harmonics*, and $n\omega_0$ is called the nth harmonic radian frequency (n is an integer). The first harmonic is usually called the *fundamental*. Later on, we shall discuss methods of finding numerical values for the coefficients, a_n and b_n.

The method for finding the forced response to a periodic forcing function is to expand the function in a Fourier series, find the response to each individual harmonic, and add the responses to obtain the total response, a procedure justified by the theorem of superposition of sources. In practice, though, the detail of this procedure is not the only application of the Fourier series. But rather the method itself is the basis for many techniques in engineering and design. For example, there is a technique used in stereo audio systems called biamplification, or simply "biamp." A block diagram of a biamp system is shown in Fig. 7.2a, with a more

278

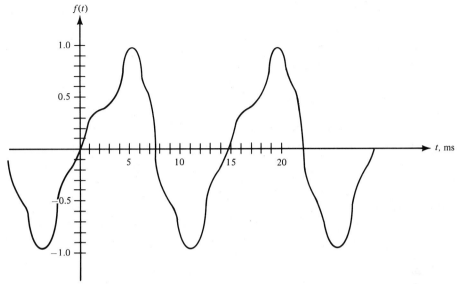

fig. 7.1 Periodic signal of the type that might occur in audio systems.

conventional system shown for comparison in Fig. 7.2b. The high-pass filter is designed to pass harmonics above approximately 150 Hz and attenuate those below 150 Hz. The low-pass filter is designed to attenuate harmonics above 150 Hz and pass those below 150 Hz. The two filters together are called a crossover network because their frequency responses cross over at 150 Hz.

In the conventional amplification system, the power amplifier must be capable of amplifying the maximum value of the signal without saturating, and the filters must be able to withstand high-power signals. In the biamp system, the filters need pass only low-power signals, and thus design and construction are made easier. Also, the two power amplifiers need not have as high a power rating as the amplifier in the conventional system because neither the high-frequency part of the signal nor the low-frequency part of the signal has a maximum value as great as that of the signal itself. It is usually more economical to construct two low-power amplifiers than it is to construct one high-power amplifier. Another advantage of the biamp system is that lower distortion results when the signals are split before the power amplification.

The biamp system is a good example of how Fourier series techniques are used in engineering design. The filters and amplifiers are designed to produce the desired response to the harmonics of the signal, in this case sending some harmonics along one path and others along another path, but the actual response to a signal such as the one in Fig. 7.1 is not always calculated, because that calculation is not necessary. The design is based on response to harmonics, but does not require the magnitude of each harmonic in detail. It is very important, however, for the design engineer to have good comprehension of Fourier series techniques. Much of the use of Fourier

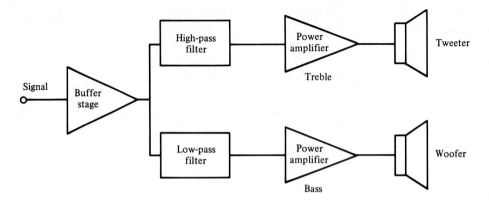

(a) Biamplification system

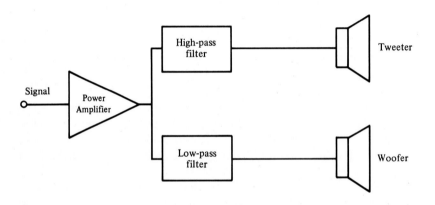

(b) Conventional amplification system

fig. 7.2 Comparison of biamplification of audio signals with conventional amplification.

series is based on qualitative understanding. For example, an experienced engineer can look at a given waveform and tell from its shape that it contains no second harmonic and that it contains a very strong third harmonic, without obtaining a number for the coefficient of the third harmonic. Just knowing that the third harmonic is very strong is often important in explaining the behavior of a system.

After you study this chapter, you too will be able to look at a waveform and make wise statements about its harmonic content. We shall stress concepts and qualitative understanding as we develop Fourier series techniques in the remainder of the chapter, hoping to give you a good basis for achieving skill in applying the techniques to the design and analysis of physical systems.

7.1 EXPANSION OF A SPECIFIC PERIODIC FUNCTION IN A FOURIER SERIES

In this section we show how to expand the half-wave rectified cosine wave shown in Fig. 7.3 in a Fourier series, just to introduce the techniques. Then in the next section we generalize the methods to any periodic waveform.

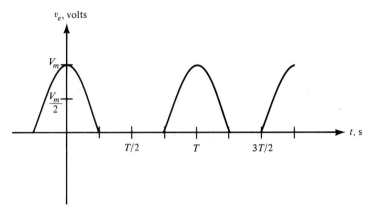

fig. 7.3   Half - wave rectified sinusoid used in the example.

Fourier's Theorem

As previously mentioned, Fourier's theorem states that any periodic waveform $f(t)$ can be expanded in an infinite series of the form

$$f(t) = a_0 + a_1 \cos \omega_0 t + a_2 \cos 2\omega_0 t + \cdots + a_n \cos n\omega_0 t + \cdots$$
$$+ b_1 \sin \omega_0 t + b_2 \sin 2\omega_0 t + \cdots + b_n \sin n\omega_0 t + \cdots \qquad (7.1)$$

if $f(t)$ is single-valued everywhere, has a finite number of discontinuities in any period, and if

$$\int_{t_0}^{t_0 + T} |f(t)| \, dt$$

exists for any t_0. The mathematical statement that $f(t)$ is periodic is

$$f(t) = f(t + T) \qquad (7.2)$$

where T is the *period* and ω_0 is related to T by

$$\omega_0 = \frac{2\pi}{T} \qquad (7.3)$$

The frequency is given by $f_0 = 1/T$. By the nth *harmonic*, we mean $a_n \cos n\omega_0 t + b_n \sin n\omega_0 t$, and the nth harmonic frequency is nf_0. When the function represents voltage or current, the term a_0 is called the dc component, since it represents the average value of the function. Usually, the $n = 1$ term is called the *fundamental*. The a's and b's, of course, are fixed by $f(t)$ (the wave shape), and are determined as described below.

Expansion of a Rectified Sinusoid

Let's expand the rectified cosine wave in Fig. 7.3 in a Fourier series—that is, find the a's and b's such that

$$v_e = a_0 + a_1 \cos \omega_0 t + a_2 \cos 2\omega_0 t + \cdots$$
$$+ \, b_1 \sin \omega_0 t + b_2 \sin 2\omega_0 t + \cdots \tag{7.4}$$

The value of a_0 can be found by integrating both sides of (7.4) over any period. An easy period is from $-T/4$ to $3T/4$. Thus

$$\int_{-T/4}^{3T/4} v_e \, dt = \int_{-T/4}^{3T/4} a_0 \, dt + \int_{-T/4}^{3T/4} a_1 \cos \omega_0 t + \int_{-T/4}^{3T/4} a_2 \cos 2\omega_0 t + \cdots$$
$$+ \int_{-T/4}^{3T/4} b_1 \sin \omega_0 t \, dt + \int_{-T/4}^{3T/4} b_2 \sin 2\omega_0 t \, dt + \cdots \tag{7.5}$$

The first term on the right-hand side of (7.5) is

$$\int_{-T/4}^{3T/4} a_0 \, dt = a_0 T \tag{7.6}$$

All the other terms on the right-hand side are zero because the integral of a sinusoid over a period or an integral number of periods is zero. For example,

$$\int_{-T/4}^{3T/4} a_2 \cos 2\omega_0 t \, dt = \int_{-T/4}^{3T/4} a_2 \cos \frac{4\pi}{T} t \, dt = \frac{a_2 T}{4\pi} \sin \frac{4\pi}{T} t \bigg]_{-T/4}^{3T/4} = 0$$

To evaluate the term on the left-hand side of (7.5), we can divide the integral into two terms:

$$\int_{-T/4}^{3T/4} v_e \, dt = \int_{-T/4}^{T/4} v_e \, dt + \int_{T/4}^{3T/4} v_e \, dt$$

From Fig. 7.3, we can see that $v_e = V_m \cos \omega_0 t$ for $-T/4 \le t \le T/4$ and $v_e = 0$ for $T/4 \le t \le 3T/4$. Hence

$$\int_{-T/4}^{3T/4} v_e \, dt = \int_{-T/4}^{T/4} V_m \cos \frac{2\pi}{T} t \, dt = V_m \frac{T}{\pi} \tag{7.7}$$

Now from (7.6) and (7.7), we get

$$a_0 = \frac{V_m}{\pi} \tag{7.8}$$

Since a_0 is equal to the integral of the function over a period divided by the period, a_0 is equal to the average value of the function, and when the function is a voltage or current, a_0 is called the *dc component*.

The other a's and b's can be found by using some very convenient relations having the property of *orthogonality*, as illustrated in the following example. We can find a_1 by multiplying both sides of (7.4) by $\cos \omega_0 t$ and integrating over any period. By so doing, we get

$$\int_{-T/4}^{3T/4} v_e \cos \omega_0 t \, dt = \int_{-T/4}^{3T/4} a_0 \cos \omega_0 t \, dt + \int_{-T/4}^{3T/4} a_1 \cos^2 \omega_0 t \, dt$$

$$+ \int_{-T/4}^{3T/4} a_2 \cos \omega_0 t \cos 2\omega_0 t \, dt + \cdots + \int_{-T/4}^{3T/4} b_1 \cos \omega_0 t \sin \omega_0 t \, dt$$

$$+ \int_{-T/4}^{3T/4} b_2 \cos \omega_0 t \sin 2\omega_0 t \, dt + \cdots \tag{7.9}$$

From integral tables, we can find that

$$\int_{-T/4}^{3T/4} a_1 \cos^2 \omega_0 t \, dt = \frac{a_1 T}{2} \tag{7.10}$$

and all the other terms on the right-hand side are zero. The general relations are that

$$\int_{t_0}^{t_0 + T} \cos m\omega_0 t \cos n\omega_0 t \, dt = \begin{cases} 0 & \text{if } m \neq n \\ \dfrac{T}{2} & \text{if } m = n \end{cases} \tag{7.11}$$

$$\int_{t_0}^{t_0 + T} \cos m\omega_0 t \sin n\omega_0 t \, dt = 0 \qquad \text{for all } m, n \tag{7.12}$$

$$\int_{t_0}^{t_0 + T} \sin m\omega_0 t \sin n\omega_0 t \, dt = \begin{cases} 0 & \text{if } m \neq n \\ \dfrac{T}{2} & \text{if } m = n \end{cases} \tag{7.13}$$

where m and n are integers and t_0 is a constant. Equations (7.11) to (7.13) are called *orthogonality* relations and $\cos m\omega_0 t$ and $\cos n\omega_0 t$ are said to be *orthogonal* when $m \neq n$ because the integral of the product of these functions over a period is zero. Likewise $\sin m\omega_0 t$ and $\sin n\omega_0 t$ are orthogonal for $m \neq n$ and $\cos m\omega_0 t$ and $\sin n\omega_0 t$ are orthogonal for all m and n. From the orthogonality relations, then, we find that all the terms on the right-hand side of (7.9) are zero except the a_1 term, which conveniently allows us to solve for a_1.

The left-hand side of (7.9) is

$$\int_{-T/4}^{3T/4} v_e \cos \omega_0 t \, dt = \int_{-T/4}^{T/4} V_m \cos^2 \omega_0 t \, dt + 0$$

$$= \frac{V_m T}{4} \tag{7.14}$$

Now a_1 is found from (7.14) and (7.10) to be

$$a_1 = \frac{V_m}{2} \tag{7.15}$$

Similarly, by multiplying both sides of (7.4) by $\cos 2\omega_0 t$ and integrating over a period, we find that all terms on the right-hand side are zero except the a_2 term, and evaluating the left-hand side gives us

$$a_2 = \frac{2V_m}{3\pi} \tag{7.16}$$

We can find the value for b_1 by multiplying both sides of (7.14) by $\sin \omega_0 t$ and integrating over a period. Then all terms except the b_1 term on the right-hand side are zero, and in this case the left-hand side is zero also. Thus $b_1 = 0$. Multiplying in turn by $\sin 2\omega_0 t$, $\sin 3\omega_0 t$, ..., and integrating, we find that all the b's are zero. Our series expansion for v_e is thus

$$v_e = \frac{V_m}{\pi} + \frac{V_m}{2} \cos \omega_0 t + \frac{2V_m}{3\pi} \cos 2\omega_0 t - \frac{2V_m}{15\pi} \cos 4\omega_0 t$$

$$+ \frac{2V_m}{35\pi} \cos 6\omega_0 t + \cdots \tag{7.17}$$

Figure 7.4 shows the sum of the first five terms, which we have called v_{e_1}. It is clear that the five terms represent the function very well, even in the region where the function should be zero. In general, the Fourier series is a slowly converging series, often requiring many more terms than five for a precise representation of a function, especially one that is characterized by a number of rather sharp changes in the function and its first derivative. The convergence is good in the case of the rectified wave because the function is sinusoidal over half the period.

It is quite remarkable that a's and b's can be found so easily that will cause the resulting sinusoids to have the proper magnitudes and phases to add up to a cosine wave over half the period and nearly cancel over the other half of the period. Were it not for the orthogonality relations, though, the evaluation of the a's and b's would be much more difficult.

PROBLEMS

7.1 Convince yourself that the same results can be obtained by integrating over any period by evaluating the integrals in (7.6), (7.7), (7.10), and (7.14) with limits 0 and T.

7.2 Get a better feel for the orthogonality relations by looking up integrals and evaluating the results for $t_0 = T/2$, $m = 1$, and $n = 2$ in (7.11), (7.12), and (7.13).

7.3 Evaluate the next term in the series in (7.17).

7.4 Add the next two nonzero terms to those given in (7.17). Then write a computer program to calculate the sum of the terms on the right-hand side of (7.17) for appropriate values of ωt, plot the curve, and compare the results with Fig. 7.4.

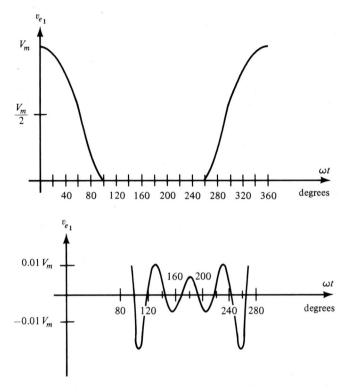

fig. 7.4 The sum of the first five terms in the Fourier series for v_e as given in (7.17). The lower graph is an expansion of the sum in the range 100° to 260°, where the sum is too small to plot in the top graph.

7.2 GENERALIZATION OF RELATIONS FOR FOURIER SERIES

In the preceding section we showed how one specific function can be represented by the sum of a Fourier series. Orthogonality relations were used to find the appropriate coefficients for each harmonic. In this section we generalize the results, giving expressions for the coefficients in terms of integrals of the product of the function and appropriate sinusoids. We also discuss symmetry properties and consistency checks.

General Expression for Coefficients

Consider the expansion of a function $f(t)$ in a Fourier series as expressed in (7.1), where $f(t)$ satisfies the requirements described in Section 7.1. To evaluate the coefficient of the nth cosine term, a_n, we multiply both sides of (7.1) by $\cos k\omega_0 t$, where k

is any integer (not zero), and integrate to obtain

$$\int_{t_0}^{t_0+T} f(t) \cos k\omega_0 t \, dt = a_0 \int_{t_0}^{t_0+T} \cos k\omega_0 t \, dt$$

$$+ a_1 \int_{t_0}^{t_0+T} \cos \omega_0 t \cos k\omega_0 t \, dt + \cdots + a_n \int_{t_0}^{t_0+T} \cos n\omega_0 t \cos k\omega_0 t \, dt + \cdots$$

$$+ b_1 \int_{t_0}^{t_0+T} \sin \omega_0 t \cos k\omega_0 t \, dt + \cdots + b_n \int_{t_0}^{t_0+T} \sin n\omega_0 t \cos k\omega_0 t \, dt + \cdots$$

$$(7.18)$$

Because of the orthogonality relations (7.11) to (7.13), all the terms on the right-hand side are zero except the a_k term. For example, if $k = 3$, then all terms on the right-hand side are zero except the a_3 term; if $k = 10$, then all terms are zero except the a_{10} term. Thus (7.18) reduces to

$$\int_{t_0}^{t_0+T} f(t) \cos k\omega_0 t \, dt = a_k \int_{t_0}^{t_0+T} \cos k\omega_0 t \cos k\omega_0 t \, dt$$

$$= \frac{a_k T}{2}$$

Solving for a_k, we get

$$a_k = \frac{2}{T} \int_{t_0}^{t_0+T} f(t) \cos k\omega_0 t \, dt \tag{7.19}$$

Similarly, multiplying both sides of (7.1) by $\sin k\omega_0 t$, integrating, and using orthogonality, we get

$$b_k = \frac{2}{T} \int_{t_0}^{t_0+T} f(t) \sin k\omega_0 t \, dt \tag{7.20}$$

For the special case of $k = 0$, we integrate both sides of (7.1) and use orthogonality to get

$$a_0 = \frac{1}{T} \int_{t_0}^{t_0+T} f(t) \, dt \tag{7.21}$$

With these general relations for the a's and b's, we need not go through the whole process of multiplying by the proper sinusoid, integrating, and using orthogonality each time, but we need only evaluate the appropriate integrals.

For example, let's expand the rectangular wave shown in Fig. 7.5 in a Fourier series. Since inspection of the waveform shows that it repeats itself every 2 seconds, $T = 2$ and $\omega_0 = 2\pi/2 = \pi$. From (7.21), with $t_0 = 0$,

$$a_0 = \frac{1}{2} \int_0^2 f(t) \, dt = \frac{1}{2} \int_0^1 1 \, dt + \frac{1}{2} \int_1^2 0 \, dt = \frac{1}{2} \tag{7.22}$$

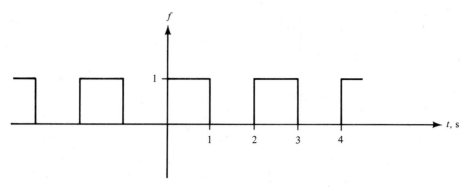

fig. 7.5   Rectangular periodic waveform.

From (7.19),

$$a_k = \frac{2}{2} \int_0^2 f(t) \cos k\pi t \; dt = \int_0^1 1 \cos k\pi t \; dt + \int_1^2 0 \cos k\pi t \; dt$$

$$a_k = \frac{\sin k\pi}{k\pi} = 0 \tag{7.23}$$

From (7.20),

$$b_k = \int_0^1 \sin k\pi t \; dt = \frac{-(\cos k\pi - 1)}{k\pi}$$

$$b_k = \begin{cases} 2/k\pi & \text{for } k \text{ odd} \\ 0 & \text{otherwise} \end{cases} \tag{7.24}$$

So, the expansion for the rectangular wave is

$$f(t) = \frac{1}{2} + \frac{2}{\pi} \sin \pi t + \frac{2}{3\pi} \sin 3\pi t + \frac{2}{5\pi} \sin 5\pi t + \cdots$$

$$f(t) = \frac{1}{2} + \frac{2}{\pi} \sum_{n=1}^{\infty} \frac{1}{n} \sin n\pi t \qquad \text{for } n \text{ odd} \tag{7.25}$$

PROBLEMS

7.5   Try your hand at expanding a periodic waveform in a Fourier series by expanding the triangular periodic waveform shown in Fig. 7.6.

7.6   Pick some value of t_0 other than zero and show that (7.19) to (7.21) give the same results for the rectangular wave as (7.22) to (7.24).

Symmetry and Shifting Properties

You may be wondering why there are no cosine terms in (7.25), the expansion of the rectangular wave, and whether there is any connection between the properties of the function and the a's all being zero. There is: the symmetry properties of the function, as explained below.

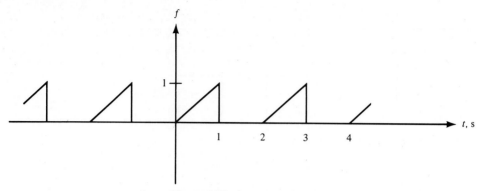

fig. 7.6   Triangular periodic waveform.

Even and Odd Functions.   Some important symmetries that a function can possess
are described by saying that a function is an *even function* or an *odd function*. The
function in Fig. 7.3 is an even function because $v(t) = v(-t)$. For example, the
function has the same value for $t = T/8$ as for $t = -T/8$. Graphically, this means
that if the function on the right-hand side of the vertical axis were rotated about the
vertical axis to the left-hand side, the right-hand side of the function would fit
exactly on top of the left-hand side of the function. Another way of saying this is
that the function has mirror symmetry about the $t = 0$ axis. Another even function
is shown in Fig. 7.7. The general definition of an even function is that $f(t)$ is even if

$$f(-t) = f(t) \qquad \text{even function}$$

A function $f(t)$ is said to be an odd function if

$$f(-t) = -f(t) \qquad \text{odd function}$$

Graphically, this means that if the right-hand side of the function is folded to the
left-hand side of the vertical axis and then rotated about the horizontal axis, the

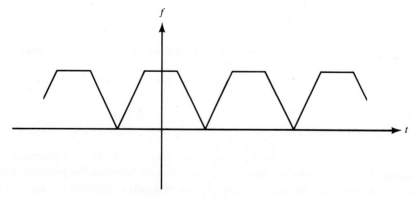

fig. 7.7   An even function.

right-hand side of the function will then fit exactly on top of the left-hand side of the function. Said another way, rotating the right-hand side of the function about the vertical axis would give the negative of the left-hand side of the function. Examples of odd functions are shown in Fig. 7.8. The function shown in Fig. 7.6 is neither even nor odd, because $f(-t) \neq f(t)$ and $f(-t) \neq -f(t)$. Some other examples are:

$f(t) = t^2$ is even because $f(-t) = (-t)^2 = t^2 = f(t)$

$f(t) = t^3$ is odd because $f(-t) = (-t)^3 = -t^3 = -f(t)$

$f(t) = 2 + t^3$ is neither even nor odd

$\cos t$ is even

$\sin t$ is odd

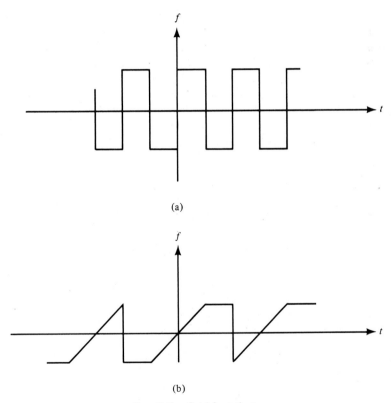

(a)

(b)

fig. 7.8 *Odd functions.*

PROBLEM

7.7 State whether the function in Fig. 7.1 is odd, even, or neither. Construct one odd function, one even function, and one function that is neither even nor odd.

We will next show that the Fourier series of an even function has no sine terms and the Fourier series of an odd function has no cosine terms. From (7.20),

$$b_k = \frac{2}{T} \int_{t_0}^{t_0+T} f(t) \sin k\omega_0 t \, dt$$

Since the expression is valid for any t_0, we choose $t_0 = -T/2$ and write

$$b_k = \frac{2}{T} \int_{-T/2}^{T/2} f(t) \sin k\omega_0 t \, dt$$

$$= \frac{2}{T} \int_{-T/2}^{0} f(t) \sin k\omega_0 t \, dt + \frac{2}{T} \int_{0}^{T/2} f(t) \sin k\omega_0 t \, dt$$

Now let's replace t by $-t$ in the first integral on the right:

$$b_k = \frac{2}{T} \int_{T/2}^{0} f(-t)\sin(-k\omega_0 t)(-dt) + \frac{2}{T} \int_{0}^{T/2} f(t)\sin k\omega_0 t \, dt$$

But $\sin(-k\omega_0 t) = -\sin k\omega_0 t$, and interchanging the limits on the integral is equivalent to changing the sign in front of the integral; thus

$$b_k = -\frac{2}{T} \int_{0}^{T/2} f(-t)\sin(k\omega_0 t) \, dt + \frac{2}{T} \int_{0}^{T/2} f(t)\sin k\omega_0 t \, dt$$

For an even function, $f(-t) = f(t)$, resulting in

$$b_k = -\frac{2}{T} \int_{0}^{T/2} f(t)\sin k\omega_0 t \, dt + \frac{2}{T} \int_{0}^{T/2} f(t)\sin k\omega_0 t \, dt$$

$$b_k = 0 \qquad \text{for an even function} \tag{7.26}$$

and we have proved that the Fourier series of an even function has no sine terms; that is, all the b's are zero.

Similarly, we can show that $a_k = 0$ when the function is odd. From (7.19),

$$a_k = \frac{2}{T} \int_{-T/2}^{0} f(t)\cos k\omega_0 t \, dt + \frac{2}{T} \int_{0}^{T/2} f(t)\cos k\omega_0 t \, dt$$

Replacing t by $-t$ in the first integral, interchanging the limits, using the facts that $\cos(-k\omega_0 t) = \cos k\omega_0 t$ and $f(-t) = -f(t)$, we get

$$a_k = -\frac{2}{T} \int_{0}^{T/2} f(t) \cos k\omega_0 t \, dt + \frac{2}{T} \int_{0}^{T/2} f(t) \cos k\omega_0 t \, dt$$

$$a_k = 0 \qquad \text{for an odd function} \tag{7.27}$$

The work required for evaluating the coefficients in a Fourier series can often be lessened by shifting the function either vertically or horizontally to make it an even or odd function. For example, consider the function $f(t)$ shown in Fig. 7.9a. Shifting it to the left by $T/4$ makes it an even function, which will have no sine terms in its

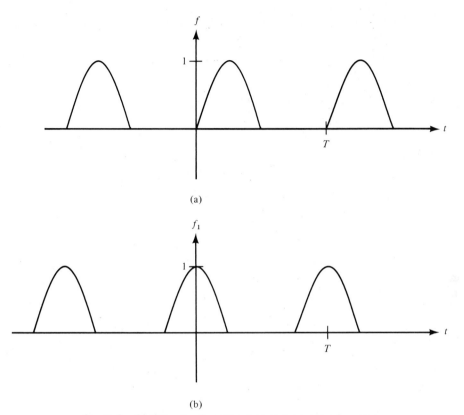

(a)

(b)

fig. 7.9   Shifting the function makes it an even function.

Fourier series expansion. The expansion for f_1 is

$$f_1(t) = \frac{1}{\pi} + \frac{1}{2} \cos \omega_0 t + \frac{2}{3\pi} \cos 2\omega_0 t - \frac{2}{15\pi} \cos 4\omega_0 t + \frac{2}{35\pi} \cos 6\omega_0 t + \cdots \quad (7.28)$$

(Note that this is the same as (7.17) with $V_m = 1$.) Now since f is the same as f_1 shifted to the right by $T/4$, we know that

$$f(t) = f_1\left(t - \frac{T}{4}\right)$$

Hence we can replace t by $t - T/4$ in (7.28) and get

$$f(t) = \frac{1}{\pi} + \frac{1}{2} \cos \omega_0\left(t - \frac{T}{4}\right) + \frac{2}{3\pi} \cos 2\omega_0\left(t - \frac{T}{4}\right) - \frac{2}{15\pi} \cos 4\omega_0\left(t - \frac{T}{4}\right) + \cdots$$

$$(7.29)$$

By shifting the function to make it an even function, we saved the work of evaluating the coefficients of the sine terms. Quite often, we really do not care whether a function is shifted or not; that is, we do not care when $t = 0$ occurs for steady-state functions. Consequently, we often shift the function to make evaluation of the Fourier series easier.

Superposition can sometimes be used in conjunction with shifting to find the Fourier series for a function. The f_1 in Fig. 7.10 can be shifted to obtain f_2 and the f_1 and f_2 can be added to obtain the full-wave rectified sine wave, f. Since $f_2(t) = f_1(t - T/2)$, the Fourier series for f_2 can be obtained by shifting (7.28):

$$f_2(t) = f_1\left(t - \frac{T}{2}\right) = \frac{1}{\pi} + \frac{1}{2}\cos\omega_0\left(t - \frac{T}{2}\right) + \frac{2}{3\pi}\cos 2\omega_0\left(t - \frac{T}{2}\right)$$

$$-\frac{2}{15\pi}\cos 4\omega_0\left(t - \frac{T}{2}\right) + \frac{2}{35\pi}\cos 6\omega_0\left(t - \frac{T}{2}\right) + \cdots$$

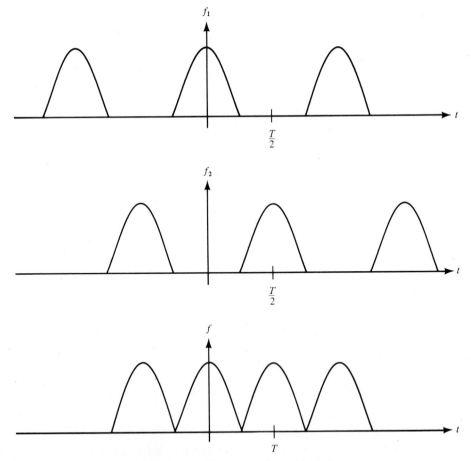

fig. 7.10   The sum of f_1 and f_2 is f.

Since

$$\frac{\omega_0 T}{2} = \frac{2\pi}{T} \frac{T}{2} = \pi$$

we have

$$f_2(t) = \frac{1}{\pi} + \frac{1}{2}\cos(\omega_0 t - \pi) + \frac{2}{3\pi}\cos(2\omega_0 t - 2\pi) - \frac{2}{15\pi}\cos(4\omega_0 t - 4\pi)$$

$$+ \frac{2}{35\pi}\cos(6\omega_0 t - 6\pi) + \cdots$$

which reduces through trigonometric identities to

$$f_2(t) = \frac{1}{\pi} - \frac{1}{2}\cos \omega_0 t + \frac{2}{3\pi}\cos 2\omega_0 t - \frac{2}{15\pi}\cos 4\omega_0 t$$

$$+ \frac{2}{35\pi}\cos 6\omega_0 t + \cdots \tag{7.30}$$

Now adding (7.28) and (7.30) gives us

$$f(t) = f_1(t) + f_2(t) = \frac{2}{\pi} + \frac{4}{3\pi}\cos 2\omega_0 t - \frac{4}{15\pi}\cos 4\omega_0 t + \frac{4}{35\pi}\cos 6\omega_0 t + \cdots \tag{7.31}$$

Half - Wave Symmetry

Another kind of symmetry is called *half-wave symmetry*. A function $f(t)$ is said to possess half-wave symmetry if

$$f\left(t + \frac{T}{2}\right) = -f(t)$$

or if

$$f\left(t - \frac{T}{2}\right) = -f(t)$$

that is, each half-wave is the same except for a difference in sign. Half-wave symmetry does not depend on the point where $t = 0$, or whether the function is shifted to the right or to the left, as does evenness or oddness of a function.

One example of a function with half-wave symmetry is the function in Fig. 7.8a; another is the one in Fig. 7.11. Notice that the function is neither even nor odd, but could be made either even or odd by shifting it appropriately. However, it does possess half-wave symmetry, regardless of whether it is shifted.

The Fourier series of a function with half-wave symmetry will have no even

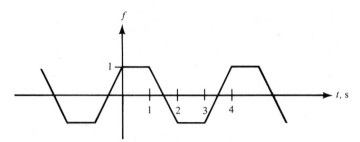

fig. 7.11   A function with half-wave symmetry.

harmonics; that is, a_2, a_4, a_6, ... and b_2, b_4, b_6, ... will each be zero. This can be shown by starting with (7.19) and $t_0 = -T/2$. Then

$$a_k = \frac{2}{T} \int_{-T/2}^{T/2} f(t)\cos k\omega_0 t \; dt$$

$$= \frac{2}{T} \int_{-T/2}^{0} f(t)\cos k\omega_0 t \; dt + \frac{2}{T} \int_{0}^{T/2} f(t)\cos k\omega_0 t \; dt$$

In the first term on the right, we let $t = x - T/2$, where x is a dummy variable:

$$a_k = \frac{2}{T} \int_{0}^{T/2} f(x - T/2)\cos k\omega_0\left(x - \frac{T}{2}\right) dx + \frac{2}{T} \int_{0}^{T/2} f(t) \cos k\omega_0 t \; dt$$

But $f(x - T/2) = -f(x)$ if f has half-wave symmetry, and $\cos k\omega_0(x - T/2) = \cos k\pi \cos k\omega_0 x$. Hence

$$a_k = -\frac{2}{T} \int_{0}^{T/2} \cos k\pi \, f(x)\cos k\omega_0 x \; dx + \frac{2}{T} \int_{0}^{T/2} f(t)\cos k\omega_0 t \; dt$$

$$a_k = \frac{2}{T} (1 - \cos k\pi) \int_{0}^{T/2} f(t)\cos k\omega_0 t \; dt$$

Thus for functions with half-wave symmetry,

$$a_k = \begin{cases} 0 & \text{if } k \text{ is even} \\ \dfrac{4}{T} \displaystyle\int_{0}^{T/2} f(t)\cos k\omega_0 t \; dt & \text{if } k \text{ is odd} \end{cases} \qquad (7.32)$$

A similar derivation leads to

$$b_k = \begin{cases} 0 & \text{if } k \text{ is even} \\ \dfrac{4}{T} \displaystyle\int_{0}^{T/2} f(t)\sin k\omega_0 t \; dt & \text{if } k \text{ is odd} \end{cases} \qquad (7.33)$$

for functions with half-wave symmetry.

The Fourier series given in (7.25) for the rectangular wave in Fig. 7.5 is an example in which the even harmonics are zero because of half-wave symmetry.

PROBLEM

7.8 State whether the Fourier series of the function in Fig. 7.11 would have sine and/or cosine terms and which harmonics, if any, would be zero. Shift f so that it becomes an even function and answer the same questions. Shift f so that it becomes an odd function and answer the same questions. Find the Fourier series of f shifted to be an odd function.

  A summary of some useful relations for Fourier series is given in Table 7.1. The next section gives an example of the use of Fourier series in analysis and design.

TABLE 7.1 Summary of Useful Relations for Fourier Series

Definition of Fourier series:

$$f(t) = a_0 + a_1 \cos \omega_0 t + a_2 \cos 2\omega_0 t + \cdots + a_n \cos n\omega_0 t + \cdots$$

$$+ b_1 \sin \omega_0 t + b_2 \sin 2\omega_0 t + \cdots + b_n \sin n\omega_0 t + \cdots$$

where

$$a_0 = \frac{2}{T} \int_{t_0}^{t_0 + T} f(t)\, dt$$

$$a_k = \frac{2}{T} \int_{t_0}^{t_0 + T} f(t)\cos k\omega_0 t\, dt$$

$$b_k = \frac{2}{T} \int_{t_0}^{t_0 + T} f(t)\sin k\omega_0 t\, dt$$

$\omega_0 = 2\pi/T$, T is the period of $f(t)$, and t_0 is arbitrary.

Even function: If $f(-t) = f(t)$, $f(t)$ is even; then $b_k = 0$.

Odd function: If $f(-t) = -f(t)$, $f(t)$ is odd; then $a_k = 0$.

Half-wave symmetry: If $f(t - T/2) = -f(t)$, $f(t)$ has half-wave symmetry; then all even harmonics are zero; that is, $a_2, a_4, a_6, \ldots = 0$; $b_2, b_4, b_6, \ldots = 0$.

An Example of the Use of Fourier Series in Design

In many engineering applications it is important to be able to select signals of a given frequency, or signals in a given band of frequencies, as discussed in Section 6.8 and illustrated at the beginning of this chapter by the example of biamplification. Another example is the tuner in a radio receiver. The signal picked up by a radio antenna consists of signals generated by thousands of transmitting stations (plus noise signals) at frequencies that literally cover the spectrum. The tuner in the receiver selects one station by allowing signals at that station frequency (actually a very narrow band of frequencies) to pass, while rejecting signals at other frequencies.

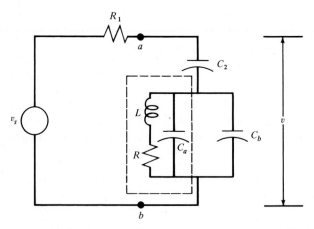

fig. 7.12 Circuit designed to reject the first harmonic and pass the third harmonic. The elements inside the dashed line represent the model of an inductor.

Let us now describe the design of a circuit that will strongly reject signals at one frequency and strongly pass signals at another frequency, with intermediate response for frequencies in between. More specifically, we shall design the circuit to reject a first harmonic and pass a third harmonic.

The diagram of the circuit is shown in Fig. 7.12, and the voltage produced by the signal generator shown in Fig. 7.13. Elements inside the dashed lines are the model of an inductor, which is represented by the combination of inductance, capacitance, and resistance shown. Our first step in designing the circuit is to establish a method of analyzing a circuit containing a source with a triangular waveform, and this will illustrate the use of Fourier series in analysis and design. By using the

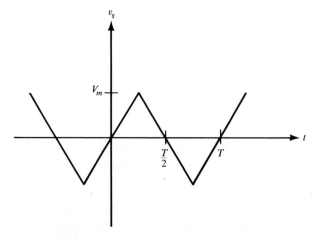

fig. 7.13 Voltage produced by the signal generator in Fig. 7.12.

techniques described in previous sections, we can write the v_s in Fig. 7.13 as

$$v_s = \frac{8V_m}{\pi^2}\left[\sin \omega_0 t - \frac{1}{9}\sin 3\omega_0 t + \frac{1}{25}\sin 5\omega_0 t - \cdots\right]\text{V} \qquad (7.34)$$

where $\omega_0 = 2\pi/T$ and T is the period, as indicated in Fig. 7.13. Now since v_s is the sum of a series of terms and since voltage sources add in series, we can represent v_s as a series combination of sources, as shown in Fig. 7.14. By superposition, we can find the response due to each of these sources separately and then add the responses to get the total response for the combination of sources. If v_1 is the voltage across $a\text{-}b$ due to v_{s1} alone, v_3 the voltage across $a\text{-}b$ due to v_{s3} alone, and so on, then by superposition,

$$v = v_1 + v_3 + v_5 + \cdots$$

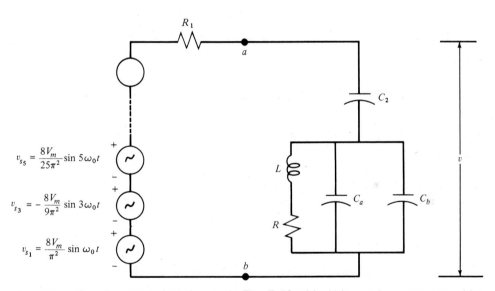

fig. 7.14 *Circuit equivalent to the one in Fig. 7.12 with each term in v_s represented by a separate voltage source.*

Since each source is sinusoidal, we can find the sinusoidal steady-state response to each of the individual components v_{s1}, v_{s3}, ... by transforming the circuit into the frequency domain and using phasor transform techniques (Chapter 4), as shown for the fundamental component in the circuit of Fig. 7.15. Letting $\mathbf{Z}$ be the impedance across $a\text{-}b$, we can write

$$\mathbf{V}_1 = \frac{\mathbf{Z}_1}{\mathbf{Z}_1 + R_1}\mathbf{V}_{s1} \qquad (7.35)$$

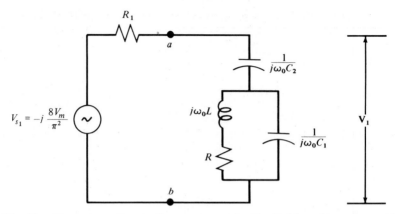

fig. 7.15   *Circuit corresponding to the one source in Fig. 7.14 and transformed into the* ω *domain with* $C_1 = C_a + C_b$.

where

$$Z_1 = \frac{1}{j\omega_0 C_2} + \frac{(R + j\omega_0 L)\dfrac{1}{j\omega_0 C_1}}{R + j\omega_0 L + \dfrac{1}{j\omega_0 C_1}}$$

which, after some algebraic manipulation, can be written as

$$Z_1 = \frac{(C_1 + C_2)}{j\omega_0 C_1 C_2} \times \frac{R + j\left(\omega_0 L - \dfrac{1}{\omega_0 (C_1 + C_2)}\right)}{R + j\left(\omega_0 L - \dfrac{1}{\omega_0 C_1}\right)} \qquad (7.36)$$

Similarly, for the source corresponding to the third harmonic, we can write

$$V_3 = \frac{Z_3}{Z_3 + R_1}\, V_{s3} \qquad (7.37)$$

$$Z_3 = \frac{(C_1 + C_2)}{j3\omega_0 C_1 C_2} \times \frac{R + j\left(3\omega_0 L - \dfrac{1}{3\omega_0 (C_1 + C_2)}\right)}{R + j\left(3\omega_0 L - \dfrac{1}{3\omega_0 C_1}\right)} \qquad (7.38)$$

From these equations, we can now see what our design objectives are in terms of selection of parameters. We want V_1 to be very small since it represents the response to the fundamental, and we want V_3 to be as large as possible since it represents the response to the third harmonic. If we choose R_1 to be large enough, then we can choose values of L, C_1, and C_2 that will simultaneously make the numerator of Z_1 small and the denominator of Z_3 small, thus making V_1 small and V_3 large. In terms of circuit behavior, this corresponds to a series resonance of L

with $C_1 \| C_2$ at ω_0, which makes $\mathbf{Z}_1$ small, and a parallel resonance of L with C_1 at $3\omega_0$, which makes $\mathbf{Z}_3$ large. The series resonance means that in (7.36),

$$\omega_0 L - \frac{1}{\omega_0(C_1 + C_2)} = 0 \qquad (7.39)$$

and the parallel resonance means that in (7.38),

$$3\omega_0 L - \frac{1}{3\omega_0 C_1} = 0 \qquad (7.40)$$

Equations (7.39) and (7.40) could have been obtained more directly by noting, as explained in Section 6.10, that the resonance frequencies are essentially independent of the resistances. Combining (7.39) and (7.40), we get

$$\frac{1}{\omega_0(C_1 + C_2)} = \frac{1}{9\omega_0 C_1}$$

which requires that

$$C_2 = 8C_1 \qquad (7.41)$$

for the two resonances to occur at ω_0 and $3\omega_0$.

To proceed with the design, we need to have some numbers for the parameters. Let's choose the fundamental frequency of the triangular wave to be 9 kHz so that

$$\omega_0 = 2\pi 9(10^3) \text{ rad/s}$$

and use typical values of

$$L = 100 \text{ mH}$$

$$R = 200 \ \Omega$$

Then, to satisfy (7.39) and (7.40), we must use

$$C_1 = 347.47 \text{ pF}$$

$$C_2 = 2{,}779.70 \text{ pF}$$

Let's try a value of

$$R_1 = 680 \text{ k}\Omega$$

and if it is not large enough, we will have to choose another value later.

Now let's calculate v for these parameters and see whether our design objectives are satisfied. Substituting the values in (7.36) and ploughing through the complex algebra gives us

$$\mathbf{Z}_1 = 253.09 - j1.12 \ \Omega \qquad (7.42)$$

and putting this value in (7.35) with $V_m = 1$ V results in

$$\mathbf{V}_1 = 301.58 \times 10^{-6} e^{-j90.25°} \text{ V} \qquad (7.43)$$

which is quite small, as we had hoped.

Substituting in values also gives us

$$Z_3 = 1.33 \times 10^6 - j20.37 \times 10^3 \ \Omega \qquad (7.44)$$

$$V_3 = 61.18 \times 10^{-3}e^{j89.74°} \ V \qquad (7.45)$$

and we note that V_3 is about 200 times larger than V_1. Calculating Z_5 and V_5 gives us

$$Z_5 = 9.64 - j13.68 \times 10^3 \ \Omega \qquad (7.46)$$

$$V_5 = 652.13 \times 10^{-6}e^{-j178.81°} \ V \qquad (7.47)$$

and we see that V_5 is about 100 times smaller than V_3. Consideration of V_7, V_9, and the following terms shows that these terms are all smaller than V_5 because the V_{s7}, V_{s9}, and so on are smaller and the Z_7, Z_9, and so on are all smaller.

Taking the inverse transforms of V_1, V_3, and V_5 gives us

$$v_1 = 301.58 \times 10^{-6} \cos(\omega_0 t - 90.25°) \ V$$

$$v_3 = 61.18 \times 10^{-3} \cos(3\omega_0 t - 89.74°) \ V$$

$$v_5 = 652.13 \times 10^{-6} \cos(5\omega_0 t - 178.81°) \ V$$

Now since

$$v = v_1 + v_3 + v_5 + \cdots$$

we get

$$v \approx 61.18 \times 10^{-3} \cos(3\omega_0 t - 89.74°) \ V \qquad (7.48)$$

since all the terms added to v_1 are at least 100 times smaller than v_1 and therefore negligible.

Equation (7.48) is a remarkable result, because it states that the response of the circuit in Fig. 7.12 to the *triangular-wave input* is a *sinusoid* with a frequency three times that of the triangular wave! According to (7.48), our design objectives have indeed been reached. The first harmonic was rejected and the third harmonic passed, and even though the amplitude of the third is nine times smaller than the amplitude of the first harmonic (see (7.34)), the response of the circuit to the third harmonic is about 200 times greater than the response to the first harmonic.

The usefulness of the Fourier series techniques in analyzing and designing circuits is apparent from this example. In practice, you will not need to go through all the detailed steps including the multiple sources shown in Fig. 7.14, but you should study these steps carefully the first time that you use these techniques so that you gain a clear understanding of the concepts and procedures involved.

It is not difficult to set up a circuit like the one in Fig. 7.12 and display the third harmonic on an oscilloscope, and it would be well worth your while to do so —and it is also fun. Find a suitable inductor and be careful in constructing a model of it. Techniques for measuring the inductance, capacitance, and resistance values needed to model the inductor will depend on the equipment that you have available. Perhaps you can get a lab instructor to help you if you have difficulty.

7.3   *CONSISTENCY CHECKS*

Instead of merely plugging into the equations to evaluate the coefficients in the Fourier series, you should get in the habit of estimating the harmonic content of a waveform and checking the results that you get from numerical evaluations of the equations. At the very least, you should look for possible symmetry in the waveform and use that information to anticipate the nature of the Fourier series expansion. But you can go further, even to the point of making a ballpark estimate of the amplitude of each harmonic. A method for making such estimates is described below. First the method is described; then a derivation of the method is given for those who are interested in it. You can use the method effectively without understanding the derivation.

* Method of Weighted Areas for Estimating Harmonic Amplitudes.   To begin with, let's look at the fundamental for a rectangular wave superimposed on that rectangular wave, as shown in Fig. 7.16. The areas that are crosshatched are the areas between the function and the harmonic, some labeled $\oplus$ and some labeled $\ominus$. In a crude sense, the amplitude of the fundamental must be such that the positive and negative areas are approximately equal, which is the basis for estimating the amplitude of the fundamental. As described below, a more accurate estimation is obtained by requiring positive and negative *weighted* areas to be equal.

An algebraic sign is given to each area according to the following rule:

$$\text{Sign of area} = [\text{sign}(f - h)][\text{sign } h] \tag{7.49}$$

where f is the value of the function and h is the value of the harmonic. For example, for area A, $f - h$ is a positive number (for example, $1 - \frac{1}{2} = \frac{1}{2}$) and the sign of h is positive. Hence area A is labeled positive. On the other hand, for area B, $f - h$ is a negative number because h is larger than f. Since h is positive, the sign of B is therefore negative. For area C, $f - h$ is a negative number (for example, $-1 - (-\frac{1}{2}) = -\frac{1}{2}$) and h is also negative, making C a negative area.

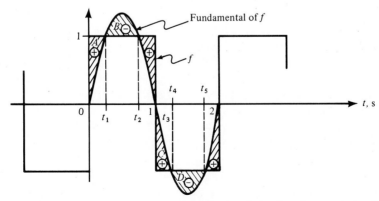

fig. 7.16   *Areas used in estimating the amplitude of the first harmonic in the Fourier series for a rectangular wave.*

You will notice in Fig. 7.16 that the total positive area is a little larger than the total negative area. The reason is that it is not just the areas that must balance, but the *weighted* areas. To see what we mean by a weighted area, look at Fig. 7.17, which shows a differential element of the area between f and its fundamental. The total area is composed of the sum of differential areas. By weighted area, we mean the sum of differential areas, each multiplied by the absolute value of the fundamental at that point. Thus the differential element shown in Fig. 7.17 would be multiplied by 0.4, which is the absolute value of the harmonic at that point. It is the sum of the weighted areas that must add to zero. In practice, we do not really try to assign a numerical value to each differential element and add up numbers, we just look at the areas between the two curves and note that the weighted area A will be reduced because the value of the fundamental is low near the left-hand side of the area. Area B, however, is not affected much by the weighting factor because the magnitude of the fundamental is almost unity over the interval which B extends.

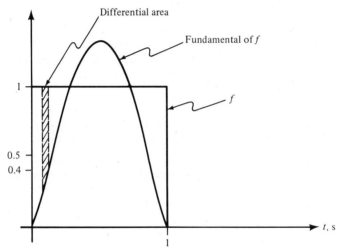

fig. 7.17   One differential area of the many that make up the area between the fundamental and f.

The method we are describing is most useful for estimating harmonic content and making consistency checks on calculated results. It could be carried far enough to get fairly good numbers for the amplitudes of harmonics, but it is usually easier to evaluate the integrals than it is to carry this method that far. Figure 7.18 shows, though, how easy it is to make an estimate or a consistency check—the two amplitudes of the sine wave shown cannot possibly be correct for the fundamental of f because in (a) there is no negative area at all, and in (b) the weighted negative area far exceeds the weighted positive areas. Note that because of symmetry, we need only consider one-half period. You can see that it is not too difficult to arrive at a

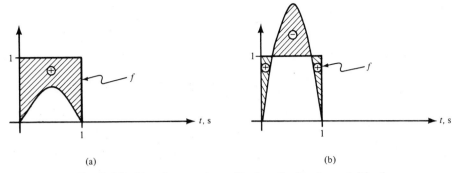

fig. 7.18 Two incorrect amplitudes of a fundamental for f.

good estimation for the proper amplitude of the fundamental by making one or two sketches to arrive at an approximate balance of the weighted areas as shown in Fig. 7.16.

Another example is shown in Fig. 7.19, where the areas are shown for the third harmonic. Again, it is easy to see how the weighted positive and negative areas balance. Figure 7.20 shows the third harmonic for a triangular wave, illustrating another point that we wish to make about this method. When the function has a dc level, the harmonics are drawn about the dc level, or the function is shifted vertically until the dc level is zero.

Note that the third harmonic in Fig. 7.20 must have the polarity shown; otherwise, the weighted positive area would exceed the weighted negative area. Also, note that if the amplitude were increased, the weighted negative areas would exceed the weighted positive areas, and vice versa if the amplitude were decreased.

An important observation to make is that the fundamental must intersect the function in order to have both positive and negative areas. In some cases, this fact alone is enough to make a good consistency check.

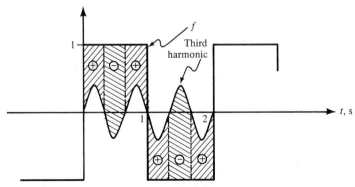

fig. 7.19 Areas used in estimating the amplitude of the third harmonic of f.

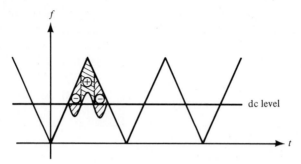

fig. 7.20   Third harmonic drawn about the dc level.

Now let's summarize the method of weighted areas.

1. Sketch the function.
2. Draw in the dc level of the function or shift the function until the dc level is zero.
3. Sketch the harmonic about the dc level (about zero amplitude if the function has no dc level).
4. Label the areas between the function and the harmonic with algebraic signs according to (7.49).
5. Estimate the weighted areas (as described above) and adjust the amplitude of the harmonic until the sum of the weighted positive areas is approximately equal to the sum of the weighted negative areas.
6. If you are making a consistency check on a calculated harmonic amplitude, sketch the harmonic with the calculated amplitude and see if the weighted areas balance.

After completing the following problems, you should be able to make quick estimates and consistency checks.

PROBLEMS

7.9   Estimate the magnitude of the fundamental of the triangular wave shown in Fig. 7.20 by using the method of weighted areas described above. Note that the fundamental must intersect the function. Calculate the magnitude of the fundamental and compare with your estimate.

7.10   Using the method of weighted areas, make consistency checks on the values of the fundamental and second harmonic in (7.17) calculated for v_e in Fig. 7.3.

7.11   Show that the weighted areas for the second harmonic of f in Fig. 7.8a cannot possibly balance unless the amplitude of the second harmonic is zero.

7.12   Using the method of weighted areas, estimate the amplitudes of the fundamental and first harmonic of the waveform of Fig. 7.6. Calculate the amplitudes and compare the result with your estimates.

7.13   Using the method of weighted areas, estimate the amplitude of the fundamentals (both sine and cosine terms) of f in Fig. 7.1. Repeat for the second harmonic. Note how easy it is to tell whether the second harmonic sine term should be positive or negative.

7.14   Sketch a periodic waveform and estimate the amplitude of the fundamental and first harmonic.

*** Derivation of Method of Weighted Areas.** Although you need not understand the details of the derivation of the method of weighted areas to use it effectively as described above, we shall give a brief outline of the derivation for those who are interested. The method is based on the equation following (7.18), which is obtained from orthogonality relations:

$$\int_{t_0}^{t_0+T} f(t)\cos k\omega_0 t\, dt = a_k \int_{t_0}^{t_0+T} \cos k\omega_0 t \cos k\omega_0 t\, dt$$

and the similar equation involving $\sin k\omega_0 t$ and b_k. The derivation will be given only for the equation shown above, but a similar derivation applies for the equation containing $\sin k\omega_0 t$ and b_k. Rearranging the equation gives us

$$\int_{t_0}^{t_0+T} f(t)\cos k\omega_0 t\, dt - \int_{t_0}^{t_0+T} a_k \cos k\omega_0 t \cos k\omega_0 t\, dt = 0$$

Combining under one integral and factoring out $\cos k\omega_0 t$ results in

$$\int_{t_0}^{t_0+T} \cos k\omega_0 t[f(t) - a_k \cos k\omega_0 t]\, dt = 0 \tag{7.50}$$

Now we can divide the integration up into intervals in which the integrand has the same algebraic sign throughout that interval:

$$\int_{t_0}^{t_1} \cos k\omega_0 t[f(t) - a_k \cos k\omega_0 t]\, dt$$

$$+ \int_{t_1}^{t_2} \cos k\omega_0 t[f(t) - a_k \cos k\omega_0 t]\, dt + \cdots$$

$$+ \int_{t_n}^{t_0+T} \cos k\omega_0 t[f(t) - a_k \cos k\omega_0 t]\, dt = 0 \tag{7.51}$$

Thus if (7.51) were the equation for b_1 (which would contain $\sin \omega_0 t$ in place of $\cos \omega_0 t$), then for the function in Fig. 7.16, t_0 would be zero and t_1, t_2, t_3, t_4, and t_5 would be as shown in the figure. For the first term, the integrand is positive, for the second it is negative, and so on. The part of the integrand in each term:

$$[f(t) - a_k \cos k\omega_0 t]\, dt$$

represents a differential area between $f(t)$ and the harmonic like the one shown in Fig. 7.17. The complete integrand

$$\cos k\omega_0 t[f(t) - a_k \cos k\omega_0 t]\, dt$$

represents a *weighted* differential area—that is, the differential area multiplied by the amplitude of the harmonic. Each term in (7.51) is equivalent to the sum of the weighted differential areas, and hence we call each term in (7.51) a *weighted area*. Thus (7.51) states that the algebraic sum of the weighted areas must be zero. The method of weighted areas is a graphical interpretation of (7.51).

7.4 FOURIER SERIES EXPANSION OF A FUNCTION FROM TABULATED DATA

Many waveforms encountered in practice are not as easily described as the rectangular and triangular kinds of waves that we have used as examples in this chapter. Consequently, the coefficients in the Fourier series for these practical waveforms are not easily obtained from the integral expressions given in (7.19) and (7.20). However, if tabulated values of the function are available, good approximate values for the coefficients can be obtained by numerical integration of the integrals in (7.19) and (7.20), as described below.

Numerical Integration

Since the approximate method for evaluating the coefficients in the Fourier series is based on a form of numerical integration, let's first discuss the method of numerical integration. Consider first the "stairstep" function shown in Fig. 7.21, and the numerical evaluation of $\int_0^6 f(x)\, dx$. Since f is constant in intervals of x from 0 to 1, from 1 to 2, and so on, we can easily evaluate the integral by breaking the integration up into a sum of integrations. Thus,

$$\int_0^6 f(x)\, dx = \int_0^1 f(x)\, dx + \int_1^2 f(x)\, dx + \int_2^3 f(x)\, dx$$

$$+ \int_3^4 f(x)\, dx + \int_4^5 f(x)\, dx + \int_5^6 f(x)\, dx$$

Putting in the appropriate values of x for each interval gives

$$\int_0^6 f(x)\, dx = \int_0^1 3\, dx + \int_1^2 5\, dx + \int_2^3 6\, dx$$

$$+ \int_3^4 7\, dx + \int_4^5 8\, dx + \int_5^6 8.5\, dx$$

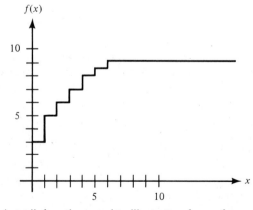

fig. 7.21 *"Stairstep" function used to illustrate a form of numerical integration.*

Now since the integrand in each term is a constant, we can quickly evaluate the integrals to get

$$\int_0^6 f(x)\, dx = 3 \cdot 1 + 5(2 - 1) + 6(3 - 2) + 7(4 - 3) + 8(5 - 4) + 8.5(6 - 5)$$

$$= 3 + 5 + 6 + 7 + 8 + 8.5$$

$$= 37.50$$

This result is equivalent to that obtained by finding the area under the curve, which can be obtained by adding up the areas of the rectangular elements shown in Fig. 7.22. The area of each element is the height, which is equal to the value of the

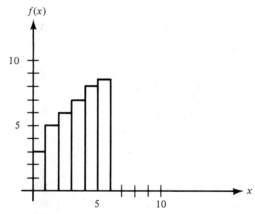

fig. 7.22   Areas equal to the integral of f in Fig. 7.21 from x = 0 to x = 6.

function, times the width, which we usually call Δx. That is,

$$dA_i = f(x_i)\, \Delta x_i$$

where dA_i is the area of the ith element. Then,

$$\int_0^6 f(x)\, dx = \sum_{i=1}^6 f(x_i)\, \Delta x_i$$

An approximate value for an integral can be found by approximating the function by a stairstep function and finding the area under the curve by adding the areas of the rectangular elements. For example, $\int_0^{x13} g(x)\, dx$ can be obtained by approximating $g(x)$ by a stairstep function, as indicated in Fig. 7.23, and writing

$$\int_0^W g(x)\, dx \approx \sum_{i=1}^N g(x_i)\, \Delta x_i \qquad (7.52)$$

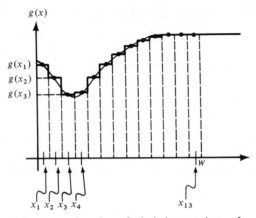

fig. 7.23 Approximation of g(x) by a stairstep function.

The smaller we make the Δx's, the better the approximation. As you may recall, the exact value of the integral is the limit:

$$\int_0^W g(x)\,dx = \lim_{\substack{\Delta x_i \to 0 \\ N \to \infty}} \sum_{i=1}^N g(x_i)\,\Delta x_i$$

We can get a good approximation in many cases with a surprisingly small number of elements. For example, let's find an approximation to

$$\int_0^{\pi/2} \cos x\,dx = 1$$

by choosing N elements of width Δx between 0 and $\pi/2$. Then

$$\Delta x = \frac{\pi}{2N}$$

and

$$\int_0^{\pi/2} \cos x\,dx \approx \sum_{i=1}^N \cos x_i\,\Delta x_i = \frac{\pi}{2N} \sum_{i=1}^N \cos x_i$$

since all the Δx_i's are equal.

For $N = 5$, we get

$$\int_0^{\pi/2} \cos x\,dx \approx \frac{\pi}{10}\left(\cos\frac{\pi}{20} + \cos\frac{3\pi}{20} + \cos\frac{5\pi}{20} + \cos\frac{7\pi}{20} + \cos\frac{9\pi}{20}\right) = 1.004124$$

(7.53)

where we have taken x_i to be in the middle of each interval, as indicated in Fig. 7.23. In this case, the result is surprisingly accurate for only 5 elements. For 10 elements, the result is 1.001029 and for 20 elements, it is 1.000257. The results are not always this good, of course, because the accuracy depends on how rapidly the function changes, but our example does indicate that we can get an approximate numerical value for a definite integral by a simple method. There are more sophisticated methods of numerical integration, but for our purposes the method we have described is adequate. The following problems will give you some practice in using this method of numerical integration.

PROBLEMS

7.15 Use numerical integration to find approximate values for $\int_0^{10} x^2 \, dx$ by using in turn 5, 10, and 20 elements and compare the results with the result obtained by standard integration.

7.16 Repeat Prob. 7.15 for $\int_0^{\pi/4} \tan x \, dx$.

Approximate Fourier Coefficients

With the expression for numerical integration (7.52) that we developed above, we can derive expressions for approximate values of Fourier coefficients from (7.19), (7.20), and (7.21). Thus, using equal Δt_i's throughout the interval, we can write

$$a_k \approx \frac{2}{T} \sum_{i=1}^{N} f(t_i)\cos k\omega_0 t_i \, \Delta t$$

$$a_k \approx \frac{2}{N} \sum_{i=1}^{N} f(t_i)\cos k\omega_0 t_i \tag{7.54}$$

$$a_0 \approx \frac{1}{N} \sum_{i=1}^{N} f(t_i) \tag{7.55}$$

$$b_k \approx \frac{2}{N} \sum_{i=1}^{N} f(t_i)\sin k\omega_0 t_i \tag{7.56}$$

where we have used

$$\Delta t = \frac{T}{N} \tag{7.57}$$

Another useful relation is that

$$t_i = \frac{2i-1}{2} \times \frac{T}{N} \tag{7.58}$$

where t_i is the ith value of t.

As an example of an approximate Fourier coefficient, consider b_1 for the rectangular wave of Fig. 7.5. The value that we found by evaluation of (7.20) is $2/\pi$, as given in (7.24). An approximate value is given by

$$b_1 \approx \frac{2}{N} \sum_{i=1}^{N} f(t_i)\sin \omega_0 t_i = \frac{2}{N} \sum_{i=1}^{N/2} 1 \sin \omega_0 t_i \tag{7.59}$$

since $f(t) = 0$ for $T/2 < t < T$. Putting (7.58) and $N = 20$ in (7.59) gives us

$$b_1 \approx \frac{1}{10} \sum_{i=1}^{10} \sin\left[\frac{(2i - 1)\pi}{20}\right] = 0.63925 \qquad (7.60)$$

Since the value in (7.24) is $2/\pi = 0.63662$, it is clear that the approximate value given in (7.60) is a reasonable approximation.

Equations (7.54) to (7.56) are easily programmed for computer calculations or for calculations on a programmable calculator, making it a relatively simple matter to obtain good approximate values of Fourier coefficients.

PROBLEM

7.17 Find approximate values for a_0, a_1, and a_2 for the function in Fig. 7.3 and compare with the values in (7.17). Try two different values of N.

Perhaps the most useful application of (7.54) to (7.56) is to find Fourier coefficients for a function such as the one in Fig. 7.1, for which no mathematical expression is available. To illustrate this application, we shall find the Fourier series for f in Fig. 7.1 from the values of the function tabulated in Table 7.2. Using (7.54) to (7.56) with

$$\Delta t = 0.5 \text{ ms} \qquad\qquad t_1 = 0.5 \text{ ms}$$

$$T = 15 \text{ ms} \qquad\qquad t_2 = 1.0 \text{ ms}$$

$$N = 30$$

TABLE 7.2 Values of the Function in Fig. 7.1

t, milliseconds	$f(t)$	t, milliseconds	$f(t)$
0.5	0.15	8.0	−0.32
1.0	0.30	8.5	−0.42
1.5	0.375	9.0	−0.50
2.0	0.40	9.5	−0.60
2.5	0.42	10.0	−0.77
3.0	0.45	10.5	−0.92
3.5	0.50	11.0	−1.00
4.0	0.57	11.5	−0.97
4.5	0.90	12.0	−0.88
5.0	1.00	12.5	−0.72
5.5	0.97	13.0	−0.40
6.0	0.87	13.5	−0.30
6.5	0.65	14.0	−0.20
7.0	0.50	14.5	−0.13
7.5	0.00	15.0	0.00

and so on, we get

$$a_0 = -0.0025 \qquad b_1 = 0.8247$$

$$a_1 = -0.0772 \qquad b_2 = -0.1950$$

$$a_2 = 0.1188 \qquad b_3 = 0.0490$$

$$a_3 = 0.0730$$

Note that we have approximated $f(t_i)$ by the value of the function at the right side of the Δt_i element, rather than at the center as in Fig. 7.23. The choice should have a small effect on the values of the coefficients.

Figure 7.24 shows a comparison of the first seven terms of the Fourier series of the function with the function itself.

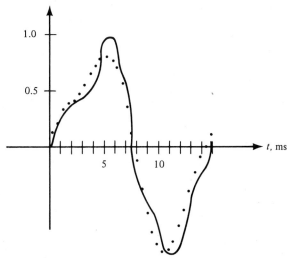

fig. 7.24 *Dots show the sum of the first seven terms of the Fourier series for the function in Fig. 7.1. The solid line is the function in Fig. 7.1.*

PROBLEM

7.18 On a piece of graph paper, sketch out a periodic function of any shape. Then make a table of values of the function similar to Table 7.2 and calculate approximate values for several of the Fourier coefficients. Sum the terms of the Fourier series and compare with the original function. You will probably find it best to use a computer or programmable calculator for the calculations.

7.5 THE EXPONENTIAL FORM OF THE FOURIER SERIES

The Fourier series, which was given in (7.1), may be expressed in a useful, tidy form known as the exponential form. First, let us write the series (7.1) in the following compact form,

$$f(t) = a_0 + \sum_{k=1}^{\infty} a_k \cos k\omega_0 t + \sum_{k=1}^{\infty} b_k \sin k\omega_0 t \qquad (7.61)$$

Since, from Euler's equations, $\cos \theta = (e^{j\theta} + e^{-j\theta})/2$ and $\sin \theta = (e^{j\theta} - e^{-j\theta})/2j$, we may write (7.61) as

$$f(t) = a_0 + \sum_{k=1}^{\infty} a_k \frac{(e^{jk\omega_0 t} + e^{-jk\omega_0 t})}{2} + \sum_{k=1}^{\infty} -jb_k \frac{(e^{jk\omega_0 t} - e^{-jk\omega_0 t})}{2} \qquad (7.62)$$

Rearranging (7.62), we obtain

$$f(t) = a_0 + \sum_{k=1}^{\infty} \frac{(a_k - jb_k)}{2} e^{jk\omega_0 t} + \sum_{k=1}^{\infty} \frac{(a_k + jb_k)}{2} e^{-jk\omega_0 t} \qquad (7.63)$$

If we change the summation from $+k$ to $-k$, we do not change the sign of a_k because a_k is the coefficient of the cosine term and $\cos(-k\omega_0 t) = \cos k\omega_0 t$, but we do change the sign of b_k because $\sin(-k\omega_0 t) = -\sin k\omega_0 t$. Therefore, we may write (7.63) as

$$f(t) = a_0 + \sum_{k=1}^{\infty} \frac{(a_k - jb_k)}{2} e^{jk\omega_0 t} + \sum_{k=-1}^{-\infty} \frac{(a_k - jb_k)}{2} e^{jk\omega_0 t} \qquad (7.64)$$

We may now include all the terms on the right-hand side of (7.64) in one summation:

$$f(t) = \sum_{k=-\infty}^{\infty} \frac{a_k - jb_k}{2} e^{jk\omega_0 t} = \sum_{k=-\infty}^{\infty} C_k e^{jk\omega_0 t} \qquad (7.65)$$

where

$$C_k = \frac{a_k - jb_k}{2} \qquad (7.66)$$

PROBLEM
7.19   Show that $C_0 = a_0$.

Let us substitute the expressions for a_k and b_k given in either (7.19) and (7.20) or Table 7.1 into (7.66) to obtain

$$C_k = \frac{1}{T} \int_{t_0}^{t_0+T} f(t) \cos k\omega_0 t\ dt - \frac{j}{T} \int_{t_0}^{t_0+T} f(t) \sin k\omega_0 t\ dt \qquad (7.67)$$

The sum of these integrals may be replaced by the integral of the sum to obtain

$$C_k = \frac{1}{T} \int_{t_0}^{t_0+T} f(t)(\cos k\omega_0 t - j \sin k\omega_0 t)\ dt \qquad (7.68)$$

Then using Euler's relationship, $e^{-j\theta} = \cos \theta - j \sin \theta$, (7.68) becomes

$$C_k = \frac{1}{T} \int_{t_0}^{t_0+T} f(t) e^{-jk\omega_0 t}\ dt \qquad (7.69)$$

Observe that all of the complex amplitudes C_k of the frequency components are given by a single expression in (7.69).

Let us determine the frequency components of the rectangular voltage pulse shown in Fig. 7.25 using (7.69) to determine the amplitude of the harmonics as an example of the use of the exponential form. Time $t = 0$ has been arbitrarily chosen at the center of the pulse. Then, using (7.69),

$$C_k = \frac{1}{T} \int_{-t_d/2}^{t_d/2} e^{-jk\omega_0 t} \, dt = \frac{1}{-jk\omega_0 T} [e^{-jk\omega_0 t_d/2} - e^{jk\omega_0 t_d/2}] \qquad (7.70)$$

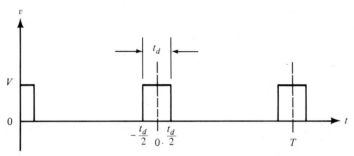

fig. 7.25   A rectangular voltage pulse having amplitude V, duration (or width) t_d , and period T.

Using Euler's relationship, $\sin \theta = (e^{j\theta} - e^{-j\theta})/2j$, (7.70) becomes

$$C_k = \frac{2}{k\omega_0 T} \sin\left(\frac{k\omega_0 t_d}{2}\right) \qquad (7.71)$$

Equation (7.71) may be cast into the familiar $(\sin x)/x$ form by multiplying both numerator and denominator by $t_d/2$ to obtain

$$C_k = \frac{t_d}{T} \frac{\sin(k\omega_0 t_d/2)}{k\omega_0 t_d/2} \qquad (7.72)$$

Substituting this expression for C_k into (7.65), we obtain the expression for the frequency components of the rectangular pulse, $f_1(t)$.

$$f_1(t) = \sum_{k=-\infty}^{\infty} \frac{t_d}{T} \frac{\sin(k\omega_0 t_d/2)}{k\omega_0 t_d/2} e^{jk\omega_0 t} \qquad (7.73)$$

These frequency components are illustrated in Fig. 7.26a, where $T = 10^{-2}$ s and $t_d = 10^{-3}$ s.

You may be disturbed because both negative and positive frequency components are shown in Fig. 7.26a. This comes about because the frequency components in the exponential form are, from (6.65), $C_k e^{jk\omega_0 t}$. Therefore, when k is negative, the exponent of e is negative and the phasor represented by this exponent rotates clockwise in contrast with the standard counterclockwise rotation of the phasor when the exponent is positive. Thus, each real frequency consists of two

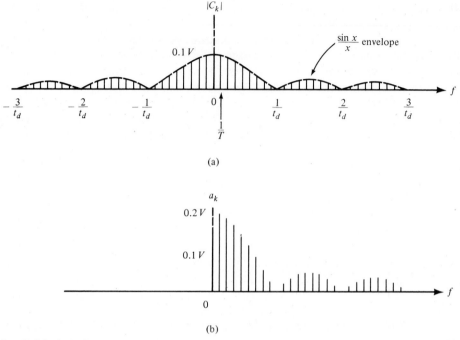

fig. 7.26   (a) Both negative and positive frequency components of a rectangular pulse having amplitude V and duration $t_d = 0.1\,T$. (b) The real-frequency spectrum of the rectangular pulse described in (a).

phasor components: one rotating counterclockwise and the other rotating clockwise. The *real* frequency at any given harmonic k is obtained by adding these two components as follows, using (7.66):

$$C_k e^{jk\omega_0 t} + C_{-k} e^{-jk\omega_0 t} = \frac{(a_k - jb_k)}{2} e^{jk\omega_0 t} + \frac{(a_k + jb_k)}{2} e^{-jk\omega_0 t}$$

$$= a_k \cos k\omega_0 t + b_k \sin k\omega_0 t$$

Since we chose $t = 0$ at the center of the pulse, the b_k's are zero and $a_k = 2C_k$ as seen from (7.66). The frequency spectrum for the rectangular pulse train using only real, positive frequencies is given in Fig. 7.26b. Observe that these real-frequency components may be obtained by folding the negative half of the frequency plot in Fig. 7.26a onto the positive half and then adding the magnitudes of each pair of components having the same frequency. Observe that $a_0 = c_0$, so the dc component is unaffected in the folding process.

PROBLEM

7.19   Change the time reference t_0 to the leading edge of the rectangular pulse in Fig. 7.25 and use the exponent form of Fourier series to determine the C_k coefficients. Can these coefficients be folded, or added, to give the magnitude of the real-frequency coefficients?

* 7.6   THE FOURIER TRANSFORM

The Fourier series is a very useful technique for obtaining the frequency components of a periodic voltage or current expressed as a function of time or, in other words, for transforming a time-domain expression into a frequency-domain expression. On the other hand, if we know the frequency components, we may plot each component as a sinusoidal function of time and add the magnitudes of all the components at many Δt intervals along the time axis to obtain the voltage or current as a function of time. This was done to show, in Fig. 7.24, the similarity of the reconstituted function of time from the first seven components to the original time function. This reconstitution of the time function from the frequency components could be called a frequency-domain to time-domain transformation, which is quite tedious because of the large number of frequency components usually required to obtain an accurate time plot. Thus the Fourier series is normally used only to transform a function of time into its frequency components, or frequency domain. In addition, the Fourier series is applicable only to repeating, or periodic, functions.

In order to see how the Fourier series technique can be applied to events that are not periodic, or occur so rarely that they may be considered as single events, let us reconsider the Fourier series in exponential form and the double-sided frequency spectrum for the rectangular pulse given in Fig. 7.26a. This spectrum is repeated in Fig. 7.27 to illustrate the effect of increasing the period T, or time between pulses. In

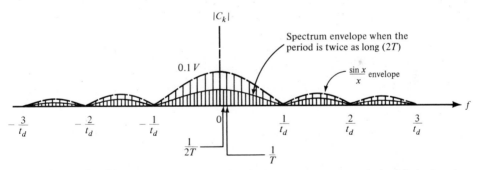

fig. 7.27   Illustration of the effect on the frequency spectrum of doubling the time between rectangular pulses.

this figure, the period T has been doubled, so the spectral lines are twice as close together and the amplitude of each component is only half as great as before the period was doubled, since $t_d/2T$, is reduced to one-half its previous value. We may now visualize that as the period T, or the distance between pulses, continues to increase, the spectral lines become more tightly packed and their amplitudes decrease, since both spacing and amplitude are inversely proportional to T. We would like T to approach ∞ in order to obtain the frequency components of a single pulse, but then the spectral amplitude would approach zero, or vanish, and we would lose the information we need. Therefore, let us modify the expression for C_k given in (7.70) and repeated below for convenience.

$$C_k = \frac{1}{T} \int_{-t_d/2}^{t_d/2} f(t)e^{-jk\omega_0 t} \, dt \tag{7.74}$$

We may replace the period T with $1/f_0 = 2\pi/\omega_0$. Also, $k\omega_0$ is the frequency of the kth spectral line, or frequency component, so we will change its designation to ω_k. Making these substitutions in (7.74), we obtain

$$C_k = \frac{\omega_0}{2\pi} \int_{-t_d/2}^{t_d/2} f(t)e^{-j\omega_k t} \, dt \tag{7.75}$$

Now, we may divide both sides of (7.75) by ω_0 to obtain

$$\frac{C_k}{\omega_0} = \frac{1}{2\pi} \int_{-t_d/2}^{t_d/2} f(t)e^{-j\omega_k t} \, dt \tag{7.76}$$

The ratio C_k/ω_0 does *not* change as $T \to \infty$ because both C_k and ω_0 are inversely proportional to frequency, as previously mentioned. We may therefore let $T \to \infty$ in (7.76) and, in this case, the space between the spectral lines disappears and ω_k becomes a continuous function ω. Also, the ratio C_k/ω_0 becomes the *spectral density*, which we shall call $F(\omega)$. Making these changes in (7.76),

$$F(\omega) = \frac{1}{2\pi} \int_{-t_d/2}^{t_d/2} f(t)e^{-j\omega t} \, dt \tag{7.77}$$

This is the Fourier transform of a discrete time function that begins at time $-t_d/2$ and ends at $t_d/2$. We can make the transform general by extending the integration limits from $-\infty$ to ∞, thus allowing the engineer the freedom to choose the limits of the actual integration most appropriately to include the nonzero values of the particular time function at hand. The general Fourier transform is then

$$F(\omega) = \frac{1}{2\pi} \int_{-\infty}^{\infty} f(t)e^{-j\omega t} \, dt \tag{7.78}$$

PROBLEM

7.20 Show that the Fourier transform of a single rectangular pulse having amplitude V and width t_d is a continuous spectral density having the form $(\sin x)/x$.

We may note from Fig. 7.27 that the lowest-frequency spectral lines produced by a train of rectangular pulses having a width t_d have maximum amplitude at zero frequency, and the line amplitudes decrease as their frequencies increase until zero amplitude is reached at frequency $f = 1/t_d$. The spectral amplitudes of the higher-frequency components are relatively small. The same may be said for the spectral density of a single pulse, as you may have observed from Prob. 7.20. Therefore, as the pulse duration t_d becomes long, the first zero-amplitude pulse occurs at a low frequency, and the significant frequency components, or frequency spectrum, are low frequencies. As t_d approaches ∞, there remains essentially one spectral line at zero frequency, or dc. This is as it should be, since an extremely long pulse may be a dc voltage switched on in the morning and switched off in the evening. On the other hand, as the pulse width narrows, the frequency spectrum broadens until an extremely narrow pulse produces a frequency spectrum amplitude that is essentially independent of amplitude over a very wide frequency range, perhaps over the entire frequency range of interest.

As an example of the relationship between the duration of a rectangular pulse and its frequency spectrum, let us consider two different pulse durations of 1 ms and 0.1 s and compare their spectral densities. These pulses and their spectral densities are sketched in Fig. 7.28 using the same time and frequency scales for both pulses. The spectral densities may be obtained by substituting the pulse durations into your solution of Prob. 7.20. It should not be surprising to see that short-duration pulses have a broad frequency spectrum, whereas long-duration pulses have a narrow frequency spectrum, since time and frequency are reciprocal relationships.

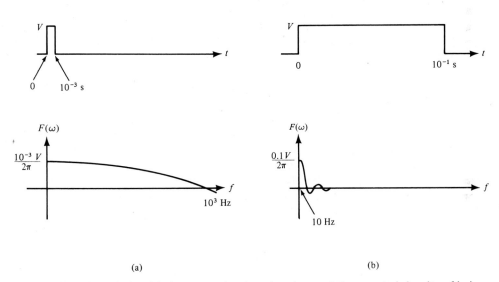

(a)   (b)

fig. 7.28   The relationship between the time duration and the spectral density of (a) a 1 - ms rectangular pulse and (b) a 0 .1 - s pulse.

The Fourier analysis has indicated that quick time changes are associated with high-frequency components. Thus, the frequency components beyond the first zero crossing are produced primarily by the sharp corners on the rectangular pulse. However, pulses are never precisely rectangular due to the time constants of practical circuits, so these higher-frequency components are less significant for pulses obtained by practical circuits.

Now we need to consider the transformation of the spectral density back into a function of time. Let us begin with the exponential Fourier series that summed the frequency components to obtain the time function. This was (7.65) repeated below for convenience.

$$f(t) = \sum_{k=-\infty}^{\infty} C_k e^{jk\omega_0 t} \tag{7.79}$$

Let us make the same changes in (7.79) that we did in (7.74), namely, $k\omega_0 = \omega_k$, and in (7.76), $C_k/\omega_0 = F(\omega)$, so $C_k = F(\omega)\omega_0$. Then (7.79) becomes

$$f(t) = \sum_{k=-\infty}^{\infty} F(\omega)\omega_0\, e^{j\omega_k t} \tag{7.80}$$

Now, letting $T \to \infty$, ω_0 becomes continuous (ω) as before, ω_0 becomes infinitesimal, or $d\omega$ and the summation sign become an integral. Then (7.80) becomes

$$f(t) = \int_{-\infty}^{\infty} F(\omega)e^{j\omega t}\, d\omega \tag{7.81}$$

This transformation from the frequency domain to the time domain is known as the inverse Fourier transform.

7.7   SUMMARY

In Chapter 4 the phasor transform was developed to transform a steady-state sinusoidal time function into the frequency domain, thus permitting the solution of ac circuits by the use of complex algebra rather than differential equations. It was also shown in Chapter 4 that the frequency-domain expressions may be readily transformed back into time-domain expressions, although this latter transform is not usually necessary. Similarly, in this chapter we showed how Fourier's series may be used to transform a nonsinusoidal periodic, or repeating, function into a sum of frequency components. The circuit response to each of these components may then be obtained by the complex algebra techniques discussed in Chapter 4. These frequency components are obtained by the general expressions given in (7.19) to (7.21). The first two of these expressions are repeated below for convenience.

$$a_k = \frac{2}{T} \int_{t_0}^{t_0+T} f(t)\cos k\omega_0 t\, dt \tag{7.82}$$

$$b_k = \frac{2}{T} \int_{t_0}^{t_0+T} f(t)\sin k\omega_0 t\, dt \tag{7.83}$$

The dc component a_0 may be obtained by letting $k = 0$ in (7.82) and then dividing the result by 2. In other words, if we designate $a_k|_{k=0}$ as a_0', then $a_0 = a_0'/2$. This may be verified by comparing (7.21) with (7.19). After determining the circuit response to each frequency component, the time response may be obtained by expressing each component as a function of time and then adding these component values at a large number of sufficiently close time intervals to permit an accurate plot of this sum as a function of time. This inverse transform is seldom taken, but it may be simply expressed as the Fourier series.

$$f(t) = a_0 + \sum_{k=1}^{\infty} a_k \cos k\omega_0 t + \sum_{k=1}^{\infty} b_k \sin k\omega_0 t \qquad (7.84)$$

The exponential form of the Fourier series presents a convenient and tidy form of time-to-frequency transformation, as given in (7.69), and is repeated below for convenience.

$$C_k = \frac{1}{T} \int_{t_0}^{t_0 + T} f(t)e^{-jk\omega_0 t} \, dt \qquad (7.85)$$

The summing of the frequency components to obtain the time function is expressed in (7.65), repeated below.

$$f(t) = \sum_{k=-\infty}^{\infty} \frac{a_k - jb_k}{2} e^{jk\omega_0 t} = \sum_{k=-\infty}^{\infty} C_k e^{jk\omega_0 t} \qquad (7.86)$$

Since the frequency-domain expression, or spectral density $f(\omega)$, is a continuous function when the time function is a single event, the transformation from the frequency domain to the time domain as well as from the time domain to the frequency domain may be accomplished by integration as indicated by the Fourier transform pair that follows.

$$F(\omega) = \frac{1}{2\pi} \int_{-\infty}^{\infty} f(t)e^{-j\omega t} \, dt \qquad (7.87)$$

$$F(t) = \int_{-\infty}^{\infty} f(\omega)e^{j\omega t} \, d\omega \qquad (7.88)$$

These expressions are the same as (7.78) and (7.81), which were developed in Section 7.6.

PROBLEMS

7.21 The type of sound generated by a steam nozzle is known as *white noise*. White noise has a spectral density that is independent of frequency. Let us assume that white noise having spectral density D is passed through an ideal filter that does not attenuate the noise components below frequency ω_1 at all but completely eliminates all the components above ω_1, as shown in Fig. 7.29. Change this spectrum into a double-sided

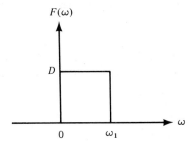

fig. 7.29 An ideal filter.

spectrum and then use the inverse Fourier transform to find the time function that would result from the frequency spectrum shown in Fig. 7.29. What effect does the cutoff frequency ω_1 have on the time response? In other words, imagine ω_1 as being variable from 0 to infinity and consider the effect of the bandwidth ω_1 on the time function.

7.22 A dc voltage V is suddenly applied to a series RC circuit so the current in the circuit is $(V/R)e^{-t/RC}$. Determine the spectral density $F(\omega)$ as a function of ω.

7.23 Determine the spectral density $F(\omega)$ of the voltage across the capacitor in the series RC circuit of Prob. 7.22. Sketch $F(\omega)$ as a function of ω.

7.24 Design a circuit, like the one in Fig. 7.12, that will reject a fundamental and pass a third harmonic:

a. First attempt to obtain an inductor and model it by measuring the appropriate L, R, and C_a as contained in the dashed lines in Fig. 7.12. Ask your instructor for help in using the available equipment to make the appropriate measurements. If no inductor is available, assume that $L_1 = 0.1$ H, $R = 50$ Ω, and $C_a = 100$ pF.

b. Write the appropriate design equations and choose values of capacitance that will result in passing the third harmonic and rejecting the fundamental of a triangular wave having a frequency of 5 kHz. You may have to use trimmer capacitors in parallel with fixed capacitors to obtain the needed values of capacitance.

c. For a sinusoidal voltage source, calculate the response of your circuit over a frequency range $f_a \leq f \leq f_b$, where f_a is one-fifth the frequency of your triangular wave and f_b is equal to the seventh harmonic. Do this for several values of R_1 to obtain a "feel" for how R_1 affects the rejection of the fundamental. You will probably want to write a computer program to make these calculations, or use a programmable calculator, if you have access to one. Make graphs of your responses as a function of frequency.

d. Calculate the response of your circuit to the triangular-wave input. Make a plot of the output voltage as a function of time.

e. If the circuit components and measuring instruments are available, measure the frequency response of your circuit to a sinusoidal generator over the same frequency range as the values you calculated in (c) above. Plot your measured values and calculated values on the same set of axes and discuss the results. Display the response of your circuit to a triangular-wave input and compare the results, both in amplitude and frequency, with the result that you calculated in (d) above. Compare your measured and calculated results and discuss factors that influence the two, including validity of the models used, accuracy of measurements, and accuracy of calculations.

f. Discuss the success of your design procedure and the usefulness of this circuit for rejecting one frequency and passing another.

Chapter 8

Laplace Transform Analysis

Introduction The Laplace transform is a powerful tool that is used extensively in electric circuit analysis. It has broad implications in much of the linear analysis in engineering at large, and in electrical engineering in particular. Because of its elegance and power, you will find it exciting to learn how to apply this method.

Since we make extensive use of special mathematical functions called step functions in the development of the Laplace transform methods, we first discuss step functions and their use in circuit theory. Then we proceed to development of the Laplace transform and its use in circuit theory.

In its broadest sense the Laplace transform involves transformation of the complex frequency domain, thus involving complex variable theory. However, we will minimize the use of complex variable theory, but this will not be detrimental at the introductory level. We shall be content here to let you look forward to extensive use of complex variable theory in advanced courses, satisfying ourselves with trying to establish the basic concepts and techniques as firmly as possible.

8.1 STEP FUNCTIONS

We have previously analyzed circuits in which we assumed that a switch was opened or closed instantaneously, resulting in instantaneous changes in voltage or current. Of course nothing physical actually ever changes instantaneously, but it is often a good approximation to assume so, and therefore it is convenient to have a mathematical function to represent such an instantaneous change. In this section, we will discuss such a mathematical function, which is called a *step function*.

If the switch in the circuit of Fig. 8.1a were closed instantaneously at $t = 3$ s, the

321

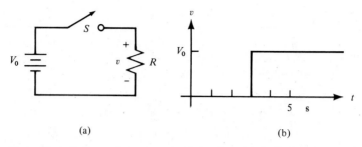

(a)   (b)

fig. 8.1   A circuit and the voltage across R in the circuit if S were closed instantaneously at $t = 3$ s.

voltage v across the resistor would increase instantaneously to V_0 at $t = 3$, as shown
in Fig. 8.1b. Since v suddenly steps up, it is called a *step function*. This is just one
example of a step function.

Mathematical Definition

A general mathematical definition of a step function is

$$u(x - x_0) = \begin{cases} 0 & \text{if } x < x_0 \\ 1 & \text{if } x > x_0 \end{cases} \tag{8.1}$$

where x is the variable and x_0 is any specific value of the variable. Note that
$u(x - x_0)$ is indeterminate at $x = x_0$, since its value changes from zero to unity at
$x = x_0$. A graph of $u(x - x_0)$ is shown in Fig. 8.2. Since $u(x - x_0) = 1$ for $x > x_0$, it
is often called a *unit* step function.

The voltage v shown in Fig. 8.1 can be written in terms of a unit step function as

$$v = V_0\, u(t - 3)$$

since

$$V_0\, u(t - 3) = \begin{cases} 0 & \text{for } t < 3 \\ V_0 & \text{for } t > 3 \end{cases}$$

Note that $u(t - t_0)$ is always dimensionless and must be multiplied by a constant
with appropriate dimensions when used in describing a physical quantity such as
voltage.

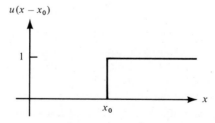

fig. 8.2   A step function.

Use of a Step Function in Circuit Analysis

As an example of how the step function is useful in circuit theory, let's analyze the same circuit that we analyzed in Chapter 3 (Fig. 3.24), assuming that the switch is closed at $t = 3$ s instead of at $t = 0$ as we did there. The circuit is shown again in Fig. 8.3a, with an alternate representation in Fig. 8.3b, where the battery and switch are represented by a voltage step-function generator.

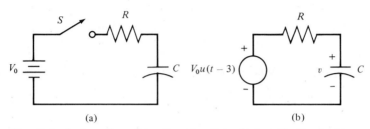

(a)   (b)

fig. 8.3   *Circuit in (a) replaced by the circuit in (b) with the battery and switch represented by the voltage step - function generator.*

The Circuit Equations.   Following the same procedure that we have used many times before, we find that the differential equation for the voltage v across C is

$$\frac{dv}{dt} + \frac{1}{RC} v = \frac{V_0\, u(t-3)}{RC} \tag{8.2}$$

which is identical to the one that we obtained in Chapter 3 except that here the driving function is a voltage step function.

The natural response is

$$v_h = Ae^{-t/RC}$$

the same as it was before, since the natural response does not depend on the driving function. The forced response, however, is now different, and since the driving function is discontinuous, we must be careful how we find the forced response. Perhaps the most straightforward way to find v_p is to find it in two intervals. For $t < 3$, the driving function is zero, and hence

$$v_p = 0 \qquad \text{for } t < 3 \tag{8.3}$$

For $t > 3$ the driving function is equal to V_0/RC, and

$$v_p = V_0 \qquad \text{for } t > 3 \tag{8.4}$$

From (8.3) and (8.4) and the definition of a step function, we see that we can write

$$v_p = V_0\, u(t-3) \tag{8.5}$$

We cannot easily substitute (8.5) into (8.2) to show that v_p does satisfy the differential equation because the derivative of $u(t-3)$ approaches infinity at $t = 3$, where

$u(t - 3)$ is discontinuous. However, $v_p = 0$ and $v_p = V_0$ does satisfy (8.2), respectively, for $t < 3$ and $t > 3$.

Combining the natural response and the forced response gives us

$$v = Ae^{-t/RC} + V_0 u(t - 3)$$

Satisfying the Initial Conditions.   The initial conditions are that

$$v(t) = 0 \qquad \text{for } t < 3$$

and

$$v(3+) = 0$$

The first condition is required because the switch is open for $t < 3$, and the second because the voltage across a capacitor cannot change instantaneously. The two conditions require, respectively, that

$$A = 0 \qquad \text{for } t < 3$$

$$Ae^{-3/RC} = -V_0 \qquad \text{for } t = 3$$

Again using the definition of a step function, we can write

$$A = -e^{3/RC}V_0\, u(t - 3)$$

The $u(t - 3)$ is required to satisfy the first condition. The second condition is also satisfied, and it does not matter that A appears to be a function of t with the $u(t - 3)$ there because $u(t - 3) = 1$ for $t > 3$, and A is just a constant for $t > 3$. The final solution is

$$v = u(t - 3)V_0(1 - e^{-(t-3)/RC}) \qquad (8.6)$$

The Nature of the Solution.   The graph of (8.6) in Fig. 8.4 for some typical values will help you understand the nature of v. The curve is the familiar exponential that we became acquainted with in Chapter 3. In fact, it is just the same curve that we obtained for the same circuit in Section 3.3, except that it is shifted to the right by 3 seconds. After a little thought, you should realize that the curve *should* be merely shifted to the right, because the only difference between this circuit and the one in Section 3.3 is the time at which S is closed. Since we can start counting time from any arbitrary point, the shape of the curve should not change when we do nothing more than delay the opening of the switch.

Shifted Functions.   Since our physical understanding makes the nature of the solution clear, the main point of this example is the mathematical function that describes the kind of curve shown in Fig. 8.4. Notice that (8.6) is equivalent to a function

$$v_1 = u(t)V_0(1 - e^{-t/RC})$$

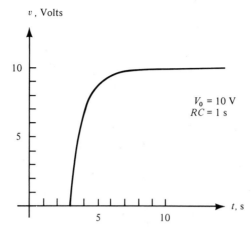

fig. 8.4   *Graph of (8.6) for some typical values.*

with every t replaced by $t - 3$, which we indicate mathematically by $v(t) = v_1 (t - 3)$. This is a hint that, in general, replacing t by $t - t_0$ ($t_0 > 0$) shifts the curve to the right, and indeed this is true. Thus, with more understanding, we could have taken our previous solution from Chapter 3 and replaced t by $t - 3$ in it to get the solution desired here. You must be wondering, though, why there was no $u(t)$ in our previous solution in Chapter 3. The answer is that we tacitly assumed that $v = 0$ for $t < 0$ and began our graphs at $t = 0$, without explicitly including $u(t)$ to indicate that $v = 0$ for $t < 0$.

Let's explore further the concept of shifting curves, because there are some tricky aspects to it. Figure 8.5 shows an example of the shifting of the very simple function shown in Fig. 8.5a, $f(x) = x$. Figure 8.5b shows the shifted function $f(x - 2) = x - 2$. You should plot a few points yourself to be sure that you see that $f(x - 2) = x - 2$ is really $f(x) = x$ shifted to the right. Note that both $f(x)$ and $f(x - 2)$ have nonzero values for $x < 0$. Figure 8.5c shows another function, $g(x) = xu(x)$, which is zero for $x < 0$, and Fig. 8.5d shows the shifted function, $g(x - 2) = (x - 2)u(x - 2)$. In this case, $g(x - 2) = 0$ for $x < 2$. The tricky point is that if a function that you are going to shift is zero for $x < 0$, then the $u(x)$ must be included to be precisely correct. A little thought should make it clear to you that $f(x + a)$ is just $f(x)$ shifted to the *left* by a units.

Using Step Functions to Represent a Pulse

To see better how step functions can be useful in circuit analysis, consider the circuit shown in Fig. 8.6. The voltage v_s can be written as the sum of two step functions, as shown in Fig. 8.7, with a corresponding equivalent circuit shown in Fig. 8.8. The circuit of Fig. 8.8 can be analyzed using the Laplace transform techniques described later in this chapter, or it can be analyzed using our current background by using the principle of superposition (Chapter 2).

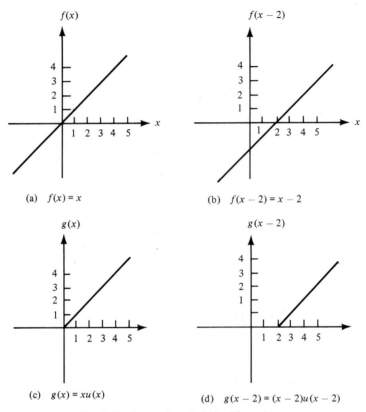

fig. 8.5   Examples of shifted functions.

PROBLEMS

8.1 If $f(x) = x^2 - 2x + 5$, write an expression for $f(x - 2)$. Plot both $f(x)$ and $f(x - 2)$, and show that $f(x - 2)$ is $f(x)$ shifted to the right by two units.

8.2 If $f(x) = x^2 - 2x + 5$, write an expression for $f(x + 2)$. Plot both expressions, and show that $f(x + 2)$ is $f(x)$ shifted to the left by two units.

8.3 Write an expression for the $f(x)$ shown in Fig. 8.9a.

8.4 Write an expression for the $f(x)$ shown in Fig. 8.9b.

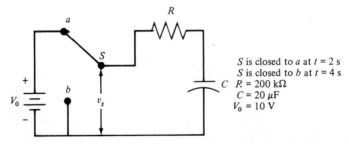

fig. 8.6   Circuit to be analyzed using step functions.

volts

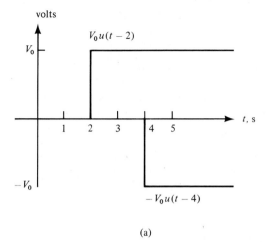

$V_0 u(t - 2)$

V_0

t, s

1   2   3   4   5

$-V_0$

$-V_0 u(t - 4)$

(a)

v_s, volts

V_0

t, s

1   2   3   4   5

(b)

fig. 8.7   *Constructing* v_s *in Fig. 8.6 from two step functions. The sum of the two step functions in (a) gives the* v_s *shown in (b).*

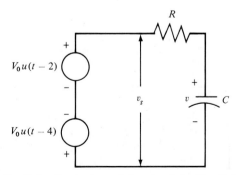

R

$V_0 u(t - 2)$

$V_0 u(t - 4)$

v_s

v C

fig. 8.8   *Equivalent of the circuit in Fig. 8.6.*

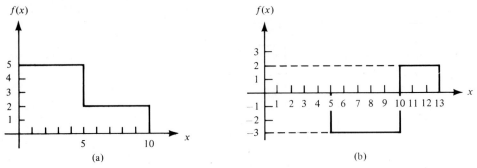

fig. 8.9 (a) See Prob. 8.3. (b) See Prob. 8.4.

Example 8.1

A coil having inductance L and resistance R_c is connected with resistances, a switch S, and a current source as shown in Fig. 8.10. The switch has been opened for an infinite time, then closed at $t = 5$ ms, and then opened again at $t = 15$ ms.

  a. Find an equation for voltage v in each of the time regions: $t < 5$ ms, 5 ms $< t < 15$ ms, and $t > 15$ ms.

  b. Using the $u(t - t_0)$ nomenclature, find a single equation for v that is valid for all values of t.

  c. Prepare a plot that shows v in the region of -10 ms $< t < 20$ ms.

  d. Show some important consistencies among the solutions to (a), (b), and (c).

Solution

a. For $t < 5$ ms (S open), the circuit comes to a steady-state condition, so the inductance L has no effect. Then it follows that

$$v = 2 \times (20\|100) = 33.33 \text{ V} \qquad \text{for } t < 5 \text{ ms}$$

For 5 ms $< t < 15$ ms, S is closed. We can write the solution for v by inspection as follows. The differential equation describing v is first order and the time constant is

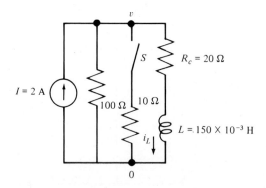

fig. 8.10 Circuit for Example 8.1.

$$\tau_1 = \frac{L}{R_1} = \frac{150 \times 10^{-3}}{20 + (10\|100)} = 5.156 \times 10^{-3} \text{ s} = 5.156 \text{ ms}$$

The particular solution in this region of time is also the value of v for $t = \infty$ if switch S had remained closed until $t = \infty$.

$$V_{p_1} = 2 \times (100\|10\|20) = 12.50 \text{ V}$$

In order to find $v(5^+)$, we first find $i(5^+)$, which is also $i(5^-)$, using the current divider concept at $t = 5^-$

$$i(5^+) = i(5^-) = \frac{100}{120} \times 2 = 1.667 \text{ A}$$

At $t = 5^+$ the current in $(10\|100) = 2 - 1.667 = 0.3333$ A. So $v(5^+) = 0.3333 \times (10\|100) = 3.030$ V. It follows that for $5 \times 10^{-3} < t < 15 \times 10^{-3}$,

$$v = 12.50 - (12.50 - 3.030)e^{-(t-5) \times 10^{-3}/(5.156 \times 10^{-3})}$$

where t is in seconds, or

$$v = 12.50 - 9.470e^{-(t-5)/5.156}$$

where t is in milliseconds and for $(5 < t < 15)$. For $t > 15$ ms, S is open and

$$\tau_2 = \frac{L}{R_2} = \frac{150 \times 10^{-3}}{120} = 1.25 \times 10^{-3} \text{ s}$$

To find $v(15^+)$ we first find $i_L(15^+)$ which is also $i_L(15^-)$. From the equation above,

$$v(15^-) = 12.50 - 9.470e^{-10/5.156} = 11.14$$

S is closed, so

$$i_L(15^-) = 2 - \frac{11.14}{(10\|100)} = 0.7746 = i(15^+)$$

$$v(15^+) = (2 - 0.7746)100 = 122.54 \qquad S \text{ open}$$

The particular solution for $t > 15$ ms is the steady solution at $t = \infty$. This particular solution is also the value of v for $t < 5$ ms.

$$v_{p_2} = 2 \times (20\|100) = 33.33 \text{ V}$$

So, for $t > 15$ ms,

$$v = 33.33 + (122.54 - 33.33)e^{-t/1.25}$$

or

$$v = 33.33 + 89.21e^{-(t-15)/1.25} \text{ V}$$

where t is in ms and where $t > 15$.

To summarize, equations for v in the three regions of time are (t in milliseconds):

For $t < 5$,

$$v = 33.33 \text{ V}$$

For $5 < t < 15$,

$$v = 12.50 - 9.470e^{-(t-5)/5.156} \text{ V}$$

For $t > 15$,

$$v = 33.33 + 89.21e^{-(t-15)/1.25} \text{ V}$$

    b. A single equation that applies through all three of these regions of time is as follows:

$$v = [u(t-(-\infty)) - u(t-5)]33.33 \qquad \text{(term 1)}$$
$$+ [u(t-5) - u(t-15)](12.5 - 9.470e^{-(t-5)/5.156}) \qquad \text{(term 2)}$$
$$+ [u(t-15)](33.33 + 89.21e^{-(t-15)/1.25}) \qquad \text{(term 3)}$$

Notice that by using the $u(t-t_0)$ nomenclature, term 1 is made to be zero for $t > 5$, the second term is zero except in the region $5 < t < 15$, and the third term is zero for $t < 15$.

    c. The plot of v as a function of time is shown in Fig. 8.11.

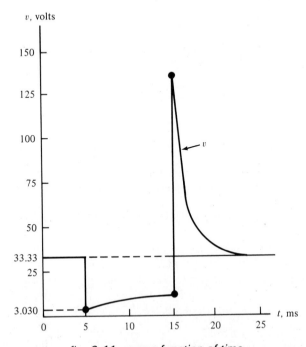

fig. 8.11   *v as a function of time.*

d. We have previously calculated $v(5^-) = 33.33$ V and $v(5^+) = 3.030$ V. The equations of parts (a) and (b) as well as the plot of part (c) show these values. Because the source current is constant and because the current in L cannot change at the instant S is closed ($t = 5$), the current of $2 - 1.667 = 0.3333$ A flows through the 100-Ω resistance at $t = 5^-$ and through $(100\|10)$ Ω at $t = 5^+$. Therefore the resistance ratio $(100\|10)/100 = 0.09091$ should for consistency be equal to the inverse ratio of the corresponding values of v or $3.030/33.33 = 0.09091$. This is one consistency check on our previous solutions.

For a similar consistency check at $t = 15$, we calculate

$$\frac{v(15^-)}{v(15^+)} = \frac{11.14}{122.54} = 0.09091$$

So we have a similar consistency check at $t = 15$.

• • •

PROBLEM

8.5 Switch S in Fig. 8.12 has been open for an infinite time, then closed at $t = 4$ ms, and then opened again at $t = 5$ ms.
a. Find a single equation of i for $0 < t < 8$ ms.
b. Prepare an appropriate plot of i.

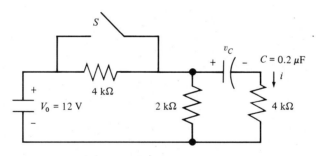

fig. 8.12 Switch closed, then opened.

8.2 BASIC ELEMENTS OF LAPLACE TRANSFORM METHODS

As an example for use in developing Laplace transform methods, we shall use the circuit shown in Fig. 8.13, which is a simplification of a "coupling circuit" commonly used in electronic circuits. First we shall start with the differential equation for v_o in Fig. 8.13 and show that the differential equation can be "transformed" to an algebraic equation in the complex frequency domain (s domain), with initial conditions included. The equations in the s domain can then be solved and a transformation made back to the time domain, thus yielding the desired solution.

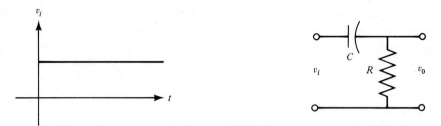

fig. 8.13   A simplification of a coupling circuit.

Using the methods of Chapter 3, we find that the differential equation for v_o in Fig. 8.13 is

$$\frac{dv_o}{dt} + \frac{1}{RC}v_o = \frac{dv_i}{dt} \tag{8.7}$$

and we shall assume that the initial condition is that

$$v_C(0) = 0 \tag{8.8}$$

where v_C is the voltage across the capacitor. Our goal is to transform (8.7) to an algebraic equation in the s domain (s is a complex frequency) and subsequently obtain an expression for v_o.

Definition of the Laplace Transform

The Laplace transform, named after the famous mathematician Pierre Simon de Laplace, is defined by

$$F(s) = \int_0^\infty f(t)e^{-st}\,dt \tag{8.9}$$

where $F(s)$ is said to be the Laplace transform of the function $f(t)$, and s is a complex quantity, usually written as $s = \sigma + j\omega$, where σ and ω are both real quantities. Observe that the Laplace transform is very much like the Fourier transform (7.78). The essential difference is that the s in the Laplace transform includes a *real part* σ, in addition to the $j\omega$ found in the Fourier transform. Therefore, the function of time $f(t)$ is transformed into the *complex frequency* domain $F(\sigma + j\omega)$ by the Laplace transform, whereas $f(t)$ is transformed into the *steady-state sinusoidal* frequency domain $F(j\omega)$ by the Fourier transform.

The Laplace transform has several advantages over the Fourier and other transform methods we have used in the past, at least in the solution of some circuit problems, because of the following features:

1. The real part of s assures convergence of the term $f(t)e^{-st}$ as t approaches infinity, unless $f(t)$ increases without bound. Thus, the upper limit of the Laplace transform integral is almost always finite (usually zero). This is not generally

true for the Fourier transform, where $e^{-j\omega t}$ is indeterminite as t approaches infinity and the limit can be obtained only if $f(t)$ approaches zero.

2. The s in the Laplace transform is essentially the same s that we used in the *operator technique* for obtaining a *complete* time solution for a differential equation in Chapter 5. Therefore, we might expect that complete solutions are obtainable by the Laplace transform technique.

3. The initial conditions are handled much more easily in the Laplace transform method than in the operator method.

The *inverse* Laplace transform is also very similar to the inverse Fourier transform (7.81). It can be shown that the function $f(t)$ can be found from $F(s)$ by

$$f(t) = \frac{1}{2\pi j} \int_{\sigma - j\infty}^{\sigma + j\infty} F(s)e^{st} \, ds \tag{8.10}$$

which is an integration in the complex s plane. Equations (8.9) and (8.10) are said to be a *transform pair*, and we use the notation that

$$F(s) = \mathscr{L}[f(t)] \tag{8.11}$$

$$f(t) = \mathscr{L}^{-1}[F(s)] \tag{8.12}$$

The script $\mathscr{L}$ in (8.11) means that $F(s)$ is the Laplace transform of $f(t)$, as given by (8.9), and $\mathscr{L}^{-1}$ means that $f(t)$ is the *inverse* Laplace transform of $F(s)$, as given by (8.10). The conditions[1] on $f(t)$ required for the Laplace transform pair are:

1. In every interval $0 \le t_1 \le t \le t_2$, $f(t)$ is bounded and has a finite number of maxima and minima and a finite number of finite discontinuities.

2. A real constant b exists such that

$$\int_0^\infty e^{-bt}|f(t)| \, dt$$

is convergent.

Since evaluation of the integral in (8.10) involves complex variable theory, we shall avoid finding inverse transforms by use of (8.10). Instead, we shall develop a table of transforms from (8.9) and use that to find inverse transforms. To begin, let's consider one of the simplest functions of time to transform, $f(t) = u(t)e^{-at}$. The $u(t)$ is present because it is implicit in (8.9) that the transform represents functions only for $t > 0$, although initial conditions are taken into account, as we shall see shortly. Evaluating (8.9) for the exponential time function gives us

$$F(s) = \int_0^\infty u(t)e^{-at}e^{-st} \, dt = \int_0^\infty u(t)e^{-(s+a)t} \, dt = -\frac{1}{s+a}e^{-(s+a)t}\bigg]_0^\infty \tag{8.13}$$

Here is a subtle point in taking the Laplace transform.

[1] C. R. Wylie, Jr., *Advanced Engineering Mathematics*, 3d ed. New York: McGraw-Hill, 1966.

1. We describe the physical system as a function of time.
2. We transform this description to a function of s by means of the Laplace transform.
3. Then after performing simplifying algebraic operations, we transform the desired function back into the time domain.

We can see in (8.13) that $F(s) = 0$ at the upper limit where $t = \infty$, providing $s + a$ has a positive real part, as it usually does. Then it follows that

$$F(s) = 0 + \frac{1}{s + a} = \frac{1}{s + a}$$

Thus we have one transform pair,

$$\mathcal{L}[u(t)e^{-at}] = \frac{1}{s + a} \tag{8.14a}$$

$$\mathcal{L}^{-1}\left[\frac{1}{s + a}\right] = e^{-at}u(t) \tag{8.14b}$$

Another transform can be obtained by evaluating (8.9) for $f(t) = u(t)$, a simple step function. If $s > 0$, it follows that

$$\mathcal{L}[u(t)] = \int_0^\infty u(t)e^{-st}\,dt = -\frac{1}{s}e^{-st}\bigg]_0^\infty = \frac{1}{s} \tag{8.15a}$$

and therefore

$$\mathcal{L}^{-1}\left[\frac{1}{s}\right] = u(t) \tag{8.15b}$$

We will generate additional transform pairs as we proceed. These pairs are listed in Table 8.2, on page 364.

With this introduction to the transform, let's now proceed to transform the differential equation (8.7) that we are using for an example and see how to use transforms to solve circuit problems.

Transformation of Differential Equations

The first step in transforming the differential equation is to show that the Laplace transform is linear; that is,

$$\mathcal{L}[f_1(t) + f_2(t)] = \mathcal{L}[f_1(t)] + \mathcal{L}[f_2(t)] = F_1(s) + F_2(s) \tag{8.16}$$

This property is easily shown from (8.9):

$$\mathcal{L}[f_1(t) + f_2(t)] = \int_0^\infty [f_1(t) + f_2(t)]e^{-st}\,dt$$

$$= \int_0^\infty f_1(t)e^{-st}\,dt + \int_0^\infty f_2(t)e^{-st}\,dt = F_1(s) + F_2(s)$$

Linearity allows us to transform the differential equation by transforming each term individually and then adding the terms.

Another property that we need is that

$$\mathscr{L}[kf(t)] = kF(s) \tag{8.17}$$

where k is a constant. This property is also shown to be true by substitution into (8.9):

$$\mathscr{L}[kf(t)] = \int_0^\infty kf(t)e^{-st}\,dt = k\int_0^\infty f(t)e^{-st}\,dt = kF(s)$$

Since the differential equation includes derivatives of functions, we also need to be able to transform derivatives. From (8.9) we can write

$$\mathscr{L}\left[\frac{df}{dt}\right] = \int_0^\infty \frac{df}{dt}e^{-st}\,dt$$

Integrating by parts gives us

$$\mathscr{L}\left[\frac{df}{dt}\right] = f(t)e^{-st}\Big|_0^\infty - \int_0^\infty (-s)f(t)e^{-st}\,dt$$

Again, if s has a positive real part, $f(t)e^{-st}\big|_{t=\infty} = 0$, then it follows that

$$\mathscr{L}\left[\frac{df}{dt}\right] = -f(0) + s\int_0^\infty f(t)e^{-st}\,dt \tag{8.18}$$

Again from (8.9), we recognize the integral of (8.18) to be $F(s)$. Then

$$\mathscr{L}\left[\frac{df}{dt}\right] = -f(0) + sF(s) \tag{8.19}$$

where $F(s) = \mathscr{L}[f(t)]$. Note that $f(0)$ is the value of $f(t)$ at $t = 0$. If there is a discontinuity at $t = 0$, the lower limit of the defining integral in (8.9) must be 0^- in order to include the discontinuity, and the corresponding term in (8.19) will be $f(0^-)$. In some cases, this will correspond to throwing a switch at $t = 0^-$ instead of at $t = 0$. This merely amounts to shifting the origin of time slightly to make discontinuities easier to handle.

Some circuit equations contain integrals resulting from relations like

$$v_C = \frac{1}{C}\int_{-\infty}^t i\,dt$$

so let's also evaluate

$$\mathscr{L}\left[\int_{-\infty}^t f(t)\,dt\right]$$

First we split the integral into two parts:

$$\int_{-\infty}^{t} f(t)\,dt = \int_{-\infty}^{0} f(t)\,dt + \int_{0}^{t} f(t)\,dt \qquad (8.20)$$

$$\quad\;(1)\qquad\qquad (2)\qquad\qquad (3)$$

Integral (2) in (8.20) is a constant because of the limits $(-\infty)$ and (0). It is of special interest to us. If $f(t)$ were the current i, then this integral has the meaning of the charge accumulated from $t = -\infty$ to time $t = 0$. For brevity, let this integral be

$$\int_{-\infty}^{0} f(t)\,dt = f^{-1}(0)$$

and

$$\mathscr{L} \int_{-\infty}^{t} f(t)\,dt = \frac{f^{-1}(0)}{s} + \int_{0}^{\infty} e^{-st}\left[\int_{0}^{t} f(t)\,dt\right] dt$$

Integrating by parts, we get

$$\int u\,dv = uv - \int v\,du$$

$$dv = e^{-st}\,dt \qquad u = \int_{0}^{t} f(t)\,dt$$

$$v = -\frac{e^{-st}}{s} \qquad du = f(t)\,dt$$

$$\mathscr{L}\left[\int_{-\infty}^{t} f(t)\,dt\right] = \frac{f^{-1}(0)}{s} - \frac{e^{-st}}{s}\int_{0}^{t} f(t)\,dt\,\Big|_{0}^{\infty} + \int_{0}^{\infty}\frac{e^{-st}}{s} f(t)\,dt \qquad (8.21)$$

$$\quad\;(1)\qquad\qquad (2)\qquad\qquad\qquad (3)\qquad\qquad\qquad (4)$$

Examine term (3). Substituting the two limits 0 and ∞ gives

$$-\frac{e^{-st}}{s}\int_{0}^{t} f(t)\,dt\,\Big|_{0}^{\infty} = -\frac{e^{-st}}{s}\int_{0}^{t} f(t)\,dt\,\Big|_{t=\infty} + \frac{1}{s}\int_{0}^{t} f(t)\,dt\,\Big|_{t=0}$$

$$\qquad\;(a)\qquad\qquad\qquad\quad (b)\qquad\qquad\qquad\quad (c)$$

The (b) term is zero in practical situations because the integral evaluated at $t = \infty$ is finite and s normally has a positive real part. The (c) term is also zero because the two limits on the integral are equal (zero).

We recognize term (4) of (8.21) to be $F(s)/s$. It follows that

$$\mathscr{L} \int_{-\infty}^{t} f(t)\,dt = \frac{f^{-1}(0)}{s} + \frac{F(s)}{s} \qquad (8.22)$$

With these derived results, let's now take the transform of (8.7), the differential equation for the circuit in Fig. 8.11. Using linearity and (8.18), we can write

$$\mathscr{L}\left[\frac{dv_o}{dt} + \frac{1}{RC} v_o = \frac{dv_i}{dt}\right]$$

$$\mathscr{L}\left[\frac{dv_o}{dt}\right] + \frac{1}{RC}\mathscr{L}[v_o] = \mathscr{L}\left[\frac{dv_i}{dt}\right]$$

From (8.19), we get

$$sV_o(s) - v_o(0) + \frac{1}{RC} V_o(s) = sV_i(s) - v_i(0) \tag{8.23}$$

where $V_o(s) = \mathscr{L}[v_o]$ and $V_i(s) = \mathscr{L}[v_i]$. Solving for $V_o(s)$ gives us

$$V_o(s) = \frac{sV_i(s) + v_o(0) - v_i(0)}{s + \dfrac{1}{RC}} \tag{8.24}$$

Now that we have an expression for $V_o(s)$, the transform of $v_o(t)$, we would like to be able to take the inverse transform to obtain an expression for $v_o(t)$.

Example of an Inverse Transformation

For a first example, let's look at the inverse of (8.24) when $v_i = u(t)$, a unit step function. For this case, we use (8.15a) to get $V_i(s) = 1/s$. Also, let's assume that the voltage across the capacitor at $t = 0$ is zero, which means that $v_o(0) = v_i(0)$. Consequently, (8.24) becomes

$$V_o(s) = \frac{1}{s + \dfrac{1}{RC}} \tag{8.25}$$

Now we can find $\mathscr{L}^{-1}[V_o(s)]$ from (8.14b), with $a = 1/RC$. Thus,

$$v_o(t) = \mathscr{L}^{-1}[V_o(s)] = \mathscr{L}^{-1}\left[\frac{1}{s + \dfrac{1}{RC}}\right] = u(t)e^{-t/RC} \tag{8.26}$$

This is the same result that we previously obtained by the methods of Chapter 3. This simple example illustrates the use of Laplace transform methods, although it does not really indicate the power and advantage of transform methods because the example is so simple. Before going on to more complex examples, let's summarize the transform method.

Summary of the Laplace Transform Method

In summarizing the use of Laplace transforms in circuit analysis we will use another simple example, the *RL* circuit shown in Fig. 8.14. The first step is to write the

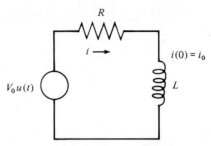

fig. 8.14   *RL circuit used in summarizing Laplace transform methods.*

differential equation for the current i:

$$L\frac{di}{dt} + Ri = V_0\,u(t)$$

Then we transform the differential equation to the s domain, using the relations derived above, to get

$$LsI(s) - Li(0) + RI(s) = \frac{V_0}{s}$$

where $I(s) = \mathcal{L}[i]$. Next, we solve for $I(s)$:

$$I(s) = \frac{(V_0/L) + si(0)}{s(s + R/L)} \tag{8.27}$$

Having an expression for $I(s)$, we next want to find the inverse transform to get an expression for $i(t)$. However, we have no transform pair for the first term in (8.27). Consequently, we use the familiar partial-fraction expansion (reviewed in Section 8.4); thus (8.27) becomes

$$I(s) = \frac{V_0/R}{s} + \frac{i(0) - V_0/R}{s + R/L} \tag{8.28}$$

From (8.14b) and (8.15b), we can find the inverse transform of $I(s)$:

$$i(t) = \left[\frac{V_0}{R} + (i(0) - V_0/R)e^{-Rt/L}\right]u(t) \tag{8.29}$$

As consistency checks, we note that as $t \to \infty$, $i(t) \to V_0/R$, as it should, and as $t \to 0$, $i(t) \to i(0)$, the initial current in the coil.

The procedure for finding $i(t)$ by Laplace transform methods is summarized in Fig. 8.15. Although the example is a simple one, the process used is the same process used in more complex cases. The differential equation is transformed into an algebraic equation in the s domain, the desired function is solved for, and the inverse transformation back to the time domain is made. The initial conditions are included as part of the transformation to the s domain. Now we shall show that Kirchhoff's

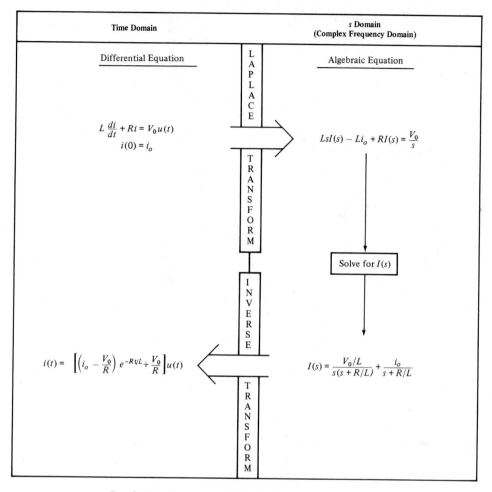

fig. 8.15 Summary of the Laplace transform method.

laws apply in the s domain, and a circuit representation in the s domain can be used to write s-domain expressions for voltages and currents directly, without writing differential equations. In the process, we will also define *impedance* in the s domain.

PROBLEM

8.6 Write the time-domain equation that includes an integral for the voltage across the inductor in Fig. 8.14. Transform the equation to the s domain, assume that $i_o = 0$, solve for the transformed voltage across the inductance, and find the time-domain equation for v by taking the inverse Laplace transform. Make consistency checks on your answer. Then check your answer by classical analysis all in the time domain.

8.3   REPRESENTATION OF CIRCUITS IN THE s DOMAIN

Since Kirchhoff's two laws are statements involving sums of voltages and currents, it is easy to show by linearity, (8.16), that Kirchhoff's laws hold in the s domain. For example, for the circuit of Fig. 8.13, Kirchhoff's voltage law states that

$$v_i = v_C + v_o$$

By taking the Laplace transform of this equation, we get

$$V_i(s) = V_C(s) + V_o(s)$$

Thus we see that Kirchhoff's voltage law is true for the transformed voltages. A simple extension can be made to show that both of Kirchhoff's laws apply to transformed voltages and currents. Consequently, after we transform the voltage and current relations like $v = L\, di/dt$, we can work directly with circuit representations in the s domain, without writing differential equations.

Transformation of Voltage - Current Relations

The three voltage-current relations for inductance, capacitance, and resistance are, respectively:

$$v_L = L\frac{di_L}{dt}$$

$$i_C = C\frac{dv_C}{dt}$$

$$v_R = Ri_R$$

Transforming each of these to the s domain gives us

$$V_L(s) = LsI_L(s) - Li(0) \qquad (8.30)$$

$$I_C(s) = CsV_C(s) - Cv(0) \qquad (8.31)$$

$$V_R(s) = RI_R(s) \qquad (8.32)$$

On the basis of these relations, we define a new quantity, impedance, as the ratio of voltage to current in the s domain. As a first step, let's suppose that $i(0)$ and $v(0)$ are both zero. Then the voltage-current relations for inductance and capacitance are just like those for resistance, leading us to define

$$Z_L(s) = \frac{V_L(s)}{I_L(s)} = sL \qquad Z_C(s) = \frac{V_C(s)}{I_C(s)} = \frac{1}{sC} \qquad \frac{V_R(s)}{I_R(s)} = R$$

as *impedances*. Thus we say that sL is the impedance of inductance, $1/sC$ is the impedance of capacitance, and R is the impedance of resistance R. We use the notation $Z(s)$ for impedances to emphasize that impedance is a function of s. You may have observed that the order of Ls and Cs in (8.30) and (8.31) was changed to sL and sC in the impedance definitions above. The purpose of this change was to point out the similarity of the s-domain impedances to the steady-state sinusoidal impedances $j\omega L$ and $1/j\omega C$ with which we are familiar and the operational impedances sL and $1/sC$ that were used in Chapter 5. The resistance R is the same in all cases. Mathematically, the order of the terms in a product makes no difference, so you will find both orders used in the work that follows.

It is extremely important to remember the limitations on these impedance concepts of $Z_L(s) = Ls$; $Z_C(s) = 1/Cs$; and $Z_R(s) = R$. The first two are valid only when $i(0)$ and $v(0)$ are equal to zero, respectively.

Now let's return to the case when $i(0)$ and $v(0)$ are not necessarily zero. Rearranging (8.30) to

$$I_L(s) = \frac{V_L(s)}{sL} + \frac{i(0)}{s}$$

we note that this voltage-current relation is the same as that for an element with an impedance of sL in parallel with a current source of magnitude $i(0)/s$, as shown in Fig. 8.16. Similarly, if (8.31) is solved explicitly for $V_C(s)$, then we see that this voltage-current relation holds also for an impedance of $1/sC$ in series with a voltage source of $v(0)/s$. See in Fig. 8.16 the s-domain equivalent of a capacitive element. This figure in its three parts summarizes the s-domain circuit equivalence of the

Time-Domain Relation	s-Domain Transformation	s-Domain Models	s-Domain Impedance
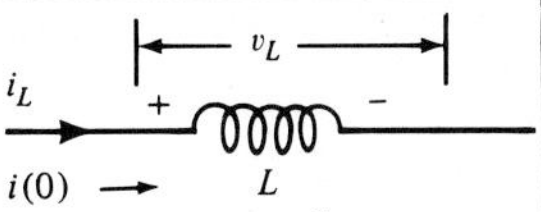 i_L v_L $i(0) \rightarrow$ L $v_L = L\frac{di_L}{dt}$	$V_L(s) = sLI_L(s) - Li(0)$	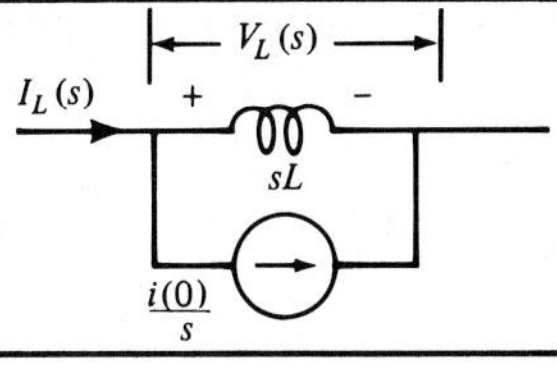 $I_L(s)$ $V_L(s)$ sL $\frac{i(0)}{s}$	sL
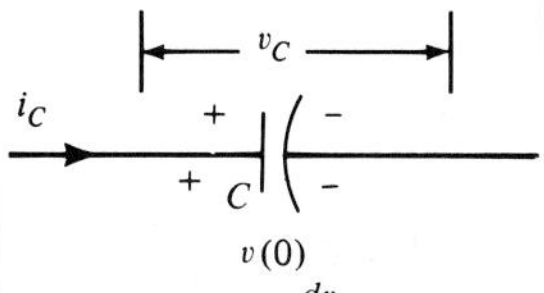 i_C v_C C $v(0)$ $i_C = C\frac{dv_C}{dt}$	$I_C(s) = sCV_C(s) - Cv(0)$	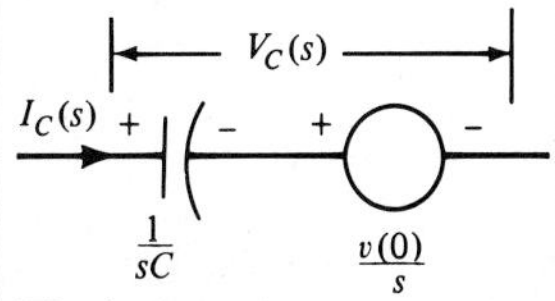 $I_C(s)$ $V_C(s)$ $\frac{1}{sC}$ $\frac{v(0)}{s}$ The independent voltage source has the same polarity as $v(0)$	$\frac{1}{sC}$
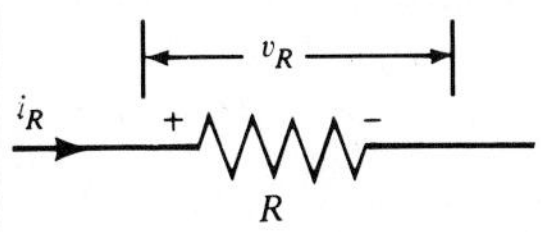 i_R v_R R $v_R = i_R R$	$V_R(s) = I_R(s)R$	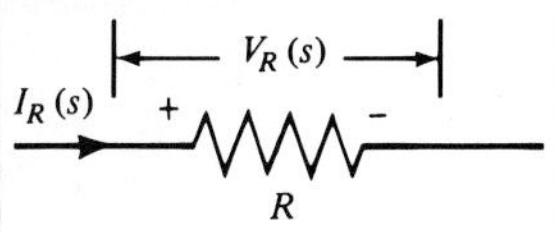 $I_R(s)$ $V_R(s)$ R	R

fig. 8.16 *Voltage-current relations for circuit elements and the transformations into the s domain.*

time-domain elements L, R, and C, respectively. These s-domain impedance representations of circuit elements are very powerful, but it is important for you to realize that impedance treated here is an s-domain quantity, and the ratio of v/i (time domain) is not impedance and has no particular conceptual value.

Finally, we should note that the $[i(0)]$ in (8.30) could have been shown in an equivalent circuit with the voltage source $V_L(s)$, and the $v(0)$ in (8.31) in a circuit with the current source $I_C(s)$, but we have chosen the representations in Fig. 8.16 because they seem to have more physical meaning. You should especially note that the independent voltage source $v(0)/s$ has the same polarity as $v(0)$, and the independent current source $i(0)/s$ has the same direction as $i(0)$.

Furthermore, we should note that although $\mathscr{L}^{-1}[I_L(s)] = i_L$, the real current $I_L(s)$ in the real inductance L is *not* the s-domain current in fictitious sL of the model of Fig. 8.16. In other words, $sL \neq V_L(s)/I_L(s)$ except where $i(0) = 0$.

Examples of Circuit Analysis in the s Domain

Figure 8.17 shows a representation in the s domain of the circuit of Fig. 8.14. The current generator i_0/s in the s domain corresponds to the initial current i_0 in the time domain. Note that the current $I_1(s)$ through the impedance sL is not the Laplace transform of the current through the coil in the time domain. $I(s)$ is the transform of the time-domain current in the coil.

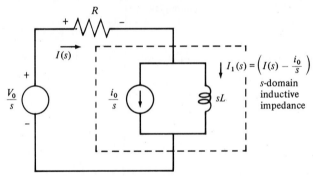

fig. 8.17   Representation in the s domain of the circuit of Fig. 8.14.

With the circuit representation in the s domain, we can use Kirchhoff's laws in the s domain to obtain the equations relating the transformed voltages and currents without ever writing the differential equations in the time domain. From Kirchhoff's voltage law,

$$\frac{V_0}{s} = RI(s) + sL\left(I(s) - \frac{i_0}{s}\right) \tag{8.33}$$

or

$$\frac{V_0}{s} = RI(s) + sLI(s) - Li_0 \tag{8.34}$$

which can be solved for $I(s)$ to get

$$I(s) = \frac{V_0/L + si_0}{s(s + R/L)}$$

which is the same equation as (8.27), the one obtained by transforming the differential equation into the s domain. This expression for $I(s)$ can now be transformed back to the time domain to obtain an expression for $i(t)$, just as we did to get (8.29). The procedure for circuit analysis using circuit representations in the s domain is illustrated for a series RL circuit in Fig. 8.18. Where capacitance and/or resistance are involved, Fig. 8.16 shows the s-domain equivalents.

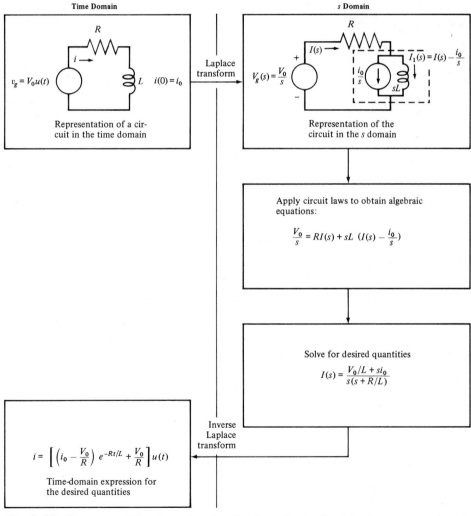

fig. 8.18 Summary of the procedure for circuit analysis using circuit representations in the s domain.

An advantage of transforming circuits to the s domain is that the circuit reduction techniques described in Chapter 2 for resistive circuits can also be used with impedances. For example, by the same techniques used for resistive circuits it is easy to show that impedances in series add as resistors in series. Thus

$$Z_t(s) = Z_1(s) + Z_2(s) + \cdots + Z_n(s)$$

where $Z_t(s)$ is the equivalent impedance of the n impedances in series. Similarly, for parallel circuits,

$$Y_t(s) = Y_1(s) + Y_2(s) + \cdots + Y_n(s)$$

where $Y_n(s)$ is the equivalent admittance of the n admittances in parallel. In addition, all the network theorems and circuit analysis techniques developed in the preceding chapters are applicable to s-domain circuit models.

Example 8.2

An inductor having resistance 3 Ω and inductance 4 H is connected with resistances, switch, and current source as shown in Fig. 8.19. The circuit has reached steady state with switch S open, then this switch is closed at $t = 0$.

a. For $t > 0$, find the nodal voltage v by using Laplace transform techniques.

b. By inspection, find the same solution for v.

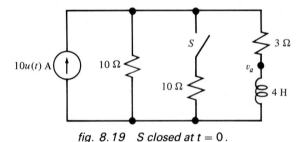

fig. 8.19   *S closed at $t = 0$.*

Solution

a. To develop the s-domain model for the time-domain model of Fig. 8.19, first find the initial current $i(0)$ in the inductor. Of course, in L the current cannot change when the switch is thrown. Use the current divider concept when the circuit is at steady state (S open).

$$i(0) = 10 \frac{\dfrac{1}{3}}{\dfrac{1}{3} + \dfrac{1}{10}} = \frac{100}{13} \text{ A}$$

Then the s-domain model of the circuit for $t > 0$ is shown in Fig. 8.20.

Now we proceed to find the s-domain voltage $V(s)$ by using impedances 10, 10, 3, and $s4$ in the patterns so familiar to us. We might proceed with a two-nodal analysis in terms of unknown nodal voltages $V_a(s)$ and $V(s)$. But to avoid si-

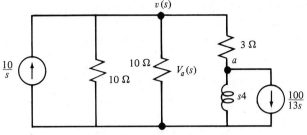

fig. 8.20   S closed at t > 0.

multaneous equations, we will choose a one-nodal analysis to find $V(s)$ directly. We remember that we write the voltage V at a single node as

$$V = \frac{\text{Sum of all separate entering currents, each due to its own forcing function alone}}{\text{Total admittance seen by } V \text{ alone}}$$

The current into node V due to the left-hand source is, of course, $10/s$. The current into $V(s)$ due to the right-hand source alone is a fraction of $-100/13s$ because of the division in the two branches, 3 and $s4$. We write

$$V(s) = \frac{\dfrac{10}{s} - \dfrac{100}{13s}\left(\dfrac{1/3}{1/3 + 1/s4}\right)}{\dfrac{1}{10} + \dfrac{1}{10} + \dfrac{1}{3 + s4}}$$

This reduces to

$$V(s) = \frac{10 - \dfrac{100}{13}\left(\dfrac{s4}{s4 + 3}\right)}{s\left(\dfrac{1}{5} + \dfrac{1}{s4 + 3}\right)}$$

Multiply numerator and denominator by $5(s4 + 3)$.

$$V(s) = \frac{5\left[s40 + 30 - s\dfrac{400}{13}\right]}{s(s4 + 8)} = \frac{11.54(s + 3.250)}{s(s + 2)}$$

Now in order to take the inverse transform, we break this equation into partial fractions:[2]

$$V(s) = \frac{18.75}{s} - \frac{7.212}{s + 2}$$

---

[2] If you are unable to perform this partial-fraction expansion without a review, you may wish to skip this step at the moment. The text material soon will treat partial fractions at some length.

By referring to the table of transforms, we write the equation for v (time domain)

$$v = (18.75 - 7.212e^{-t/2})\, u(t)$$

b.   The solution "by inspection" in this case is much simpler compared to the Laplace method. We see that this is a first-order situation and the time constant $\tau = 4/[3 + (10\|10)] = 4/8 = 1/2$. The initial value $v(0)$ is found by noting that $i(0) = 100/13$. Then we can write $v(0)$ as

$$v(0) = \left(10 - \frac{100}{13}\right)(10\|10) = 11.54 \text{ V}$$

The final value $(t = \infty)$ or the steady value after S has been closed sufficiently long is

$$v(\infty) = 10(10\|10\|3) = 18.75$$

So

$$v = 18.75 - (18.75 - 11.54)e^{-t/2}$$
$$= 18.75 - 7.210e^{-t/2}$$

● ● ●

This solution, of course, agrees with the solution by Laplace transform.
    This example suggests that where the problem is based on a first-order situation, solution by inspection is perhaps "the best way to go." The "payoff" of the transform approach is in much more complex situations where "solution by inspection" is of little or no value.

PROBLEM

8.7   Using s-domain impedances and an appropriate s-domain model, write the s-domain equation for the voltage $V_C(s)$ across the capacitor in the circuit of Fig. 8.21, assuming that (a) the initial conditions are zero and (b) the initial voltage across the capacitor is 3 V.

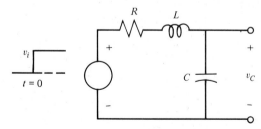

fig. 8.21   The model of a series RLC circuit with a step - voltage input.

Forced and Natural Responses

In Chapters 3 and 5 we found that the general solution to the differential equation consists of the two components: *complementary function* and *particular solution.* As explained there, the complementary function is called the *natural response* and the particular solution is called the *forced response.* In all of our examples so far, the natural response has been one or two exponentially decaying terms. Because it is often true that the natural response consists of exponentially decaying terms, the natural response is sometimes called the *transient response.* The natural response, in rather rare situations where there is *feedback,* may include one or more exponentially increasing terms. Because in most situations the natural response decays to zero, the forced response that remains is ofttimes also called the *steady-state response.*

Independently of whether the solution is obtained by classical solution of differential equations or through the Laplace transform methods, the response, of course, will contain both the forced and natural components. In the example of Fig. 8.18, the forced response is

$$i_{\text{forced}} = \frac{V_0}{R} u(t)$$

and the natural response is

$$i_{\text{natural}} = \left(i_0 - \frac{V_0}{R} \right) e^{-Rt/L} u(t)$$

Because the forced response is merely a function that satisfies the differential equation, and this differential equation is not influenced by the initial conditions, then the forced response is completely independent of the initial conditions. Only the circuit and the forcing function determine the forced response. On the other hand, the natural response (solution to the homogeneous differential equation) contains the arbitrary constant (see Chapter 3), which in turn is directly dependent upon the initial conditions as well as both the forcing function and the circuit.

In many situations we are interested in the forced response only. In these cases we can greatly simplify our analyses by assuming that initial currents in inductances and initial voltages across capacitances are all equal to zero. This will lead to a correct forced response for all initial conditions and a correct natural response only when the initial conditions are zero.

To illustrate this point, consider the series *RL* circuit of Fig. 8.18 where $i(0) = 0$.

$$\frac{V_0}{s} = (R + sL)I(s)$$

Solving for $I(s)$,

$$I(s) = \frac{V_0/L}{s\left(s + \dfrac{R}{L} \right)} = \frac{V_0/R}{s} - \frac{V_0/R}{s + \dfrac{R}{L}}$$

Taking the inverse Laplace transform,

$$i = \frac{V_0}{R}(1 - e^{-Rt/L})u(t)$$

Here

$$i_{\text{forced}} = \frac{V_0}{R}u(t)$$

$$i_{\text{natural}} = -\frac{V_0}{R}e^{-Rt/L}u(t)$$

This result is consistent with the results that we obtained above. The two forced responses are identical but the two natural responses, of course, differ in magnitude because, in the former case, $i(0) = i_0$.

So this brings us to the following important point. If you are interested only in the forced response, you can simplify the analysis by letting the initial conditions go to zero, thus reducing the s-domain models of capacitances and inductances to their impedances, as shown in Fig. 8.16.

8.4   METHODS FOR FINDING INVERSE TRANSFORMS

In order to illustrate some methods of finding inverse Laplace transforms of second-order functions, let us consider the following equation,

$$V_0(s) = \frac{sV_i(s)/R_2 C_1}{s^2 + (1/R_1C_1 + 1/R_2 C_2 + 1/R_2 C_1)s + 1/R_1C_1R_2 C_2} \tag{8.35}$$

First, we will let v_i be a unit step function and use the following values for substitution into (8.35):

$$R_1 = 10 \text{ k}\Omega \qquad C_1 = 10^{-8} \text{ F}$$
$$R_2 = 1 \text{ k}\Omega \qquad C_2 = 10^{-6} \text{ F}$$

Substituting $V_i = 1/s$ and the numbers above into (8.35) gives us

$$V_0(s) = \frac{10^5}{s^2 + 1.11(10^5)s + 10^7} \tag{8.36}$$

The first step in finding the inverse is to factor the denominator, which can be done by setting the denominator equal to zero and using the quadratic formula to obtain the two roots. Thus,

$$s_{1,2} = -\frac{1.11(10^5)}{2} \pm \sqrt{\left[\frac{1.11(10^5)}{2}\right]^2 - 10^7} = -110{,}909.84 \ -90.16$$

Therefore

$$V_0(s) = \frac{10^5}{(s + 110,909.84)(s + 90.16)} \tag{8.37}$$

Note that if $-110,909.84$ is a root of the quadratic, that the corresponding factor is $(s + 110,909.84)$. As a check on the factoring, it is a good idea to multiply out the denominator in (8.37) to check with the original expression in (8.36).

The next step is to expand (8.37) in partial fractions. That is, to write

$$\frac{10^5}{(s + 110,909.84)(s + 90.16)} = \frac{K_1}{s + 110,909.84} + \frac{K_2}{s + 90.16} \tag{8.38}$$

and solve for the values of K_1 and K_2 that make the equation true. You should have learned partial fraction expansions in algebra, but we will review the procedure here and in the next subsection to help you over any difficulties. To find K_1 we multiply both sides of (8.38) by $(s + 110,909.84)$ to get

$$\frac{10^5}{(s + 90.16)} = K_1 + \frac{(s + 110,909.84)}{(s + 90.16)} K_2$$

and then let $s = -110,909.84$ to get

$$K_1 = \frac{10^5}{(-110,909.84 + 90.16)} = -0.9024$$

Similarly, we multiply both sides of (8.38) by $(s + 90.16)$ and then let $s = -90.16$ to get

$$K_2 = \frac{10^5}{(-90.16 + 110,909.84)} = 0.9024$$

So that we can write

$$V_0(s) = \frac{-0.9024}{(s + 110,909.84)} + \frac{0.9024}{(s + 90.16)} \tag{8.39}$$

With $V_0(s)$ written in this form, the inverse can easily be found from the appropriate transform in Table 8.2 (see page 364) and an application of the principle of linearity.

$$v_0(t) = u(t)(0.9024)[-e^{-110,909.84t} + e^{-90.16t}] \tag{8.40}$$

We may recall that the solution for the circuit response by the operator method given in Chapter 5 appeared to follow the same general procedure as the Laplace transform method. However, the complete solution is obtained by taking the inverse transform, while the task of evaluating arbitrary constants was required in the operator method.

Finding the inverse transform of (8.36) is an example of a general procedure in which the function is expanded in a partial-fraction expansion and the inverse found term by term, as justified by the linearity of the Laplace transform. The general relations for this procedure are developed in the next subsection.

Partial - Fraction Expansions

In general, we would like to have relations for finding the inverse transform of

$$F(s) = \frac{P(s)}{Q(s)} \tag{8.41}$$

where $P(s)$ and $Q(s)$ are each polynomials in s. First, let's consider the case where $P(s)$ and $Q(s)$ are factored and there are no repeated roots. That is, we have $F(s)$ in the form

$$F(s) = \frac{(s - z_1)(s - z_2)(s - z_3) \ldots (s - z_m)}{(s - p_1)(s - p_2)(s - p_3) \ldots (s - p_n)} \tag{8.42}$$

where the z's are called the *zeros* of $F(s)$ and the p's are called the *poles* of $F(s)$, because when $s = z_1$ or $s = z_2$, $\ldots$, $F(s) = 0$, and when $s = p_1$ or $s = p_2$, $\ldots$ $F(s)$ becomes infinite. The idea of the partial-fraction expansion is to write $F(s)$ as the sum of terms, the inverse of each of which can be found from a table of inverses (such as Table 8.1 on page 363 or more extensive ones found in mathematical handbooks).

The first step in expanding (8.42) is to ensure that $m < n$—that is, that the polynomial in the numerator is of lower order than the polynomial in the denominator. If m is not less than n, we can make it so by long division. For example, if

$$F(s) = \frac{s^3 + 2s^2 + 3s + 1}{s^2 + 6s + 2} \tag{8.43}$$

we divide the denominator into the numerator by long division to get

$$
\begin{array}{r}
s - 4 \\
s^2 + 6s + 2 \overline{)s^3 + 2s^2 + 3s + 1} \\
\underline{s^3 + 6s^2 + 2s} \\
-4s^2 + s + 1 \\
\underline{-4s^2 - 24s - 8} \\
25s + 9
\end{array}
$$

$$F(s) = s - 4 + \frac{25s + 9}{s^2 + 6s + 2} \tag{8.44}$$

Now the second term is a ratio of polynomials in which $m < n$, and therefore it can be expanded in partial fractions.

Now let's return to (8.42) and assume that $m < n$. We expand $F(s)$ as

$$F(s) = \frac{K_1}{s - p_1} + \frac{K_2}{s - p_2} + \frac{K_3}{s - p_3} + \cdots + \frac{K_i}{s - p_i} + \cdots + \frac{K_n}{s - p_n} \tag{8.45}$$

To find K_1 we multiply both sides by $(s - p_1)$ and then let $s = p_1$. This makes all terms on the right-hand side except K_1 equal to zero, and we get

$$K_1 = [(s - p_1)F(s)]_{s=p_1} = \left[\frac{(s - z_1)(s - z_2)(s - z_3) \dots (s - z_m)}{(s - p_2)(s - p_3) \dots (s - p_n)}\right]_{s=p_1} \qquad (8.46)$$

$$K_1 = \frac{(p_1 - z_1)(p_1 - z_2)(p_1 - z_3) \dots (p_1 - z_m)}{(p_1 - p_2)(p_1 - p_3) \dots (p_1 - p_n)} \qquad (8.47)$$

For example, suppose that

$$F(s) = \frac{(s + 2)(s + 4)}{(s + 1)(s + 3)(s + 5)}$$

Then

$$F(s) = \frac{K_1}{s + 1} + \frac{K_2}{s + 3} + \frac{K_3}{s + 5}$$

and

$$K_1 = \left[\frac{(s + 2)(s + 4)}{(s + 3)(s + 5)}\right]_{s=-1} = \left[\frac{(-1 + 2)(-1 + 4)}{(-1 + 3)(-1 + 5)}\right] = \frac{3}{8}$$

The procedure to find any of the K's in (8.45) is the same. The general relation for the ith K is

$$K_i = [(s - p_i)F(s)]_{s=p_i} \qquad \text{(no repeated roots)} \qquad (8.48)$$

When repeated roots are present, more steps are necessary. For example, if

$$F(s) = \frac{(s + 3)}{(s + 1)(s + 2)^3} \qquad (8.49)$$

the repeated roots $(s + 2)^3$ requires the expansion to be

$$F(s) = \frac{K_1}{(s + 2)^3} + \frac{K_2}{(s + 2)^2} + \frac{K_3}{(s + 2)} + \frac{K_4}{(s + 1)} \qquad (8.50)$$

We can find K_1 by multiplying both sides by $(s + 2)^3$ and letting $s = -2$, because this makes all the terms on the right-hand side equal to zero except the K_1 term. Thus

$$[(s + 2)^3 F(s)]_{s=-2} = K_1 + \left[\frac{K_2(s + 2)^3}{(s + 2)^2}\right]_{s=-2}$$

$$+ \left[\frac{K_3(s + 2)^3}{s + 2}\right]_{s=-2} + \left[\frac{K_4(s + 2)^3}{s + 1}\right]_{s=-2}$$

$$K_1 = \frac{-2 + 3}{-2 + 1} = -1$$

We cannot find K_2 by multiplying both sides by $(s + 2)^2$ and letting $s = -2$ because this does not make the K_1 term approach zero. To get around this difficulty, we multiply both sides by $(s + 2)^3$, take the derivative of both sides with respect to s, and then let $s = -2$. Doing so, we get

$$\left[\frac{d}{ds}\frac{(s+3)}{(s+1)}\right]_{s=-2} = \left[\frac{d}{ds}K_1\right]_{s=-2} + \left[\frac{d}{ds}(s+2)K_2\right]_{s=-2}$$

$$+ \left[\frac{d}{ds}(s+2)^2K_3\right]_{s=-2} + \left[\frac{d}{ds}\frac{(s+2)^3}{(s+1)}K_4\right]_{s=-2}$$

$$\left[\frac{(s+1)-(s+3)}{(s+1)^2}\right]_{s=-2} = 0 + K_2 + 0 + 0$$

$$K_2 = -2$$

Similarly, we find K_3 by multiplying by $(s + 2)^3$ and taking the second derivative. Thus

$$\left[\frac{d^2}{ds^2}\frac{(s+3)}{(s+1)}\right]_{s=-2} = 0 + 0 + \left[\frac{d^2}{ds^2}(s+2)^2K_3\right]_{s=-2} + 0$$

$$-4 = 2K_3$$

$$K_3 = -2$$

Note that we got $2K_3$ on the right-hand side, and not just K_3. Since K_4 is for an unrepeated root in (8.50), we can use (8.48) to find K_4, and we get $K_4 = 2$. Therefore,

$$F(s) = -\frac{1}{(s+2)^3} - \frac{2}{(s+2)^2} - \frac{2}{(s+2)} + \frac{2}{(s+1)} \tag{8.51}$$

In general, for an $F(s)$ of the form

$$F(s) = \frac{P(s)}{(s-p_1)^r(s-p_2)\dots(s-p_n)}$$

the expansion will be of the form

$$F(s) = \frac{K_1}{(s-p_1)^r} + \frac{K_2}{(s-p_1)^{r-1}} + \frac{K_3}{(s-p_1)^{r-2}}$$

$$+ \dots + \frac{K_r}{s-p_1} + \dots + \frac{K_n}{s-p_n} \tag{8.52}$$

and

$$K_r = \frac{1}{(r-1)!}\left[\frac{d^{(r-1)}}{ds^{(r-1)}}[(s-p_1)^rF(s)]\right]_{s=p_1} \tag{8.53}$$

$$K_{r-1} = \frac{1}{(r-2)!}\left[\frac{d^{(r-2)}}{ds^{(r-2)}}[(s-p_1)^rF(s)]\right]_{s=p_1} \tag{8.54}$$

$$K_{r-3} = \frac{1}{(r-3)!}\left[\frac{d^{(r-3)}}{ds^{(r-3)}}[(s-p_1)^rF(s)]\right]_{s=p_1} \tag{8.55}$$

$$\vdots$$

Also

$$K_i = [(s - p_i)F(s)]_{s = p_i} \qquad \text{for } i > r \tag{8.56}$$

The same pattern is followed when more than one repeated root is present.

This technique of taking the derivative to find the partial-fraction coefficients is effective, but rather tedious. Fortunately, a repeated root to the third degree or higher is rather rare. If the repeated root is to the second degree, a simpler technique should be followed as illustrated below. Let

$$F(s) = \frac{s + 3}{(s + 1)(s + 2)^2} = \frac{a_1}{(s + 2)^2} + \frac{a_2}{s + 2} + \frac{a_3}{s + 1}$$

$$a_1 = \frac{-2 + 3}{-2 + 1} = -1 \qquad \text{and} \qquad a_3 = \frac{-1 + 3}{(-1 + 2)^2} = 2$$

Once we have found a_1 and a_3, there is a simple technique for writing a_2 at a glance, as follows. Think of clearing fractions on the right-hand side of the equal sign by multiplying a_1 by $(s + 1)$, a_2 by $(s + 1)(s + 2)$, and a_3 by $(s + 2)^2$. Then equate the coefficient of s^2 on the right to the coefficient of s^2 on the left. Thus we see that $(a_2 + a_3) = 0$ or $a_2 = -a_3 = -2$. So in this manner, one could, without any writing, have seen at a glance that $a_2 = -a_3$. Notice that in clearing fractions in this case, the highest-degree term in s was s^2 and the coefficient on the left-hand side was 0. In Prob. 8.8, the coefficient of s^2 on the left is 1. In another situation, the highest-degree term in s might be 3, 4, and so on.

PROBLEMS

8.8 Expand $(s + 3)(s + 4)/(s + 2)^2(s + 1)$ into partial fractions. Recombine the partial fractions to prove that you have found the correct expansion.

8.9 Expand $(s^2 + 4s + 3)/(s^2 + 9s + 20)$ in partial fractions. Then recombine the partial fractions and show that the original function is obtained.

8.10 Expand $(s + 1)/(s + 2)^2(s + 3)$ in partial fractions and check your answer by recombining the partial fractions.

8.11 Expand $1/(6s^2 + 7s + 2)$ in partial fractions and check your answer by recombining the partial fractions.

Example 8.3
At the time S is closed, for reasons not explained here, the voltage across the capacitance is 180 V in Fig. 8.22.

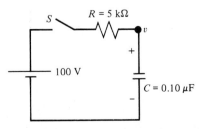

fig. 8.22 C was charged before S was closed.

a.   By Laplace transform methods, find the equation of the voltage v for time after S is closed.

b.   Also find v by classical methods of writing and solving the appropriate differential equation.

c.   Write the "solution by inspection."

Solution

a.   The s-domain model is shown in Fig. 8.23. Use the nodal-superposition method to find $V(s)$:

$$V(s) = \frac{\dfrac{100}{s5,000} + \dfrac{180s}{s \times 10^7}}{\dfrac{1}{5,000} + \dfrac{s}{10^7}}$$

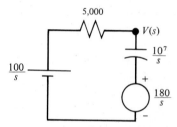

fig. 8.23   The s - domain model of the time - domain model shown in Fig. 8. 22.

Multiply numerator and denominator by $s10^7$.

$$V(s) = \frac{180\left(s + \dfrac{100 \times 10^7}{5,000 \times 180}\right)}{s\left(s + \dfrac{10^7}{5,000}\right)} = \frac{180(s + 1,111)}{s(s + 2,000)}$$

$$V(s) = \frac{180(s + 1,111)}{s(s + 2,000)} = \frac{100}{s} + \frac{80}{(s + 2,000)}$$

The inverse transform of $V(s)$ leads to the desired function of time

$$v = [100 + 80e^{-2,000t}]\, u(t)$$

b.   To find the differential equation, we note that in the time-domain model of Fig. 8.22,

$$100 = iR + v \qquad \text{and} \qquad i = C\frac{dv}{dt}$$

So

$$100 = RC\frac{dv}{dt} + v \qquad \text{or} \qquad \frac{dv}{dt} + \frac{v}{RC} = \frac{100}{RC}$$

or

$$\frac{dv}{dt} + 2{,}000v = 2 \times 10^5$$

To solve this equation, we note that the forced response is $v_p = (2 \times 10^5)/2{,}000 = 100$ V and the natural response is $v_h = Ke^{-2{,}000t}$, so

$$v = 100 + Ke^{-2{,}000t}$$

But, at $t = 0$,

$$v = 180 = K + 100$$

so

$$K = 80 \quad \text{and} \quad v = 100 + 80e^{-2{,}000t}$$

c.   By inspection, the time constant $\tau = RC = 5{,}000 \times 10^{-7} = 5 \times 10^{-4}$. The final value of v is 100 V and the initial value is 180. It follows that

$$v = 100 + 80e^{-t/\tau} = 100 + 80e^{-2{,}000t} \text{ V}$$

● ● ●

PROBLEM

8.12   Switch S has been opened for a long enough time that the circuit of Fig. 8.24 reaches steady state. Then S is closed at $t = 0$.

a.   By Laplace transform methods, find v for $t > 0$.
b.   By classical differential equations, find v.
c.   By inspection, find v.

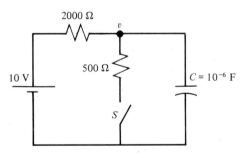

fig. 8.24   S closed at $t = 0$.

More Transform Relations

Now that we have the partial-fraction expansion, we need some more transform pairs to enable us to find inverse transforms. In particular, we need to know the inverse transform of terms like $1/(s + a)^n$. To get this, we will develop two more useful relations. First, by integration, we can show that

$$\mathscr{L}[t^n u(t)] = \int_0^\infty t^n e^{-st}\, dt = \frac{n!}{s^{n+1}} \qquad n \text{ is a positive integer} \qquad (8.57)$$

Complex Frequency Shift.   Next, we will show that multiplying by e^{-at} in the time domain corresponds to replacing s by $s + a$ in the s domain, or in other words, corresponds to a complex frequency shift. By definition,

$$\mathscr{L}[e^{-at}f(t)] = \int_0^\infty e^{-at}f(t)e^{-st}\, dt$$

Combining exponentials gives us

$$\mathscr{L}[e^{-at}f(t)] = \int_0^\infty f(t)e^{-(s+a)t}\, dt = F(s + a) \qquad (8.58)$$

where

$$F(s) = \mathscr{L}[f(t)]$$

As an example of (8.58), consider $\mathscr{L}[e^{-at}u(t)]$. We know from previous work that $\mathscr{L}[u(t)] = 1/s$. According to (8.58), we replace every s in $F(s)$ by $s + a$ to get $\mathscr{L}[e^{-at}f(t)]$. Therefore,

$$\mathscr{L}[e^{-at}u(t)] = \frac{1}{s + a}$$

which is the same as (8.14a), our previous result obtained by direct integration.

Using (8.58) in conjunction with (8.57), we can now write

$$\mathscr{L}[e^{-at}t^n u(t)] = \frac{n!}{(s + a)^{n+1}} \qquad (8.59)$$

A more convenient form of this relation is

$$\mathscr{L}^{-1}\left[\frac{1}{(s + a)^n}\right] = \frac{e^{-at}t^{n-1}}{(n - 1)!}\, u(t) \qquad (8.60)$$

This result will prove to be very useful in analysis using Laplace transforms.

Time Shift.   In order to find the response of systems to pulses like the one in Fig. 8.7, we need another result called the shifting theorem, the shift occurring in the time domain, in contrast to the shift in the s domain involved in (8.58). As we pointed out in Section 8.1, $f(t - a)$ describes the function $f(t)$ shifted to the right by a units. Now we need an additional concept, that of $u(t - a)f(t - a)$, which describes a function shifted to the right a units and cut off to the left of $t = a$; that is, because of the $u(t - a)$, the product $u(t - a)f(t - a)$ is zero for $t < a$. Figure 8.25 shows an example, with a comparison of $f(t - a)$ and $u(t - a)f(t - a)$.

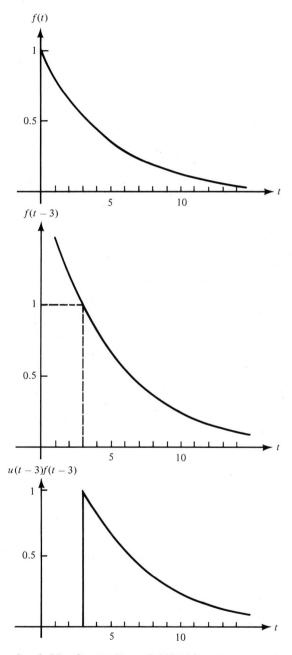

fig. 8.25   Comparison of shifted functions.

With a clear understanding of what $u(t - a)f(t - a)$ means, let's derive the Laplace transform of that quantity. Directly from the definition, we get

$$\mathscr{L}[u(t - a)f(t - a)] = \int_0^\infty u(t - a)f(t - a)e^{-st}\, dt$$

Since $u(t - a)f(t - a) = 0$ for $t < a$, we can change the lower limit on the integral and drop the $u(t - a)$; then multiply and divide the integrand by e^{-as} to get

$$\mathscr{L}[u(t - a)f(t - a)] = \int_a^\infty f(t - a)e^{-s(t - a)}e^{-as}\, dt$$

Next we define a new variable of integration to be

$$x = (t - a)$$

and get

$$\mathscr{L}[u(t - a)f(t - a)] = e^{-as}\int_0^\infty f(x)e^{-sx}\, dx = e^{-as}F(s) \tag{8.61}$$

where $F(s) = \mathscr{L}[f(t)]$.

Equation (8.61) shows that a shift in the time domain corresponds to multiplication by an exponential in the s domain, sort of the opposite of (8.58). This result can be used to find the Laplace transform of the pulse shown in Fig. 8.26. As shown in Fig. 8.26, the pulse can be considered to be the sum of two step functions, one of

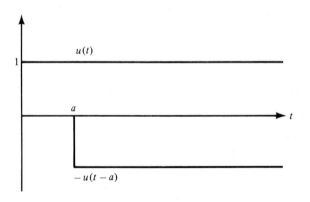

fig. 8.26   A pulse formed by the sum of two step functions.

which is shifted to the right and is negative. Using (8.61), we find the Laplace transform of the pulse to be

$$\mathscr{L}[u(t) - u(t - a)] = \frac{1}{s} - \frac{e^{-as}}{s} = \frac{1}{s}(1 - e^{-as}) \qquad (8.62)$$

Using (8.61), we can also find the Laplace transform of functions like the one shown in Fig. 8.27. Writing $f(t)$ as the sum of shifted step functions, we get

$$f(t) = u(t) - 2u(t - T) + 2u(t - 2T) - 2u(t - 3T) + \cdots \qquad (8.63)$$

Using (8.61), we find that

$$F(s) = \frac{1}{s} - \frac{2e^{-Ts}}{s} + \frac{2e^{-2Ts}}{s} - \frac{2e^{-3Ts}}{s} + \cdots$$

which can be written as

$$F(s) = \frac{2}{s}(1 - e^{-Ts} + e^{-2Ts} - e^{-3Ts} + e^{-4Ts} - \cdots) - \frac{1}{s}$$

From the relation

$$\frac{1}{1 + x} = 1 - x + x^2 - x^3 + x^4 + \cdots$$

we can write

$$F(s) = \frac{2}{s}\left(\frac{1}{1 + e^{-sT}}\right) - \frac{1}{s} = \frac{1}{s}\left(\frac{1 - e^{-sT}}{1 + e^{-sT}}\right) \qquad (8.64)$$

This result will allow us to find the response of circuits to inputs with waveforms like the one in Fig. 8.27.

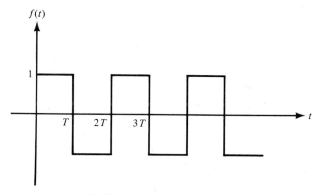

fig. 8.27   A rectangular waveform.

Complex Roots.   Another important case that we need to consider is that of complex roots in the denominator of $F(s)$. For example, we would like to find the inverse transform of a function like

$$F(s) = \frac{1}{s^2 + 4s + 13} = \frac{1}{(s + 2 + j3)(s + 2 - j3)} \tag{8.65}$$

Although the procedure is the same procedure involving partial-fraction expansion that we developed in the previous section, there are some important points that we wish to emphasize. Expanding $F(s)$ in partial fractions, we get

$$F(s) = \frac{1}{(s + 2 + j3)(s + 2 - j3)} = \frac{K_1}{s + 2 + j3} + \frac{K_2}{s + 2 - j3}$$

Since there are no repeated roots, the K's can be found from (8.48). Thus,

$$K_1 = [(s + 2 + j3)F(s)]_{s = -2 - j3} = \frac{1}{-j6}$$

$$K_2 = [(s + 2 - j3)F(s)]_{s = -2 + j3} = \frac{1}{j6}$$

Therefore,

$$F(s) = \frac{1}{-j6(s + 2 + j3)} + \frac{1}{j6(s + 2 - j3)} \tag{8.66}$$

Now (8.14) applies even when a is complex, so we can invert (8.66) using (8.14) to get

$$f(t) = \frac{u(t)e^{-(2 + j3)t}}{-j6} + \frac{u(t)e^{-(2 - j3)t}}{j6} \tag{8.67}$$

Factoring out some terms and rearranging gives us

$$f(t) = \frac{u(t)e^{-2t}}{3} \left[\frac{e^{j3t} - e^{-j3t}}{2j} \right]$$

which by Euler's relation can be written as

$$f(t) = \frac{u(t)e^{-2t}}{3} \sin 3t \tag{8.68}$$

We now have another transform pair which will be very useful. In general, if we let the real part of the complex root be α and the j-part be β, we can see by comparing (8.65) and (8.68) that

$$\mathcal{L}^{-1} \left[\frac{\beta}{(s + \alpha)^2 + \beta^2} \right] = u(t)e^{-\alpha t} \sin \beta t \tag{8.69}$$

And by a similar procedure, we can also show that

$$\mathcal{L}^{-1} \left[\frac{s + \alpha}{(s + \alpha)^2 + \beta^2} \right] = u(t)e^{-\alpha t} \cos \beta t \tag{8.70}$$

Equations (8.69) and (8.70) are useful, but an extension of these is even more useful.

From the linearity of the Laplace transform, we can write

$$\mathscr{L}^{-1}\left[\frac{A(s + \alpha) + B\beta}{(s + \alpha)^2 + \beta^2}\right] = u(t)e^{-\alpha t}(B \sin \beta t + A \cos \beta t) \qquad (8.71)$$

The sine and cosine terms of (8.71) can be added to a single sine or single cosine term by the application of some trigonometric identities. The details of this manipulation are not shown here because in Chapter 4 you learned how to add very simply any two sinusoids. It will be very easy for you to show that

$$\beta \sin \beta t + A \cos \beta t = \sqrt{A^2 + B^2} \sin \left(\beta t + \tan^{-1}\frac{A}{B}\right)$$

It follows that

$$\mathscr{L}^{-1}\left[\frac{A(s + \alpha) + B\beta}{(s + \alpha)^2 + \beta^2}\right] = u(t)e^{-\alpha t}\sqrt{A^2 + B^2} \sin (\beta t + \gamma) \qquad (8.72)$$

where $\gamma = \tan^{-1}(A/B)$.

You will remember from your trigonometry that there are two principle values of γ for each value of (A/B).

1. If A is + and B is +, then γ is in the first quadrant.
2. If A is + and B is −, then γ is in the second quadrant.
3. If A is − and B is −, then γ is in the third quadrant.
4. If A is − and B is +, then γ is in the fourth quadrant.

The time functions in (8.69), (8.70), and (8.72) are damped sinusoids because they decay with time according to $e^{-\alpha t}$. The quantity α therefore represents the damping, and it is easy to see from the $F(s)$ whether the $f(t)$ will be a damped or an undamped sinusoid ($\alpha = 0$ for undamped). In the next section, we will examine the correlation between the properties of $F(s)$ and $f(t)$ in detail.

This development has demonstrated that complex roots in the denominator of $F(s)$ lead to an $f(t)$ that is an exponentially damped sinusoid (sine alone, cosine alone, or sum of both); see (8.69), (8.70), and (8.72). Equations (8.65) to (8.72) inclusive seem to imply that in order to find $f(t)$ in these *complex-root* situations, you must (1) factor the quadratic, (2) apply partial fractions, then use the transforms of (8.69), (8.70), and (8.72). Here is a shorter route. As an example, let

$$F(s) = \frac{2s + 5}{s^2 + 4s + 20}$$

Do not factor, but arrange the denominator in the form of $(s + \alpha)^2 + \beta^2$, as in (8.72); that is,

$$F(s) = \frac{2s + 5}{s^2 + 4s + 20} = \frac{2s + 5}{(s + 2)^2 + 4^2} \qquad (8.73)$$

where $\alpha = 2$ and $\beta = 4$. Now arrange the numerator so that A and B of (8.72) are identified.

$$F(s) = \frac{2s + 5}{s^2 + 4s + 20} = \frac{2(s + 2) + 4/4}{(s + 2)^2 + 4^2}$$

where $A = 2$ and $B = 1/4$. Using the transform of (8.72), it follows that

$$f(t) = u(t)2.016e^{-2t} \sin (4t + 82.87°)$$

It is important for you to know that when $F(s)$ describes real physical systems that complex roots always occur in conjugate pairs and that the corresponding K's in the partial-fraction expansion always occur in conjugate pairs. For example, in (8.66), $K_2 = K_1^*$. This information can be used both to check results and to simplify calculations.

* Initial-Value and Final-Value Theorems

We will conclude our development of Laplace transform relations with a discussion of two more useful theorems, the initial-value theorem and the final-value theorem, proofs of which can be found in many books.[3]

The initial-value theorem states that the initial value of the function in the time domain can be found from its Laplace transform according to

$$\lim_{t \to 0^+} f(t) = \lim_{s \to \infty} [sF(s)] \tag{8.74}$$

where the limit $t \to 0^+$ is taken from the right. For example, consider $f(t) = e^{-at}$ which has a transform $F(s) = 1/(s + a)$. According to (8.74),

$$f(0^+) = \lim_{s \to \infty} \left(\frac{s}{s + a} \right) = 1$$

which we know is correct from $f(t) = e^{-at}$. This theorem is important because it tells us that the characteristics of the time-domain response for small values of t is related to the s-domain characteristics for large values of s. We shall see how this information is useful in the next section.

The final-value theorem states that

$$\lim_{t \to \infty} f(t) = \lim_{s \to 0} [sF(s)] \qquad \begin{array}{l} \text{when } F(s) \text{ has only poles} \\ \text{in the left-half plane} \end{array} \tag{8.75}$$

According to this theorem, the characteristics of $f(t)$ for very large t are related to the characteristics of $F(s)$ for very small s. The restriction that $F(s)$ has poles only in the left half-plane is necessary because if $F(s)$ has a pole in the right half-plane, $f(t)$ will contain a growing exponential and the $\lim_{t \to \infty} f(t)$ will not exist. By left half-plane, we mean values of s for which the real part of s is negative, as explained in detail in the next section.

[3] For example, C. R. Wylie, Jr., *Advanced Engineering Mathematics*, 3d ed. New York: McGraw-Hill, 1966.

As an example of application of the final-value theorem, consider again $f(t) = e^{-at}$. Then

$$\lim_{t \to \infty} f(t) = \lim_{s \to 0} \left[\frac{s}{s+a} \right] = 0$$

Note that $F(s)$ in this case has only one pole at $s = -a$, which does lie in the left half-plane because the real part of s is negative.

Summary of Transform Relations

After working our way through the development of the basic Laplace transform methods, it is time now to summarize the relations for use in applying them in analysis. Table 8.1 gives a summary of the relations that we have developed and Table 8.2 lists the transform pairs. The delta function $\delta(t)$ is explained in Section 8.6.

TABLE 8.1 Laplace Transform Relations

$$f(s) = \mathscr{L}[f(t)] = \int_0^t f(t)e^{-st}\,dt \qquad\qquad \mathscr{L}\left[\int_{-\infty}^t f(t)dt\right] = \frac{F(s)}{s} + \frac{1}{s}\int_{-\infty}^0 f(t)\,dt$$

$$f(t) = \mathscr{L}^{-1}[F(s)] \qquad\qquad \mathscr{L}[u(t)e^{-at}f(t)] = F(s+a)$$

$$f(t) = \frac{1}{2\pi j}\int_{\sigma-j\infty}^{\sigma+j\infty} F(s)e^{st}\,ds$$

$$\mathscr{L}[f_1(t) + f_2(t)] = F_1(s) + F_2(s) \qquad\qquad \mathscr{L}[u(t-a)f(t-a)] = e^{-as}F(s)$$

$$\mathscr{L}[kf(t)] = kF(s) \qquad\qquad \lim_{t \to 0^+} f(t) = \lim_{s \to \infty}[sF(s)]$$

$$\mathscr{L}\left[\frac{df}{dt}\right] = sF(s) - f(0) \qquad\qquad \lim_{t \to \infty} f(t) = \lim_{s \to 0}[sF(s)]$$

when the poles of $F(s)$ are in the left half-plane only

$$\mathscr{L}\left[\frac{d^n f}{dt^n}\right] = s^n F(s) - s^{n-1}f(0)$$

$$-s^{n-2}\left.\frac{df}{dt}\right|_{t=0} \cdots -\left.\frac{d^{n-1}f}{dt^{n-1}}\right|_{t=0}$$

TABLE 8.2   *Laplace Transform Pairs*

$F(s)$	$f(t)$
1	$\delta(t)$
$\dfrac{1}{s}$	$u(t)$
$\dfrac{1}{s+a}$	$u(t)e^{-at}$
$\dfrac{1}{s^n}$	$u(t)\dfrac{t^{n-1}}{(n-1)!}$
$\left(\dfrac{1}{s+a}\right)^n$	$u(t)\dfrac{t^{n-1}}{(n-1)!}e^{-at}$
$\dfrac{B\beta}{s^2+\beta^2}$	$u(t)B\sin\beta t$
$\dfrac{As}{s^2+\beta^2}$	$u(t)A\cos\beta t$
$\dfrac{B\beta}{(s+\alpha)^2+\beta^2}$	$u(t)Be^{-\alpha t}\sin\beta t$
$\dfrac{A(s+\alpha)}{(s+\alpha)^2+\beta^2}$	$u(t)Ae^{-\alpha t}\cos\beta t$
$\dfrac{A(s+\alpha)+B\beta}{(s+\alpha)^2+\beta^2}$	$u(t)\sqrt{A^2+B^2}e^{-\alpha t}\sin\left(\beta t+\tan^{-1}\dfrac{A}{B}\right)$

More extensive tables of Laplace transform pairs can be found in mathematical handbooks.

PROBLEMS

8.13 Using only $\mathscr{L}[u(t)t]$ and (8.58), find $\mathscr{L}[u(t)te^{-at}]$.

Answer: $1/(s+a)^2$

8.14 If $f(t)=e^{-0.2t}$ in Fig. 8.25, find the Laplace transform of the bottom curve in the figure, and then the inverse Laplace transform.

Answer: $\dfrac{e^{-3s}}{s+0.2}$

8.15 Find the Laplace transform of the $f(t)$ shown in Fig. 8.28.

Answer: $\dfrac{1}{s}[e^{-s}-2e^{-2s}+e^{-3s}]$

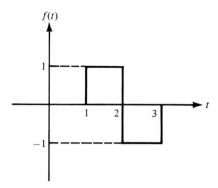

fig. 8.28 Waveform for Prob. 8.15.

8.16 Find the inverse Laplace transform of $8/(s^2 + 2s + 10)$.

Answer: $\frac{8}{3}u(t)e^{-t} \sin 3t$

8.17 Find the inverse Laplace transform of $s/(s^2 + 4s + 8)$.

Answer: $u(t)\sqrt{2}e^{-2t} \sin(2t + 135°)$

8.18 If $F(s) = 1/(s^2 + 4s + 10)$, find $f(0^+)$ without finding $\mathscr{L}^{-1}[F(s)]$.
Answer: $f(0^+) = 0$

8.19 If $F(s) = 1/(s^2 + 4s + 10)$, find $\lim_{t \to \infty} f(t)$ without finding $\mathscr{L}^{-1}[F(s)]$. Be sure and check to see whether $F(s)$ has only poles in the left half-plane.
Answer: $\lim_{t \to \infty} f(t) = 0$

*8.5 GRAPHICAL REPRESENTATIONS IN THE FREQUENCY DOMAIN

There are some important correspondences between the properties of a function $F(s)$ in the s domain and the properties of its inverse $f(t)$ in the time domain that allow us to manipulate parameters in the s domain to achieve desirable results in the time domain. In this section we introduce some of the basic ideas involved in this correspondence. You will learn more about these techniques in advanced courses in circuit theory and in courses on systems.

In the s domain, $F(s)$ is characterized by its *zeros* and *poles*, the values of s that make $F(s)$ zero and infinite, respectively. For example, if

$$F(s) = \frac{s + 3}{(s + 1)(s + 2)} \tag{8.76}$$

then $F(s)$ has a zero at $s = -3$ and poles at $s = -1$ and $s = -2$. It will now be made evident that the poles determine the nature of the time-domain response, whereas the zeros affect only the *amplitude* (and phase of sinusoids), but not the nature of the response. We can see this from (8.76). Expanding in partial fractions, we get

$$F(s) = \frac{s + 3}{(s + 1)(s + 2)} = \frac{2}{s + 1} - \frac{1}{s + 2}$$

and

$$f(t) = \mathscr{L}^{-1}[F(s)] = 2e^{-t} - e^{-2t} \tag{8.77}$$

In (8.77) the exponentials resulted from the poles at $s = -1$ and $s = -2$, whereas the zeros affected only the coefficients in front of the exponentials. To make this absolutely clear, let's change the zeros in (8.76), but not the poles and see what $f(t)$ is like. Let

$$F(s) = \frac{s + 4}{(s + 1)(s + 2)} = \frac{3}{s + 1} - \frac{2}{s + 2} \tag{8.78}$$

then

$$f(t) = 3e^{-t} - 2e^{-2t} \tag{8.79}$$

and only the coefficients in front of the exponentials are changed.

Let's consider another example:

$$F(s) = \frac{s + 3}{(s^2 + 1)(s + 2)} = \frac{As}{s^2 + 1} + \frac{B}{s^2 + 1} + \frac{1/5}{s + 2} \tag{8.80}$$

$F(s)$ was expanded as above in order to allow both sine and cosine terms as given in Table 8.2. From the coefficient of s^2, $A + \frac{1}{5} = 0$, $A = -\frac{1}{5}$. From the coefficient of s, $2A + B = 1$, $B = \frac{7}{5}$.

$$F(s) = \frac{(-1/5)s}{s^2 + 1} + \frac{7/5}{s^2 + 1} + \frac{1/5}{s + 2}$$

$$f(t) = (-\tfrac{1}{5} \cos t + \tfrac{7}{5} \sin t + \tfrac{1}{5}e^{-2t})u(t) \tag{8.81}$$

Here again, each of the poles determined the nature of a term in the time response, the pair of imaginary poles coming from $(s^2 + 1)$ producing sinusoids and the single pole at $s = -2$ producing a decaying exponential. The zero at $s = -3$ affected the magnitude of each term, but not the nature of the response; that is, there is no e^{-3t} term present.

Because of the correlation between the poles and zeros and the time response, it is customary to show the poles and zeros in the complex s plane and infer from them the nature of the time response. For example, Fig. 8.29 shows the location of the poles (crosses) and the zeros (circles) of $F(s)$ in (8.80). The plane is the complex s plane with $s = \sigma + j\omega$. We know from looking at Fig. 8.29 that $f(t)$ will have an undamped sinusoid (because of the poles at $\pm j1$) and one decaying exponential (because of the pole at $s = -2$).

Figure 8.30 shows two sets of poles corresponding to a damped sine wave. Since $\beta_1 = \beta_2$, the frequency of oscillation of $f_1(t)$ and $f_2(t)$ are equal, but since

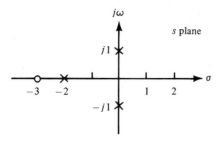

fig. 8.29   Location of the poles (crosses) and zeros (circles) of F(s) in (8.80).

$|\alpha_2| > |\alpha_1|$, the decay of $f_2(t)$ will be greater than that of $f_1(t)$. From these examples you can begin to see how the pole-zero patterns in the s plane can be used to characterize the corresponding time-domain response. Figure 8.31 shows some more characteristic pole-zero patterns and the corresponding time-domain response. One important characteristic to note is that poles in the right half-plane correspond to growing exponentials, which in physical systems often means instabilities. These instabilities occur only when feedback is applied to a system. Since the magnitude of a growing exponential approaches infinity as time approaches infinity, the growing exponential can describe a physical system only for a limited time. Usually the growing exponential is a description of the *model* of a system, which does not include the nonlinearities of the system that always cause saturation and prevent any parameters from becoming infinite. Nevertheless, poles in the right half-plane can correspond to instabilities in real systems up to the point where something happens (like a shaft breaking) to limit the response, and hence important design

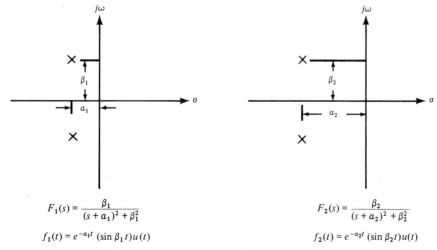

$$F_1(s) = \frac{\beta_1}{(s + a_1)^2 + \beta_1^2}$$

$$f_1(t) = e^{-a_1 t} (\sin \beta_1 t) u(t)$$

$$F_2(s) = \frac{\beta_2}{(s + a_2)^2 + \beta_2^2}$$

$$f_2(t) = e^{-a_2 t} (\sin \beta_2 t) u(t)$$

fig. 8.30   Two sets of poles in the s plane corresponding to damped oscillation in the time domain.

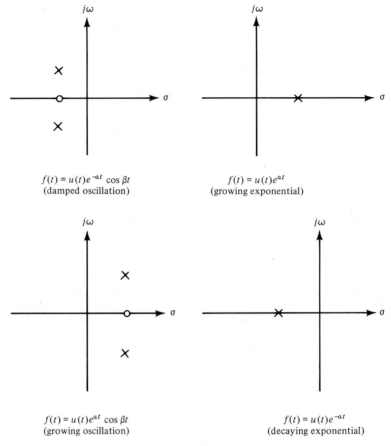

$$f(t) = u(t)e^{-at} \cos \beta t$$
(damped oscillation)

$$f(t) = u(t)e^{at}$$
(growing exponential)

$$f(t) = u(t)e^{at} \cos \beta t$$
(growing oscillation)

$$f(t) = u(t)e^{-at}$$
(decaying exponential)

fig. 8.31   More pole-zero patterns.

considerations are often based on keeping the poles in the left half-plane. In other systems, like oscillators, the poles are intentionally balanced on the imaginary axis to produce the desired effect.

Examples of Graphical Interpretation

A simple example of how time-domain response can be inferred from pole-zero patterns is the response of the circuit shown in Fig. 8.32. The voltage $V_L(s)$ is given by

$$V_L(s) = \frac{1}{s} \frac{Ls}{R + Ls} = \frac{1}{s + R/L} \tag{8.82}$$

which has one pole at $s = -R/L$, as shown in Fig. 8.33. Because there is only one pole present, and because that pole lies in the left half-plane, we know that v_L will be

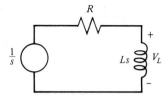

fig. 8.32 Circuit used to illustrate inference of time - domain response from pole - zero patterns.

a decaying exponential. Furthermore, if the pole is located far to the left, we know that the decay will be very fast; if the pole is located nearer the origin, we know that the decay will be slower. Consequently, since the pole is at $-R/L$, we know that increasing R moves the pole further to the left, causing the decay to be faster. On the other hand, increasing L moves the pole to the right and slows the decay. Furthermore, we can see that the pole can never move to the right half-plane for any values of R and L because changing R and L can never change the sign of R/L. This is a simple example of how pole-zero patterns can be correlated with the time-domain response, but it has the elements of procedures used with more complicated systems. In courses on feedback systems you will learn how to determine the movement of the poles as the gain of the system is changed, thus predicting stability and response characteristics. This method is called the "root locus" method, *root* meaning the pole and *locus* meaning the path taken by the pole as some parameter changes.

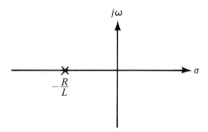

fig. 8.33 Pole - zero pattern of $V_L(s)$ of Fig. 8.32.

As a second example, let's look at the response to the circuit of Fig. 8.34. This is a typical coupling circuit in which C_2 blocks the dc component of v_i but allows the time-varying component to pass to the output v_o. The capacitance C_1 is the stray capacitance of the circuit wires to the chassis or ground, plus the unwanted but inherent capacitance across other circuit elements. In order to find the time response of the output voltage v_o, let us first write the s-domain response using the voltage divider technique.

$$V_o(s) = V_i(s) \frac{(R_1/sC_1)/(R_1 + 1/sC_1)}{(R_1/sC_1)/(R_1 + 1/sC_1) + R_2 + 1/sC_2}$$

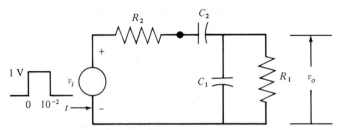

fig. 8.34 A coupling circuit. C_2 couples the time-varying signal and blocks the dc component. C_1 is shunt capacitance in the circuit.

After some algebraic reduction, we get

$$V_o(s) = \frac{sV_i(s)/R_2 C_1}{s^2 + (1/R_1 C_1 + 1/R_2 C_2 + 1/R_2 C_1)s + 1/R_1 C_1 R_2 C_2} \tag{8.83}$$

To make the example specific, let us choose the following values for the circuit of Fig. 8.34.

$$R_1 = 1 \text{ k}\Omega \qquad\qquad C_1 = 10^{-10} \text{ F}$$
$$R_2 = 20 \text{ k}\Omega \qquad\qquad C_2 = 0.5 \times 10^{-6} \text{ F}$$

You may recall that the Laplace transform of the 10-ms input pulse of unit amplitude is

$$V_i(s) = \frac{1}{s}(1 - e^{-s/100})$$

Putting these values in (8.83) results in

$$V_o(s) = \frac{0.5(1 - e^{-s/100}) \times 10^6}{(s + 95.24)[s + 10.50(10^6)]}$$

Expanding in partial fractions, we get

$$V_o(s) = 47.62(10^{-3})(1 - e^{-s/100})\left[\frac{1}{s + 95.24} - \frac{1}{s + 10.50(10^6)}\right] \tag{8.84}$$

Now we would like to correlate the pole-zero pattern of $V_o(s)$ with the time-domain response $v_o(t)$. In making a pole-zero diagram (Fig. 8.35), one strong characteristic of (8.84) is immediately obvious: The two poles are so widely separated that it is difficult to place them on one diagram. A corresponding characteristic will be present in $v_o(t)$, as we shall see.

Without taking the inverse of (8.84), we can see from this equation and/or the diagram of Fig. 8.35 that the term in v_o due to p_2 (-10.5×10^{-6}) will decay very rapidly compared to the term due to p_1 (-95.24). The two terms have equal magnitudes at $t = 0$.

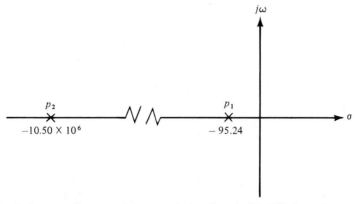

fig. 8.35   Pole - zero diagram of the term in brackets in (8.84) (not drawn to scale).

To further clarify the situation, let us take the inverse by using the shifting theorem:

$$v_o(t) = 47.62 \ u(t)[e^{-95.24t} - e^{-10.5(10^6)t}]$$

$$- \ 47.62u(t - 0.01)[e^{-95.24(t - 0.01)} - e^{-10.5(10^6)(t - 0.01)}] \ \text{mV} \quad (8.85)$$

First let us look at the nature of $v_o(t)$ for $t < 0.01$. Here the $u(t - 0.01)$ makes the second term zero. After 1 μs, the $e^{-10.5(10^6)t}$ term is less than $2.75(10^{-5})$, which is completely negligible compared to the first term, which is 0.99990 at $t = 1$ μs. Yet at $t = 0$ both $e^{-95.24t}$ and $e^{-10.5(10^6)t}$ are unity, so $v(0) = 0$. Thus the $e^{-10.5(10^6)t}$ term affects v_o for small values of t, but after 1 μs its influence is negligible and v_o for larger values of t is determined solely by the $e^{-95.24t}$ term (for $t < 0.01$ s). This behavior is just what we predicted from the pole-zero diagram, because $e^{-10.5(10^6)t}$ comes from p_2 and $e^{-95.24t}$ comes from p_1.

This example should help you see how the time-domain behavior can be inferred from the pole positions in the s plane. Figure 8.36 shows $v_o(t)$ and its two components plotted on a microsecond scale near $t = 0$, which shows the "rise time" characteristics of v_o—that is, how fast v_o rises in response to the step-function input. Figure 8.37 shows v_o plotted on a millisecond scale, where the rise time characteristics do not show up at all, but the influence of p_1 is very apparent, causing what is called "sag," as shown on the diagram. Figure 8.38 shows the characteristics on a microsecond scale near $t = 10$ ms (10,000 μs), where the input pulse has switched abruptly to zero.

There are some other important observations that we can make about the response of this circuit. By changing the circuit parameters we can move the poles around and thus change the response. By moving p_1 further to the right, we can reduce the sag. By moving p_2 further to the left, we can shorten the rise time. The next question is, what circuit parameters do we change to move the poles appropriately? The answer can be obtained by combining physical interpretation with the mathematical relations. From the circuit in Fig. 8.34, we can see that at $t = 0$, both

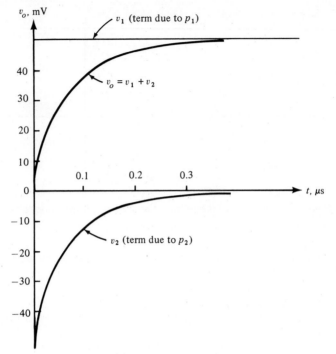

fig. 8.36 v_o for $t < 0.5 \, \mu s$.

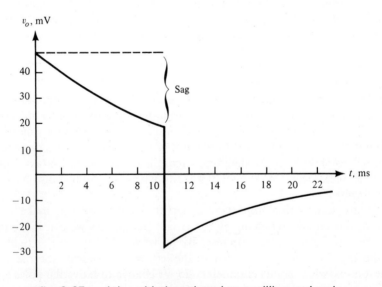

fig. 8.37 $v_o(t)$, with time plotted on a millisecond scale.

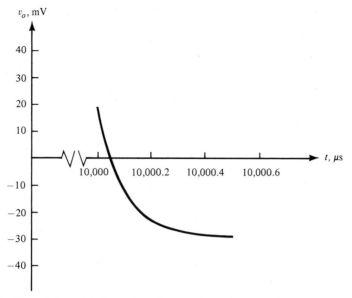

fig. 8.38 $v_o(t)$, with time plotted on a microsecond scale near $t = 10$ ms.

C_2 and C_1 look like short circuits, but since $C_2 \gg C_1$, we would expect C_2 to be nearly like a short circuit for a longer time than C_1. Therefore, for very small values of t we can approximate C_2 by a short circuit, with the result that the response for small t is determined by C_1, R_1, and R_2, with a time constant of $1/R_p C$, where $R_p = R_1 \| R_2$. Since

$$\frac{1}{R_p C_1} = 10.50(10^6)$$

we see that our reasoning appears correct, because this is just the value of p_2. Therefore we see that the shunt capacitance C_1 is the dominant factor in controlling the rise time, and we could shorten the rise time by decreasing C_1, which would move p_2 to the left. We could also change the rise time by changing R_1 and R_2, but this would also affect the sag, as we will show.

Again looking at the circuit in Fig. 8.34, we can see that for larger values of t, C_1 will probably have become charged before C_2 because C_1 is so much smaller than C_2. Hence we can ignore C_1 for larger t, in which case the response would be determined by $(R_1 + R_2)$ in series with C_2, with a time constant of

$$\frac{1}{(R_1 + R_2)C_2} = 95.24$$

which is the value of p_1. Therefore we can control the sag by adjusting C_2. By increasing C_2, we can move p_1 to the right, thus decreasing the sag.

These examples, though relatively simple, illustrate the principal concepts used in inferring time-domain response from pole-zero characteristics. You will learn more advanced techniques in other courses.

PROBLEMS

8.20 Compare the characteristics of $f_1(t)$ and $f_2(t)$ if $F_1(s)$ and $F_2(s)$ have the pole-zero diagrams shown in Fig. 8.39.

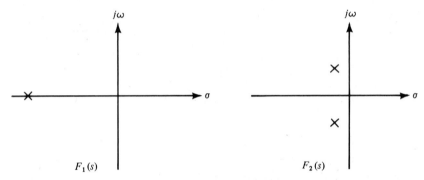

fig. 8.39   Pole - zero diagrams for $F_1(s)$ and $F_2(s)$ (see Prob. 8.19).

8.21 Make a pole-zero diagram of the $F(s)$ in (8.76) and from it explain the characteristics of $f(t)$.

8.22 Derive an expression for the $v_o(t)$ in Fig. 8.34 when $R_1 = 1$ kΩ, $R_2 = 20$ kΩ, $C_1 = 10^{-10}$ F, $C_2 = 2(10^{-6})$ F, and $V_i = (1 - e^{-s/100})/s$, plot $v_o(t)$ versus t on a millisecond scale, and show whether the sag was increased or decreased from the example given above. Make a pole-zero diagram for this case and show how the change in the pole location for these parameters correlates with the change in the sag.

8.23 Repeat Prob. 8.22 except use $C_1 = 10^{-9}$ F and $C_2 = 0.5(10^{-6})$ F and instead of finding the change in sag from the example, find how the rise time characteristics changed and correlate that change with the change in pole position.

*8.6 TRANSFER FUNCTIONS

Definition

A definition that naturally follows from the development of Laplace transform analysis is that of *transfer function*, which is defined as the ratio of an output to an input in the s domain when initial conditions are all zero. The two quantities in the ratio might be voltages or currents or mechanical quantities like torque.

An example of a transfer function is the expression for $V_o(s)$ in (8.83). From (8.83) we can write

$$\frac{V_o(s)}{V_i(s)} + \frac{s/R_2 C_1}{s^2 + (1/R_1 C_1 + 1/R_2 C_2 + 1/R_2 C_1)s + 1/R_1 C_1 R_2 C_2}$$

We call $V_o(s)/V_i(s)$ a transfer function $H(s)$, so that

$$H(s) = \frac{s/R_2 C_1}{s^2 + (1/R_1 C_1 + 1/R_2 C_2 + 1/R_2 C_1)s + 1/R_1 C_1 R_2 C_2} \qquad (8.86)$$

The transfer function characterizes a circuit or system. Thus we can represent, describe, or model the circuit in Fig. 8.34 by the transfer function, as shown in Fig. 8.40. Knowing $H(s)$, we can find where initial conditions are zero, the output for any $V_o(s)$ by multiplying the transfer function by the input:

$$V_o(s) = H(s)V_i(s) \qquad (8.87)$$

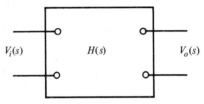

fig. 8.40   The circuit of Fig. 8.34 represented by a transfer function.

The transfer function is a convenient way to represent a system because, knowing the poles of the transfer function, we can qualitatively predict much about the response to a given input. For example, with the values for circuit components that we used in the last section, $H(s)$ becomes

$$H(s) = \frac{0.5s \times 10^6}{(s + 95.24)(s + 10.5(10^6))} \qquad (8.88)$$

which has a pole-zero diagram as shown in Fig. 8.41. The response to a step function input $1/s$ can be inferred by adding the pole of the step function to the diagram in Fig. 8.41, which cancels the zero at the origin and leaves just p_1 and p_2. Thus we know that the time-domain response is the sum of the two decaying exponentials produced by p_1 and p_2 so that $v_o = A_1 e^{-95.24t} + A_2 e^{-10.5 \times 10^6 t}$.

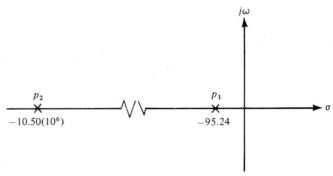

fig. 8.41   Pole - zero diagram of H(s) in (8.86) (not drawn to scale).

Impulse Response

A valuable interpretation of the transfer function is that it is the Laplace transform of the time-domain response to an impulse input. An impulse function $\delta(t - a)$ is defined by the following properties:

$$\int_b^c f(t)\delta(t - a)\, dt = f(a) \qquad \text{where } b < a < c \qquad (8.89)$$

$$\delta(t - a) = 0 \qquad \text{for } t \neq a \qquad (8.90)$$

The impulse function can be thought of in terms of

$$\delta(t) = \lim_{h \to 0} f(t) \qquad (8.91)$$

where $f(t)$ is as shown in Fig. 8.42. The function in (8.89) has a unit area with a height that approaches infinity and a width that approaches zero. This concept explains (8.89), because the product $f(t)\delta(t - a)$ is zero when $t \neq a$ since $\delta(t - a) = 0$ when $t \neq a$; in the limit as $t \to a$, the area of the product $f(t)\delta(t - a)$ becomes $f(a)$, which is equal to the integral. Although (8.91) is not a rigorous mathematical definition of the impulse function, it does have conceptual value. The rigorous mathematical derivation of the impulse function and its properties are difficult, and lie beyond the scope of this book.

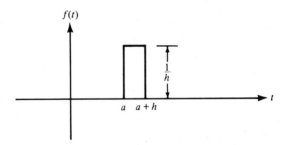

fig. 8.42   Function used to obtain a conceptualization of the impulse function.

The Laplace transform of $\delta(t - a)$ is found by applying the definition:

$$\mathscr{L}[\delta(t - a)] = \int_0^\infty \delta(t - a)e^{-st}\, dt$$

From (8.89),

$$\mathscr{L}[\delta(t - a)] = e^{-as} \qquad (8.92)$$

or when $a = 0$,

$$\mathscr{L}[\delta(t)] = 1 \qquad (8.93)$$

With 8.93, we can find the response to a unit impulse function input by letting $V_i(s) = 1$ in (8.87):

$$V_o(s) = H(s) \qquad \text{for } V_i(s) = 1 \qquad\qquad (8.94)$$

which says that $H(s)$ is the Laplace transform of the impulse response of the circuit. Transfer functions are used widely in electronic and systems analysis and design.

8.7 AN EXAMPLE OF CIRCUIT ANALYSIS

As we developed Laplace transform methods in this chapter, we used examples to illustrate the various steps, such as transforming the circuit to the s domain and finding inverse transforms. In this section, we use the methods that we have developed to analyze a moderately complex circuit so that you can see a typical procedure from the beginning to end. We hope that this will help you to solidify your own ability to analyze and design circuits using Laplace transform methods.

Figure 8.43 shows a model of a battery, a resistor, an inductor, and a capacitor. R_2 and L represent the inductor, C represents the capacitor, R_1 represents a resistor, and V_A represents the battery. In this case, the internal resistance of the battery is low enough to be neglected and the capacitor can be adequately represented by the capacitance C. The switch S has been in position 1 for a long time and is switched to position 2 at $t = 0$. We want to find an expression for the voltage v across C.

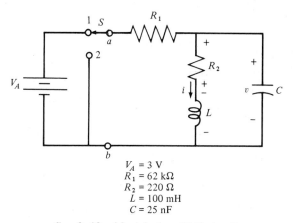

$$V_A = 3 \text{ V}$$
$$R_1 = 62 \text{ k}\Omega$$
$$R_2 = 220 \ \Omega$$
$$L = 100 \text{ mH}$$
$$C = 25 \text{ nF}$$

fig. 8.43   Model of an RLC circuit.

The first step is to transform the circuit to the s domain, which requires that we know $i(0)$, the current through L at $t = 0$ and $v(0)$, the voltage across C at $t = 0$. Both are easy to find, assuming that S has been in position 1 long enough that the natural response has died away, leaving only the forced response. Then just before the switch is thrown, C looks like an open circuit and L looks like a short circuit,

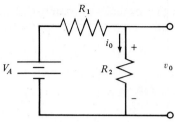

fig. 8.44   *Circuit equivalent to the one in Fig. 8.43 after the natural response has died away.*

resulting in the equivalent circuit shown in Fig. 8.44. We easily find that

$$i_0 = \frac{V_A}{R_1 + R_2}$$

$$v_0 = \frac{R_2}{R_1 + R_2} V_A$$

Since the voltage across a capacitance cannot change instantaneously and the current through an inductance cannot change instantaneously, we know that $i(0) = i_0$ and $v(0) = v_0$ and therefore

$$i(0) = \frac{V_A}{R_1 + R_2} \tag{8.95}$$

$$v(0) = \frac{R_2}{R_1 + R_2} V_A \tag{8.96}$$

Now using the representations given in Fig. 8.16, we can transform the circuit of Fig. 8.43 to the s domain, getting the representation shown in Fig. 8.45. There are many ways to obtain an expression for $V(s)$ in terms of the circuit parameters. Either loop or nodal equations are straightforward and easily applied. Let us choose nodal equations and change the initial capacitor voltage $v(0)/s$ to an equivalent

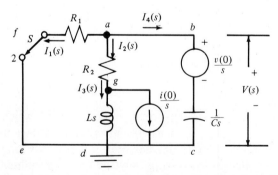

fig. 8.45   *Representation in the s domain of the circuit in Fig. 8.43.*

current source as shown in Fig. 8.46. The y-matrix $[\mathbf{Y}]$ and the I-matrix $[\mathbf{I}]$ may be written by inspection where the two unknown nodal voltages are at a and g.

$$[\mathbf{I}] = \begin{bmatrix} v(0)C \\ \\ -\dfrac{i(0)}{s} \end{bmatrix} \qquad [\mathbf{Y}] = \begin{bmatrix} sC + \dfrac{1}{R_2} + \dfrac{1}{R_1} & -\dfrac{1}{R_2} \\ \\ -\dfrac{1}{R_2} & \dfrac{1}{R_2} + \dfrac{1}{sL} \end{bmatrix}$$

Solving for $V(s)$, the node voltage V_a,

$$V(s) = \frac{\begin{vmatrix} v(0)C & -\dfrac{1}{R_2} \\ \\ -\dfrac{i(0)}{s} & \dfrac{1}{R_2} + \dfrac{1}{sL} \end{vmatrix}}{|\mathbf{Y}|} \tag{8.97}$$

which yields

$$V(s) = \frac{v(0)C\left(\dfrac{1}{R_2} + \dfrac{1}{sL}\right) - \dfrac{i(0)}{R_2 s}}{\left(sC + \dfrac{1}{R_2} + \dfrac{1}{R_1}\right)\left(\dfrac{1}{R_2} + \dfrac{1}{sL}\right) - \dfrac{1}{R_2^2}} \tag{8.98}$$

Multiplying the numerator and denominator of (8.98) by sR_2/C and performing the indicated multiplication in the denominator yields

$$V(s) = \frac{v(0)\left(s + \dfrac{R_2}{L}\right) - \dfrac{i(0)}{C}}{s^2 + \left(\dfrac{R_2}{L} + \dfrac{1}{R_1 C}\right)s + \dfrac{1}{LC}\left(1 + \dfrac{R_2}{R_1}\right)} \tag{8.99}$$

Since (8.99) is moderately complex and required a number of algebraic steps to derive, we would be foolish to proceed further without making some checks for

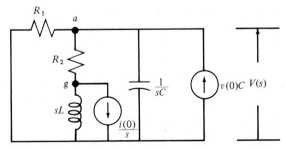

fig. 8.46   The circuit of Fig. 8.45 with the voltage source changed to a current source.

consistency. First, let's make a simple check for correctness of dimensions. Since s has the dimensions of frequency, the reciprocal of time, all denominator terms must have the dimensions of frequency squared, or $1/t^2$. Also, all numerator terms must have the dimensions of volts times frequency. You should verify for yourself that this is the case.

Although our dimensional consistency check gives us some confidence in (8.99), it does not guarantee that it is correct. We can gain more confidence in its validity by making more consistency checks. For example, from Fig. 8.45, we can see that if $C \to \infty$, $1/Cs \to 0$ and the capacitance approaches a short circuit because the impedance $1/Cs$ approaches zero. In that case, $V(s)$ must approach $v(0)/s$. The expression for $V(s)$ must be consistent with this. Let's check it. From (8.99) we see that

$$\lim_{C \to \infty} V(s) = \frac{v(0)}{s}$$

and it is consistent. A similar check can be made by letting $R_1 \to 0$. From the diagram, we see that $V(s)$ must be zero when R_1 is shorted. By multiplying numerator and denominator in (8.99) by R_1 and letting $R_1 \to 0$, we see that $V(s) \to 0$ as $R_1 \to 0$, which is consistent. By now, we can have a lot of confidence in the validity of (8.99) because it is consistent with a number of special cases that we know must be true. You can probably think of some more checks to make that will increase confidence even more, but let's go on.

At this point we must decide whether to put the values for the components given in Fig. 8.43 into (8.99), or leave the expression in general terms. If we put in numbers now, we lose information about how v is affected by the various components. On the other hand, putting in the numbers now would result in a simpler expression. If our problem were a design problem, such as choosing a given set of parameters to produce a specified v, we would not put in numbers at this point. However, since we merely want to find the v produced by the specified components, let's insert the numbers. We get

$$V(s) = \frac{0.01061(s + 2,200) - 1,928.64}{s^2 + 2,845s + 401.4 \times 10^6} \tag{8.100}$$

Now we rearrange terms in (8.100) to correspond to the transform of (8.72):

$$V(s) = \frac{0.01061(s + 1,423) - 0.09536 \times 1.998 \times 10^4}{(s + 1,423)^2 + (1.998 \times 10^4)^2} \tag{8.101}$$

that is, $\alpha = 1,423$, $\beta = 1.998 \times 10^4$, $A = 0.01061$, and $B = -0.09536$. Hence

$$v(t) = u(t)0.0959e^{-1,423t} \sin (1.998 \times 10^4 t + 179.36°) \tag{8.102}$$

Although (8.102) is the desired expression for the voltage across the capacitor, we should not stop here. A practicing engineer would have obtained (8.102) for some purpose such as designing a device, or explaining the behavior of a device, or troubleshooting. Most purposes for which (8.102) would be derived would also require some physical interpretation and qualitative understanding of the nature of the voltage. In addition, it is always important to make consistency checks on expressions like (8.102). Consequently, we shall proceed with some consistency checks and some physical interpretation before ending the discussion.

One of the first consistency checks to be made is to see whether (8.102) reduces to the number given by (8.96) at $t = 0$. A quick calculation shows that we get $v(0) = 10.6$ mV from (8.96) and $v(0) = 10.7$ mV from (8.102), a good check. Another check is to note from Fig. 8.43 that v must approach zero as t approaches infinity because the only forcing functions are the initial current in L and the initial charge in C. As current flows in the circuit, all the energy will be dissipated in the resistance, requiring that the currents and voltages must finally approach zero. Equation (8.102) is consistent with this reasoning because $v \to 0$ as $t \to \infty$. In some cases, where v is expressed in terms of the components (such as R_1, R_2, L, and C), we could make further checks by letting some of the components approach zero and infinity, as we did in connection with (8.99).

The graph of v shown in Fig. 8.47 is a great help in obtaining qualitative understanding and physical interpretation. The waveform is called a decaying sinusoid because it has the form of a sinusoid with an amplitude that decays with time. In (8.102), the exponential causes the amplitude of the sine term to decay with time. The qualitative nature of v is evident from the graph; beginning with $v(0)$, v oscillates and decays with time until it eventually reaches zero.

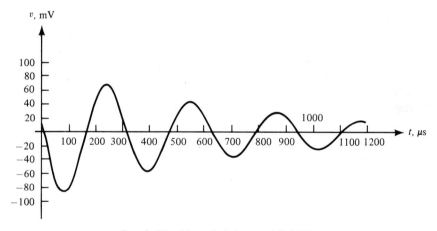

fig. 8.47   Plot of v(t) ; see (8.102) .

The nature of v can be related to the physical device in various ways—or, said in another way, various physical interpretations of v are possible. The ability to make physical interpretations is enhanced by experience, and you will probably find it difficult at first, but you should try to develop this ability because it is very important to practicing engineers. Let's restrict ourselves to a very simple interpretation here. When the switch S is moved to position 2, the inductance opposes any change in the current through it, in effect trying to keep the current flowing. Thus, at

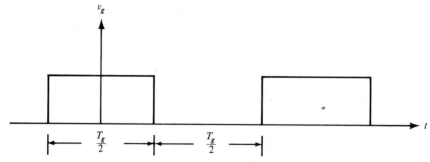

fig. 8.48   *Generator voltage used to simulate the circuit of Fig. 8.43.*

$t = 0$, $i(0)$ is like a forcing function. The charge in the capacitance also acts as a forcing function, and the voltage across the capacitance tends toward what it would be if C were an open circuit, as explained in Sections 3.3 and 3.6. In this case, that voltage would be zero, since there is no independent voltage source in the circuit. The oscillatory nature of v is caused by the interaction between L and C, as the capacitance charges and discharges, with the inductance opposing the change in current. Hence v begins at $v(0)$, oscillates, and decays toward zero. Oscillation will occur only for certain combinations of R, L, and C in a circuit, and it cannot occur unless both inductance and capacitance are present.

In this circuit, an expression showing the combinations of R, L, and C that will produce oscillation can be found from the roots of the quadratic in (8.99), as we have previously learned. When oscillation occurs, the response is said to be *under-damped* (complex conjugate roots). When no oscillation occurs, the response is *overdamped* (two distinct real roots), or *critically damped* (two equal real roots), as we have previously discussed.

We will conclude this example by considering how v can be displayed on an oscilloscope in a laboratory measurement. Closing the switch manually in the circuit of Fig. 8.43 would produce a v that would die away in a few milliseconds, allowing only a brief glimpse if displayed on an ordinary oscilloscope. However, an adequate approximation to v could be displayed on an ordinary oscilloscope by replacing V_A and S by a voltage source that produces an appropriate rectangular waveform, as shown in Fig. 8.48. If $T_g/2$ is long compared to the time it takes v to decay to a negligible value, then the response of the circuit to the voltage in Fig. 8.48 during one period will be the same as the response of the circuit in Fig. 8.43. Thus using a generator with the waveform of Fig. 8.48 would produce v repetitively, thereby allowing display on an ordinary oscilloscope.

PROBLEMS

8.24   Find a time-domain expression for i_3 in Fig. 8.49 when

$$I_0 = 1 \text{ A} \qquad C = 100 \ \mu\text{F} \qquad R_2 = 75 \ \Omega$$

$$R_1 = 47 \text{ k}\Omega \qquad L = 100 \text{ mH} \qquad i_2(0) = 0$$

$$v(0) = 0$$

Graph i_3 and discuss the characteristics of the waveform.

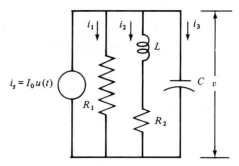

fig. 8.49   Circuit for Prob. 8.24.

8.25   Repeat Prob. 8.24 for

$$I_0 = 10 \text{ mA} \qquad C = 1 \text{ } \mu\text{F} \qquad R_2 = 2 \text{ k}\Omega$$

$$R_1 = 60 \text{ k}\Omega \qquad L = 100 \text{ mH} \qquad i_2(0) = 5 \text{ mA}$$

$$v(0) = 10 \text{ V}$$

8.26   This problem, though it involves a rather sophisticated circuit design, is intended to provide you with some fun in producing interesting patterns on an oscilloscope. We call the circuit a "spiralgraph." If you are like our students, after you get the basic circuit to work you will soon find some simple variations that will produce even more interesting displays.

   The circuit diagram of the spiralgraph is shown in Fig. 8.50. v_i is a rectangular wave having a period long compared to the important time constants of the circuit.

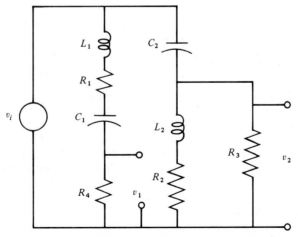

fig. 8.50   Diagram of the spiralgraph. v_i is a rectangular wave.

The voltage v_1 is connected to the vertical input of an oscilloscope and v_2 is connected to the horizontal input of the same oscilloscope. The design problem is to select circuit components so that v_1 and v_2 will be underdamped with the same frequency of oscillation and the same damping rate, but 90° out of phase. If the amplitude of the two voltages applied to the deflection plates of the cathode-ray tube are made equal by adjusting the gain of the vertical input amplifier of the oscilloscope, the resulting pattern on the oscilloscope will be a plot of the desired spiral.

a. L_1 and R_1 is a model of one inductor and L_2 and R_2 a model of a second inductor. Procure suitable inductors and measure L_1, R_1 and L_2, R_2. Inductances in the 70–100-mH range will be suitable, but you can make other values work too. You will see the design trade-offs later.

b. Transform the circuit of Fig. 8.50 to the s domain and write expressions for $V_1(s)$ and $V_2(s)$. Take the inverse transforms to get expressions of the form

$$v_1(t) = b_1 e^{-\alpha_1 t} \sin \omega_1 t$$

$$v_2(t) = b_2 e^{-\alpha_2 t} \sin (\omega_2 t - \phi)$$

where b_1, b_2, α_1, α_2, ω_1, ω_2, and ϕ are all constants related to the circuit parameters. The goal is to choose the circuit parameters so that

$$\alpha_1 = \alpha_2 \tag{1}$$

$$\omega_1 = \omega_2 \tag{2}$$

$$\phi = \frac{\pi}{2} \tag{3}$$

$$\frac{1}{\alpha} \geq 2T \tag{4}$$

where $\omega = 2\pi/T$ and T is the period of oscillation. The last requirement is to ensure that the wave does not damp out too fast.

The design procedure is complicated because you must satisfy four conditions simultaneously. There is no standard way to do this, so you must use ingenuity and perserverance. This is a very challenging problem, but don't give up because it will give you valuable experience in engineering design.

Show your work carefully and be sure that your component values satisfy (1) – (4). With the component values that you selected, plot $v_1(t)$ and $v_2(t)$.

c. Construct a breadboard circuit and display v_1 and v_2 on the two traces of a dual-trace oscilloscope. Adjust the gain of the horizontal and vertical amplifiers appropriately and connect v_2 to the horizontal input and v_1 to the vertical input of an oscilloscope. Make a sketch of the resulting display.

d. Explain why the display has the form that it does, comparing expected results with observed results.

8.27 Devise a laboratory method for measuring both the resistance and inductance of an inductor whose resistance is in the range of 10 to 50 Ω and whose inductance is in the range of 0.05 to 0.15 H. You must (1) drive your test circuit with a square wave of voltage, (2) use a capacitor in series with the coil such that the resultant circuit is oscillatory, and (3) adjust the frequency of the square wave such that its period is about 10 times the time constant of your circuit.

Use the Laplace transform as the tool for analysis, and show the following:

a. Circuit diagram of connections that includes the circuit, the oscilloscope for measurement, and the square-wave generator.

b. Expected scope image.

c. Formulas for calculating R and L of the coil in terms of values measured from scope image and the value of the capacitance.

8.28 A coil having a resistance of 50 Ω and an inductance of 0.050 H is connected in series with a capacitor having a capacitance of 10^{-7} F and negligible resistance. A certain increase of the temperature of the coil winding increases the coil resistance to 57 Ω. Does this change of coil resistance change the natural frequency of oscillation? If so, by what percent?

8.29 The coil and capacitor of Prob. 8.28 are connected in the circuit of Fig. 8.51. The switch S has been closed for a long time, then opened at $t = 0$.

    a. Working in the time domain only (Laplace and operator techniques are not allowed), find the differential equation where v is the dependent variable. Of course, no other variable current or voltage is allowed.

    b. Find the same differential equation using operator techniques. Of course, no initial conditions will appear in the differential equation.

    c. If we had to solve this differential equation, then we would have to know the initial values: $v(0)$ and $dv/dt\,|_{t=0}$. Find these two.

    d. Take the Laplace transform of this differential equation and solve for $V(s)$. Of course, this equation will contain $V(0)$ and $dv/dt\,|_{t=0}$.

    e. Show an s-domain model of the time-domain model of Fig. 8.51. Of course, the s-domain model shows the initial values of $v(0)$ and $i_L(0)$. From this model, solve for $V(s)$ and show that it agrees with that of part (d).

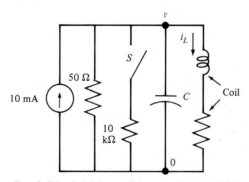

fig. 8.51   Initial conditions for Prob. 8.28.

8.30 Where the forcing function is a sinusoid, as shown in Fig. 8.52, we are required to find v_o for $t > 0$. The switch S is closed at $t = 0$, and v_o will include a transient (natural) component as well as steady-state (forced) components. We are required to find v_o by both "solution by inspection" and also by Laplace transform. The solution to this

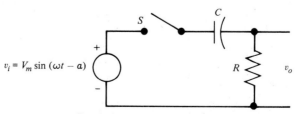

fig. 8.52   S closed at $t = 0$.

problem should demonstrate rather clearly that the "payoff" for Laplace transform is not favorable for many simple situations.

a. Find v_o (total time-domain value) using "solution by inspection." Note that we can readily write $\mathbf{V}_o$, the phasor equivalent of the steady-state part of v_o. Then we can immediately write the total solution after evaluating $v_o(0)$.

b. Find v_o by Laplace transform techniques. Reduce the steady-state part to a single sine term to match the steady-state part of the solution in part (a). In the table of transforms, you will find the transform of a sine term and of a cosine term.

Chapter 9

Coupled Circuits

Introduction Two coils wound on a cylindrical form or otherwise adjacent to each other constitute a magnetically coupled circuit (see Fig. 9.1). The current in one coil alone (Fig. 9.2) produces magnetic flux in or around that coil, and this magnetic flux leads to the self-inductance of the coil, or the time rate of change of this flux induces a voltage in the coil. When the current in one coil, such as coil #1 having N_1 turns (see Fig. 9.1) produces a magnetic flux in the other coil, #2, having N_2 turns, then we say the two coils are magnetically coupled.

Magnetically coupled circuits are used in a variety of special situations. In one situation a particular impedance $\mathbf{Z}_2$ might be connected to the v_2, i_2 terminals of the coupled circuit of Fig. 9.1. Then the apparent impedance appearing at the i_1, v_1 terminals can be adjusted to some value widely different from $\mathbf{Z}_2$ by the selection of coil turns N_1, coil turns N_2, and other coupled-circuit parameters. This procedure is called *impedance matching* and is widely used in communications systems. In another situation, v_2 can be made to be very much larger or very much smaller than v_1. This is called *voltage transformation*. The 60-Hz power delivered to the typical residence is at two voltages: 110 volts and 220

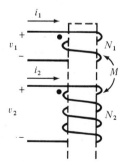

fig. 9.1   Coupled circuit.

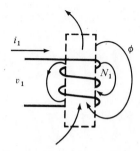

fig. 9.2 A coil having inductance L_1.

volts. The electric power delivered to your city might be at a voltage of 350,000 volts. So power voltages are transformed up and down by coupled circuits called *transformers.*

By using the magnetically coupled circuit, a signal can be transferred from one circuit to another with no conducting path (no wire) between the two. This isolation is advantageous in some situations.

Interesting and useful resonance features can be achieved by using a magnetically coupled circuit.

9.1   TWO ADJACENT COILS FORM A COUPLED CIRCUIT

Before we get involved in the details of coupled coils, let us review some ideas relative to one coil alone. In Fig. 9.2,

$$v_1 = i_1 R + \frac{d(N\phi)}{dt} = i_1 R + \frac{d\lambda}{dt} \tag{9.1}$$

where φ is magnetic flux in webers, N is turns, $\lambda = N\phi$, and R is the resistance of the coil. Here λ is called the flux linkages and is equal to the summation of all of the possible $N\phi$ products. If all of the flux were constrained to the inside of the cylindrical form on which the coil is wound, then λ would be equal to the product of that flux and the total turns. But of course some of the magnetic flux will link only part of the total turns.

You will remember that the induced voltage in the coil is given by e:

$$e = \frac{d\lambda}{dt} = L \frac{di}{dt} \tag{9.2}$$

Notice that the induced voltage can be written in terms of either $d(N\phi)/dt$, $d\lambda/dt$, or $L \, di/dt$. This leads immediately to the circuit parameter L, as follows:

$$L = \frac{d\lambda}{dt} \bigg/ \frac{di}{dt} = \frac{d\lambda}{di} \tag{9.3}$$

where L is inductance in henrys.

Now, in many situations (air path for ϕ), λ is proportional to i and it follows that

$$L = \frac{\lambda}{I} \tag{9.4}$$

Equation (9.4) gives a technique for calculating inductance L. But for a coil wound on air (or nonmagnetic form), $N\phi$ or λ is difficult to calculate with accuracy.

Let us consider the situation of Fig. 9.3, where each air gap length (g, $2g$, $5g$) is small enough compared to other pertinent dimensions that we can assume that the

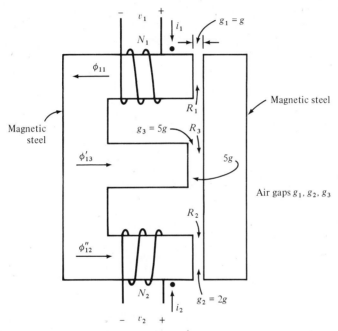

fig. 9.3   Coupled circuit.

magnetic flux does not "fringe." To further simplify our analysis, assume that the only significant magnetic reluctances are in the respective air gaps. These approximations may be reasonable for some situations and lead to a simple illustration of the concepts of self-inductance and mutual inductance. Let us proceed to find the inductance L_1 of winding N_1. Here we ignore the presence of N_2. We will assume a current I_1 in N_1 and proceed to find λ_{11}, the flux linkages in N_1 due to I_1. This system consists of magnetic steel members assembled with the three gaps: g, $2g$, and $5g$. Then the flux ϕ_{11} separates into two paths so that

$$\phi_{11} = \phi_{12} + \phi_{13} \tag{9.5}$$

If the iron reluctance is sufficiently small and the air gaps sufficiently long, all of the significant magnetic reluctances are in the three air gaps. Then where $\mathscr{R}_1$, $\mathscr{R}_2$, and

$\mathcal{R}_3$ are the respective air gap reluctances,

$$\phi_{11} = \frac{N_1 I_1}{\mathcal{R}_1 + \dfrac{\mathcal{R}_2 \mathcal{R}_3}{\mathcal{R}_2 + \mathcal{R}_3}} \tag{9.6}$$

because reluctance $\mathcal{R}_1$ is in series with the parallel combination of $\mathcal{R}_2$ and $\mathcal{R}_3$. Each of these reluctances can be written

$$\mathcal{R}_1 = \frac{g}{A\mu} \qquad \mathcal{R}_2 = \frac{5g}{A\mu} \qquad \mathcal{R}_3 = \frac{2g}{A\mu} \tag{9.7}$$

where g, $2g$, and $5g$ are the respective air gap lengths, A is the cross-sectional area of each gap, and μ is the permeability of free space ($4\pi \times 10^{-7}$) all in MKS units: g in meters and A in square meters. It follows that

$$\phi_{11} = \frac{N_1 I_1}{\dfrac{17g}{7A\mu}} = \frac{7N_1 I_1 A\mu}{17g} \tag{9.8}$$

Further,

$$L_1 = \frac{N_1 \phi_{11}}{I_1} = \frac{\lambda_{11}}{I_1} = \frac{7N_1^2 A\mu}{17g} \ \text{H} \tag{9.9}$$

It is important to notice that inductance is proportional to the square of the turns.

PROBLEMS

9.1   Calculate the numerical value of L_1 in Fig. 9.3 if $N_1 = 1,000$ turns, $A = 6 \times 6 \ \text{cm}^2$, and $g = 1$ mm.

9.2   Develop the formula for the self-inductance of N_2 of Fig. 12.3, and then the numerical value of L_2 if $N_2 = 2,000$ turns.

9.2   TWO CURRENTS IN COUPLED CIRCUIT

If both currents i_1 and i_2 are present in the coils of any magnetically coupled arrangement as in Fig. 9.1, then it follows that

$$v_1 = R_1 i_1 + \frac{d\lambda_{11}}{dt} + \frac{d\lambda_{12}}{dt} \tag{9.10}$$

where R_1 is the resistance of the coil winding, λ_{11} are the flux linkages in N_1 due to i_1, and λ_{12} are the flux linkages in N_1 due to i_2. We have previously shown how we can express the induced voltage in e_1, due to i_1 alone either as $d\lambda_{11}/dt$ or $L_1 \ di_1/dt$. But also we can express the voltage in N_1 due to i_2 alone as either $d\lambda_{12}/dt$ or as $M_{12} \ di_2/dt$, where M_{12} is the mutual inductance from coil N_2 to coil N_1. Therefore, (9.10) may be written in the alternate form,

$$v_1 = R_1 i_1 + L_1 \frac{di_1}{dt} + M_{12} \frac{di_2}{dt} \tag{9.11}$$

The induced voltages e_{11} and e_{12} are

$$e_{11} = \frac{d\lambda_{11}}{dt} = L_1 \frac{di_1}{dt} \qquad e_{12} = \frac{d\lambda_{12}}{dt} = M_{12} \frac{di_2}{dt} \qquad (9.12)$$

It follows that the mutual inductance M_{12} is given by

$$M_{12} = \frac{d\lambda_{12}}{di_2} \qquad (9.13)$$

where λ_{12} are the flux linkages in N_1 due to i_2. Again, where ϕ_{12} and i_2 are proportional,

$$M_{12} = \frac{\lambda_{12}}{I_2} \qquad (9.14)$$

This equation tells us that we can find the mutual inductance M_{12} by first assuming a current i_2 in N_2, then from this $N_2 i_2$, find the flux linkages $N_1 \phi_{12} = \lambda_{12}$ due to i_2 alone. Let us apply this concept to the situation of Fig. 9.3.

PROBLEM

9.3 For the coupled circuit of Fig. 9.3, find M_{12} in terms of N_1, N_2, g, A, and μ.

Answer: $M_{12} = 5N_1 N_2 A\mu/17g$ H

The other mutual inductance M_{21} is

$$M_{21} = \frac{\lambda_{21}}{i_1} \qquad (9.15)$$

where λ_{21} are the linkages in N_2 due to i_1 alone.

PROBLEM

9.4 For the coupled circuit of Fig. 9.3, find M_{21}. It is not obvious at this point, but $M_{21} = M_{12}$. This will serve as a validation on your solutions to subsequent problems.

9.3 POLARITY MARKS

Notice the two dots associated with Fig. 9.1. These are called polarity marks. Where the manner in which the two windings are wound in respect to each other is not obvious, these marks are needed. Polarity marks have the meaning of when the current i_1 and i_2 are both assigned directions into the respective polarity marks (or dots), the two currents will produce magnetic flux in the same direction in either coil.

PROBLEM

9.5 In the coupled circuit of Fig. 9.3, one polarity mark is shown. See the right-hand upper terminal. Establish whether the corresponding polarity mark on N_2 should be on the lower-right terminal as shown or not.

9.4 VOLTAGE EQUATIONS

Let us refer to (9.11). It is customary to think of M_{12} (or M_{21}) as a positive parameter, just as we think of L_1 and R_1 as positive parameters. Any one or more of the terms in this equation might have a negative sign, depending on directions assigned to voltages and currents. With v_1 and i_1 assigned the directions indicated in Fig. 9.1, it is obvious that the sign preceding $R_1 i_1$ and $L_1\, di_1/dt$ must each have a positive sign. On the other hand, the sign preceding the mutual inductance term ($M_{12}\, di_2/dt$) might have either a positive or negative sign preceding it, depending on the direction assigned to i_2. The correct sign can be determined easily by considering the polarity marks and the directions the currents are assigned. So to remind us that the mutually induced voltage might be preceded by a positive or negative sign, the two voltage equations for the coupled circuit are written as follows:

$$v_1 = i_1 R_1 + L_1 \frac{di_1}{dt} \pm M_{12} \frac{di_2}{dt} \tag{9.16}$$

$$v_2 = i_2 R_2 + L_2 \frac{di_2}{dt} \pm M_{21} \frac{di_1}{dt} \tag{9.17}$$

where R_1 and R_2 are the resistances of coil 1 and coil 2, respectively.

PROBLEM

9.6 In Fig. 9.4, (a) and (b) are identical coupled circuits. Note polarity marks. The coupled circuit of (c) is different from the other two. See polarity marks. Write the equivalent of (9.16) for each of these situations. You must eatablish the correct sign preceding each of the three terms in this equation. Carefully note the assigned directions of currents and polarities of voltages in all three cases.

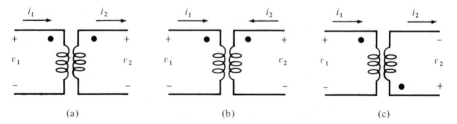

|(a)|(b)|(c)|

fig. 9.4   Polarity and sign of M.

9.5 TWO MUTUAL INDUCTANCES ($M_{12} = M_{21}$)

In calculating the two mutual inductances (M_{12} and M_{21}) of the arrangement of Fig. 9.3, you found that these two mutual inductances were equal to each other in this special case. But these two mutual inductances are equal in general. We will prove this in these next few paragraphs.

A model for the coupled circuit is given in Fig. 9.5. The proof that $M_{12} = M_{21} = M$ will be based on energy considerations. In a particular coupled-

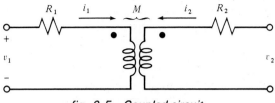

fig. 9.5   Coupled circuit.

circuit configuration, the energy stored in the magnetic field is fixed exclusively by the magnitude and direction of the two currents because these two currents will determine the magnetic density or magnetic intensity at all points in space. Obviously, the magnetic stored energy will be zero when both currents are zero. We will calculate stored energy by changing the currents from zero to some fixed value in a definite pattern. While doing this, we will integrate the power that flows into the magnetic field. The voltage applied to winding 1 is

$$v_1 = i_1 R_1 + L_1 \frac{di_1}{dt} + M \frac{di_2}{dt} \tag{9.18}$$

Part of this applied voltage v_1 is consumed in iR drop. The other part is the total induced voltage e_1 in that coil. Thus

$$v_1 = i_1 R_1 + e_1 \tag{9.19}$$

The induced voltage is, of course,

$$e_1 = L_1 \frac{di_1}{dt} + M_{12} \frac{di_2}{dt} \tag{9.20}$$

Now assume a situation where we do not allow current to flow in coil 2, but we apply a voltage to coil 1. We will calculate the power p_1 that flows into the magnetic field. This is the total power into the coil less the power into the resistance of the coil, and is given by

$$p_1 = e_1 i_1 \tag{9.21}$$

Now we can calculate the energy in the magnetic field by integrating

$$w_1 = \int_0^{t_1} p_1 \, dt = \int_0^{t_1} e_1 i_1 \, dt = \int_0^{t_1} L \frac{di_1}{dt} i_1 \, dt = L_1 \int_0^{I_1} i_1 \, di_1 = L_1 \frac{I_1^2}{2} \tag{9.22}$$

where the current in N_1 is I_1 at t_1.

Equation (9.22) gives the stored energy in the magnetic field by virtue of changing i_1 from 0 to I_1 and under the conditions that i_2 was kept at 0. Now to continue with our hypothetical experiment, assume that we hold i_1 constant at this value of I_1 while we apply a voltage at coil 2 in order to increase i_2 from 0 to I_2. This might be a difficult experiment to perform in the laboratory and would require that we apply a special voltage at coil 1 in order to hold i_1 constant at this value of I_1 because there will be an induced voltage in coil 1 as we change i_2. Now in the process of increasing i_2 from 0 to I_2, there will be power flow into both coil 1 and coil 2. The power flow into coil 1 is a consequence of the mutually induced voltage in this coil times the steady current I_1. In coil 2, both i_2 and v_2 are varying during this period, yielding a power flow into this coil. This situation is demonstrated as follows:

$$w_2 = \int_0^{t_1} I_1 v_{12} \, dt + \int_0^{t_1} i_2 v_2 \, dt \qquad (9.23)$$

The first integral in this equation yields the energy that flows into coil 1; the second integral is the energy into coil 2. In this equation, v_{12} is the induced voltage in coil 1 due to the time rate of change of the current in coil 2. There is no induced voltage in coil 2 due to i_1 because this current is not varying with respect to time. So we are able to write

$$w_2 = I_1 \int_0^{I_2} M_{12} \frac{di_2}{dt} \, dt + \int_0^{I_2} i_2 L_2 \frac{di_2}{dt} \, dt \qquad (9.24)$$

It follows that

$$w_2 = I_1 M_{12} \int_0^{I_2} di_2 + L_2 \int_0^{I_2} i_2 \, di_2 \qquad (9.25)$$

When the integrations are performed, we get

$$w_2 = I_1 I_2 M_{12} + \frac{L_2 I_2^2}{2} \qquad (9.26)$$

Equation (9.26) gives the energy stored in the magnetic field as we changed i_2 from 0 to I_2, while keeping i_1 constant at I_1. Then combining (9.22) and (9.26) to get the total energy in the magnetic field, we have

$$w_{\text{total}} = w_1 + w_2 = \frac{L_1 I_1^2}{2} + I_1 I_2 M_{12} + \frac{L_2 I_2^2}{2} \qquad (9.27)$$

Now it is important to remember that the total energy stored in the magnetic field is fixed exclusively by the parameters L_1, L_2, M_{12}, and the two currents I_1 and I_2, independent of how these two currents were changed from 0 to their values of I_1 and I_2 because these two currents in a particular configuration determine the magnetic field intensity at all points in the region of the coils.

Before taking the next step in our proof that $M_{12} = M_{21}$, it is important that you refer back to (9.24) and note the manner in which M_{12} became involved. It came into the picture here in obtaining the induced voltage in coil 1 due to a time-rated change of current in coil 2.

Now we should reverse this experiment by first holding I_1 at 0 and applying a voltage v_2 to change i_2 from 0 to I_2, then in the second step, hold i_2 constant at its value of I_2 while we forced i_1 to increase from 0 to I_1. Without developing the detail similar to (9.22) to (9.27), it is rather obvious that the total energy for this situation would be

$$w'_{\text{total}} = \frac{L_1 I_1^2}{2} + I_1 I_2 M_{21} + \frac{L_2 I_2^2}{2} \tag{9.28}$$

Obviously the total energy of (9.27) and (9.28) have to be equal because the two currents I_1 and I_2 are identically equal in the two situations. Therefore it follows that the two mutual inductances are equal:

$$M_{12} = M_{21} \tag{9.29}$$

Because these two mutual inductances are equal, there is no point in the double subscript, and we speak of a coupled circuit having one mutual inductance (no subscripts).

Equations (9.27) and (9.28) are correct but can be incorrectly applied. It is important to note the manner in which the development leading to these equations began. See (9.18). In this equation, note the plus sign in front of the mutual term. From our previous considerations of polarities on the coupled circuit, this sign was determined by the fact that both i_1 and i_2 were assigned directions into the "dots." Or, in other words, the two currents contributed in an additive direction to the total magnetic flux or field intensity. Had one of these currents been assigned in the opposite direction, then the mutual term through these equations would have been negative rather than positive. This in no way would invalidate the proof that the two mutual inductances are equal, but it does put us on guard to the extent that the total energy given by (9.27) or (9.28) holds only when the currents are assigned in the directions indicated in Fig. 9.5. To gain further insight into this situation, solve Prob. 9.7.

PROBLEM

9.7 Prove that (9.27) is valid only where i_1 and i_2 carry their own signs, and positive values of these two currents are taken when they flow into the respective "dots." In other words, what is the correct expression for the total stored energy if i_1 flows into its dot and i_2 leaves its respective dot?

9.6 COEFFICIENT OF COUPLING

The coefficient of coupling carries an important concept relative to a magnetically coupled circuit. This coefficient can be defined in terms of the fraction of the magnetic flux produced by the current in coil 1 that links coil 2, and of course also the fraction of the flux that links coil 1 but is produced by the current in coil 2. Suppose that we have two windings where the turns of one winding are wrapped extremely close to the turns of another winding, as illustrated in Fig. 9.6. In this kind of situation, nearly all of the magnetic flux that links coil 1 will also link coil 2.

Consider another situation where there are two coils of unequal turns, but assume that whatever flux is produced links all of the turns of both coils. This condition cannot be achieved precisely, but it might be approximated fairly well in

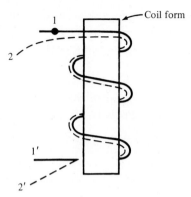

fig. 9.6 Maximum coupling.

some situations. Now we wish to write an expression for the inductance L_1 of the N_1 winding. We think of a current I_1 in this winding producing flux ϕ_1. Then the inductance of this winding is

$$L_1 = \frac{N_1\phi_1}{I_1} \tag{9.30}$$

The flux ϕ_1 is produced by I_1 and is given by

$$\phi_1 = \frac{N_1 I_1}{\mathcal{R}} \tag{9.31}$$

where $\mathcal{R}$ is the magnetic reluctance of the path taken by the flux ϕ_1. When these two equations are combined, we get

$$L_1 = \frac{N_1^2 I_1}{\mathcal{R} I_1} = \frac{N_1^2}{\mathcal{R}} \tag{9.32}$$

It is important to note that the inductance is proportional to the turns squared and inversely proportional to the magnetic reluctance of the flux path. In iron-core compared to air-core devices, we greatly increase the inductance by decreasing the magnetic reluctance $\mathcal{R}$.

In a similar fashion, we calculate the inductance of winding 2.

$$L_2 = \frac{N_2^2}{\mathcal{R}} \tag{9.33}$$

The mutual inductance might be calculated in terms of the flux linkages in coil 2 due to the current I_1.

$$M = \frac{N_2\phi_1}{I_1} = \frac{N_2 N_1 I_1}{\mathcal{R} I_1} = \frac{N_1 N_2}{\mathcal{R}} \tag{9.34}$$

The value of this exercise lies in the relationship between M^2 and $L_1 L_2$.

$$M^2 = \frac{N_1^2 N_2^2}{\mathcal{R}^2} = L_1 L_2 \qquad \text{perfect coupling} \tag{9.35}$$

Of course, perfect coupling cannot be achieved in the real world, so M^2 will always be less than $L_1 L_2$. In other words,

$$M^2 = k^2 L_1 L_2 \qquad \text{or} \qquad k = \frac{M}{\sqrt{L_1 L_2}} \tag{9.36}$$

where $k < 1$ and is called the coefficient of coupling. For iron-core transformers (two coils on an iron-core form), k, the coefficient of coupling, may be as high as 0.99. For air-core coupled circuits the coefficient of coupling k might be in the range of 0.3 to 0.8.

PROBLEM

9.8   Refer to the situation of Fig. 9.3, where it is assumed that all of the magnetic flux is constrained to the steel members and the associated air gaps. The coefficient of coupling between N_1 and N_2 is less than 1 because of the flux in the center leg ϕ_{13}.
a.   Calculate the coefficient of coupling of this configuration.
b.   Increase the gap $5g$ to $20g$. Now calculate the coefficient of coupling.

9.7   CURRENT AND VOLTAGE RELATIONS IN A COUPLED CIRCUIT

The coupled-circuit model of Fig. 9.3 would ordinarily be used in a situation where there is current i_1 and i_2 in the two coils as indicated. However, to improve our insight, let us consider the special case where impressed voltage v_1 is present, along with the consequent current i_1, but there is no closed path for i_2. For this situation there will be an induced voltage in v_1 due to its self-inductance, and also an induced voltage in v_2 due to the mutual inductance. Let us also assume that we are justified in neglecting the resistance in coil 1. Then it follows that

$$v_1 = L_1 \frac{di_1}{dt} \tag{9.37}$$

and

$$v_2 = M \frac{di_1}{dt} \tag{9.38}$$

We should be justified in debating whether the $M \, di_1/dt$ should be preceded by a minus sign or not. This sign is correct as it is, but if v_2 had been defined in the opposite direction, then the minus sign should appear in (9.38). How do we know this? We must reconsider the polarity marks again, perhaps. Earlier we defined the polarity marks such that the two currents flowing into the dots would produce magnetic flux in the same direction. It also follows that the two induced voltages due to one current alone (i_1 in this case) are such that at the instant the dot is positive at one coil, the dot will be positive at the other coil. For this reason we note that the sign in (9.38) is correct.

When the derivative of the current is eliminated between (9.37) and (9.38), we get

$$v_2 = \frac{M}{L_1} v_1 \tag{9.39}$$

Eliminate M using (9.36).

$$v_2 = k \sqrt{\frac{L_2}{L_1}} v_1 \tag{9.40}$$

Notice from (9.40) that we can make v_2 larger than v_1 by making L_2 sufficiently large compared to L_1. Simply, this amounts to increasing the turns on winding 2 compared to those on winding 1. This is called the transformer action of the coupled circuit. The iron-core transformers used everywhere in power systems are based on this principle. Voltage levels are raised and lowered where needed by means of iron-core coupled circuits having coefficients of coupling significantly greater than 0.9. Where the two coils are wound on one iron core having some particular reluctance $\mathscr{R}$ and the coefficient of coupling is assumed to be 1, then by using (9.32), (9.33), and (9.40), we get

$$v_2 = \frac{N_2}{N_1} v_1 \tag{9.41}$$

For iron-core transformers, (9.41) is not precise but only contains an error of perhaps 2 percent. On the other hand, this equation may have very little validity in air-core-coupled circuits where the coefficient of coupling k is significantly less than 1.

9.8 CIRCUIT EQUATIONS FOR A COUPLED CIRCUIT

Assume that voltages v_1 and v_2 are applied to the respective coils of a coupled circuit, as represented in Fig. 9.7. Each of the two currents i_1 and i_2 are arbitrarily assigned directions into the "dots." The respective coils have resistances R_1 and R_2. The respective self-inductances are L_1 and L_2. When Kirchhoff's voltage law is applied to each of the coil sides, we get

$$v_1 = i_1 R_1 + L_i \frac{di_1}{dt} + M \frac{di_2}{dt} \tag{9.42}$$

$$v_2 = M \frac{di_1}{dt} + i_2 R_2 + L_2 \frac{di_2}{dt} \tag{9.43}$$

All of the signs appearing in these two equations are positive because of the particular assigned polarities of the voltages and the particular assigned directions of the currents. So in writing equations such as these, one must be very careful to relate the signs to the manner in which the voltage polarities and current directions were assigned.

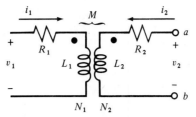

fig. 9.7 Coupled circuit.

Example 9.1

a. Assign or define v_1 differently than is shown in Fig. 9.7. Let the lower terminal of N_1 be positive rather than negative while i_1, i_2, and v_2 are left as shown. Now write equations similar to (9.42) and (9.43).

Solution

$$-v_1 = i_1 R_1 + L_1 \frac{di_1}{dt} + M \frac{di_2}{dt}$$

$$v_2 = M \frac{di_1}{dt} + i_2 R_2 + L_2 \frac{di_2}{dt}$$

b. In respect to Fig. 9.7, assign i_1 to leave (rather than enter) the dot of N_1. Leave v_1, v_2, and i_2 as shown in Fig. 9.7. Write the two equations.

Solution

$$v_1 = -i_1 R_1 - L \frac{di_1}{dt} + M \frac{di_2}{dt}$$

$$v_2 = -M \frac{di_1}{dt} + i_2 R_2 + L_2 \frac{di_2}{dt}$$

c. Assign v_1 with lower terminal positive, and assign i_1 to leave the dot of N_1. Leave i_2 and v_2 as shown. Write the equations.

Solution

$$v_1 = i_1 R_1 + L \frac{di_1}{dt} - M \frac{di_2}{dt}$$

$$v_2 = -M \frac{di_1}{dt} + i_2 R_2 + L_2 \frac{di_2}{dt}$$

• • •

PROBLEMS

9.9 a. For the circuit of Fig. 9.7, leave i_1, v_1, and v_2 as indicated, but assign the direction of i_2 so that it flows away from its dot. For these conditions, write voltage equations corresponding to (9.42) and (9.43).

b. Compared to Fig. 9.7, reverse the polarity assignment on v_2. Now write the voltage equations. Find a technique that validates or gives you confidence in your solution.

9.10 To the circuit of Fig. 9.7 add a series combination of R, L, and C to the a-b terminals. Now apply Kirchhoff's voltage law to this group, eliminating voltage v_2 so that your remaining equations involve only i_1, i_2, and v_1. Validate by some technique.

Equations (9.42) and (9.43), of course, are in the time domain. Where $i_1(0^+) = i_2(0^+) = 0$, the corresponding equations in the s domain are

$$V_1 = (R_1 + sL_1)I_1 + sMI_2 \qquad (9.44)$$

$$V_2 = sMI_1 + (R_2 + sL_2)I_2 \qquad (9.45)$$

If $i_1(0^+) \neq 0$ and/or $i_2(0^+) \neq 0$, then these equations must be modified as indicated in Fig. 8.14. The corresponding equations for steady sinusoidal voltages are

$$\mathbf{V}_1 = (R_1 + j\omega L_1)\mathbf{I}_1 + j\omega M\mathbf{I}_2 \qquad (9.46)$$

$$\mathbf{V}_2 = jM\omega\mathbf{I}_1 + (R_2 + j\omega L_2)\mathbf{I}_2 \qquad (9.47)$$

In the usual situation, a voltage is applied to one winding of the coupled circuit and a load impedance Z_L is connected to the other winding as shown in Fig. 9.8.

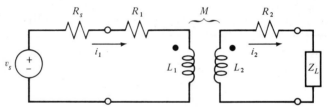

fig. 9.8 An inductively coupled circuit, or transformer, used to couple a load Z_L to a voltage source v_s .

The winding to which the voltage is applied is known as the *primary* winding, and the winding to which the load is connected is known as the *secondary* winding. This type of coupling is useful not only because of its dc isolation, but also because it provides the opportunity for maximum power transfer from the source to the load. Let us write the loop equations for the circuit of Fig. 9.8 and investigate the conditions for maximum power transfer.

$$V_s = (R_s + R_1 + sL_1)I_1 - sMi_2 \qquad (9.48)$$

$$0 = -sMI_1 + (R_2 + Z_L + sL_2)I_2 \qquad (9.49)$$

We may determine the input impedance on the primary side of the transformer by solving for i_1 from these equations, as follows:

$$i_1 = \frac{v_s(R_2 + Z_L + sL_2)}{(R_1 + R_s + sL_1)(R_2 + Z_L + sL_2) - (sM)^2} \qquad (9.50)$$

The impedance seen by v_s is v_s/i_1. From (9.50), we obtain

$$Z_i = \frac{V_s}{I_1} = (R_1 + R_s + sL_1) - \frac{(sM)^2}{(R_2 + Z_L + sL_2)} \qquad \text{(s-domain)}$$

or

$$Z_i = (R_1 + R_s + j\omega L_1) + \frac{(\omega M)^2}{R_2 + Z + j\omega L_2} \qquad \text{(}\omega\text{-domain)} \qquad (9.51)$$

Observe that the first term, in parentheses, on the right-hand side of (9.51) is the self-impedance on the primary side, commonly designated as Z_{11} written in s-domain form. Therefore the second term, which could be written as $+Z_{12}^2/Z_{22}$ is the ω-domain impedance coupled into the primary as a result of the secondary current $I_2(j\omega)$, where Z_{22} is the self-impedance of the secondary, and Z_{12} is called the mutual impedance. Equation (9.51) leads us to the new model or equivalent shown in Fig. 9.9. The impedance Z_{11} of Fig. 9.9 is easily recognized as the primary impedance (including R_s) of the circuit in Fig. 9.8. The impedance Z' is a fictitious quantity required to make the two circuits equivalent. Nevertheless, the power in Z' is the actual power in $(Z_L + R_2)$ and where Z_L is in series with R_2 when maximum power is in $(Z_L + R_2)$, there will be maximum power in Z_L, the load. Where the designer can use capacitors, or perhaps inductors, in the secondary circuit and also adjust ωM, Z' can be adjusted for maximum power in Z'. Then, for these conditions, we may see from (9.51) that maximum power is transferred to the load Z_L when the impedance Z' is the complex conjugate of the self-impedance of the primary side. Or the real part of Z' should be equal to $R_s + R_1$, and the j part of this term should be the conjugate of $j\omega L_1$.

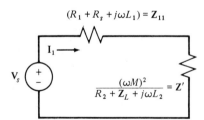

$$(R_1 + R_s + j\omega L_1) = Z_{11}$$

$$\frac{(\omega M)^2}{R_2 + Z_L + j\omega L_2} = Z'$$

fig. 9.9 Equivalent of the circuit in Fig. 9.8 [see (9.51)].

Example 9.2
Let us consider as a second example a broadband coupling and impedance transforming circuit using inductive coupling. Such a circuit might be used to couple audio amplifying devices or to couple a loudspeaker to an audio amplifier. In this case, the load will be considered to be resistive, as shown in Fig. 9.10. Let us assume that the primary winding resistance R_1 is small compared to R_s and $R_2 \ll R_L$—or alternatively, these resistances may be included in, or lumped with, R_s and R_L. Then

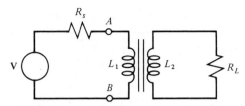

fig. 9.10 A broadband inductively (transformer) coupled circuit with a resistive load R_L.

using (9.51) we may determine the impedance Z_p looking into the transformer primary at the terminals A-B.

$$Z_p = j\omega L_1 + \frac{(\omega M)^2}{j\omega L_2 + R_L} \tag{9.52}$$

The lines between the coils in Fig. 9.10 indicate that an iron core is used to couple the coils, so the coefficient of coupling k may approach 1. Let us use the approximation that $k = 1$. Then, putting (9.52) over a common denominator, we obtain

$$Z_p = \frac{j\omega L_1(j\omega L_2 + R_L) + \omega^2 L_1 L_2}{j\omega L_2 + R_L} = \frac{j\omega L_1 R_L}{j\omega L_2 + R_L} \tag{9.53}$$

Since this impedance is a product of two impedances over a sum of two impedances, it may be more revealing to look for a parallel combination $Y_p = 1/Z_p$.

$$Y_p = \frac{j\omega L_2 + R_L}{j\omega L_1 R_L} = \frac{L_2}{L_1} G_L + \frac{1}{j\omega L_1} \tag{9.54}$$

However, using (9.32) and (9.33) we can see that

$$\frac{L_2}{L_1} = \frac{N_2^2/\mathcal{R}}{N_1^2/\mathcal{R}} = \frac{N_2^2}{N_1^2} \tag{9.55}$$

since the two windings have the same reluctance $\mathcal{R}$ in their magnetic path. Using the relationship of (9.55) in (9.54),

$$Y_p = \left(\frac{N_2}{N_1}\right)^2 G_L + \frac{1}{j\omega L_1} \tag{9.56}$$

Thus, a model of the coupling circuit of Fig. 9.10, as viewed from the primary side, is given in Fig. 9.11. You may note that the secondary load resistance has been transformed by the square of the primary-to-secondary turns ratio and appears to be in parallel with the primary inductive reactance. We say, in this case, that the secondary load resistance has been *referred to* the primary side. If we wish to have the impedance of the transformed load appear essentially as a resistance for maximum power transfer, or uniform frequency response, the inductive reactance of the primary must be large in comparison with the resistance in parallel with it. The lower edge of the passband occurs at the frequency where ωL_1 is equal to the

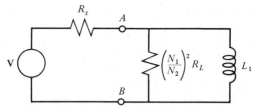

fig. 9.11   A simplified model for the coupling circuit of Fig. 9.10 with $k \to 1$.

parallel combination of $(N_1/N_2)^2 R_L$ and R_s. The upper edge of the passband is determined by the stray capacitance and the coupling coefficient.

• • •

PROBLEMS

9.11 Show that a general load impedance Z_L will transform to the primary side of a transformer by a turns-ratio-squared multiplication.

9.12 In the circuit of Fig. 9.10, $R_s = 2$ kΩ and $R_L = 8\ \Omega$. Determine the turns ratio N_1/N_2 and a suitable value for L_1 if maximum power transfer is desired for the audio-frequency range above 50 Hz. What is the power delivered to the load if v_s is a 100-V sinusoid?

9.9   IDEAL TRANSFORMERS

Iron-core transformers are closely coupled; the coefficient of coupling might be 0.99. These transformers are used extensively in raising and lowering voltages wherever significant amounts of power are involved such as in electric utilities, industry, commercial installations, and residences.

Assume that we are interested in the sinusoidal steady-state situation. Then (9.46) and (9.47) apply. Assume that the V_2 terminals deliver I_2 current to a load Z_L. Ordinarily this closely coupled iron-core circuit (transformer) is such that ωL_1 and ωL_2 are significantly greater than their respective resistances R_1 and R_2. So, where

$$\mathbf{V}_2 = -\mathbf{I}_2 Z_L \tag{9.57}$$

it follows that

$$\mathbf{V}_1 = j\omega L_1 \mathbf{I}_1 + j\omega M \mathbf{I}_2 \tag{9.58}$$

and

$$0 = j\omega M \mathbf{I}_1 + (\mathbf{Z}_L + j\omega L_2)\mathbf{I}_2 \tag{9.59}$$

It also happens that in quite a large number of these transformer applications, Z_L is significantly smaller than ωL_2. Then if we neglect Z_L in (9.59), it follows that

$$\frac{\mathbf{I}_1}{\mathbf{I}_2} = -\frac{j\omega L_2}{j\omega M} = -\frac{L_2}{M} = -\frac{L_2}{\sqrt{L_1 L_2}} = -\sqrt{\frac{L_2}{L_1}} = -\frac{N_2}{N_1} \tag{9.60}$$

A minus sign in (9.60) should not be disturbing; if i_2 of Fig. 9.7 had been assigned the opposite direction, then the negative sign of (9.60) would not be present. Notice that $\mathbf{I}_1$ and $\mathbf{I}_2$ are complex representing the respective phasor currents. The two currents are in phase (or 180° out of phase), and their ratio is equal to the inverse ratio of the respective turns N_1 and N_2. This is an approximate equation and is sufficiently accurate for many situations, but must be used with caution. The approximation that was made is that $\omega L_2 \gg R_L$, which is the same requirement we placed on ωL_1 compared with the transformed R_L in Example 9.2. Thus the current through L_1, known as the magnetizing current in the primary, is neglected. The total primary current is the transformed secondary current given by (9.60) plus the magnetizing current.

To further pursue this *ideal* transform, let us use (9.60) to eliminate I_2 from (9.58). Again neglect Z_L. Then where $a = N_1/N_2$,

$$\mathbf{V}_1 = (j\omega L_1 - j\omega M a)\mathbf{I}_1 \tag{9.61}$$

In (9.59) let $\mathbf{I}_2 \mathbf{Z}_L = -\mathbf{V}_2$ and then eliminate $\mathbf{I}_2$.

$$\mathbf{V}_2 = (j\omega M - j\omega L_2 a)\mathbf{I}_1 \tag{9.62}$$

Now the ratio of $\mathbf{V}_1$ to $\mathbf{V}_2$ is of interest.

$$\frac{\mathbf{V}_1}{\mathbf{V}_2} = \frac{L_1 - aM}{M - aL_2} = \sqrt{\frac{L_1}{L_2}} \left(\frac{\sqrt{L_1} - a\sqrt{L_2}}{\sqrt{L_1} - a\sqrt{L_2}} \right) = \sqrt{\frac{L_1}{L_2}} = \frac{N_1}{N_2} = a \tag{9.63}$$

Then in summary, for an *ideal* transformer,

$$\frac{\mathbf{V}_1}{\mathbf{V}_2} = \frac{N_1}{N_2} \quad \text{and} \quad \frac{\mathbf{I}_1}{\mathbf{I}_2} = \frac{N_2}{N_1} \tag{9.64}$$

Equations (9.64) are extremely simple and illuminating, but have restricted applications. In many situations, they can yield accuracy having only approximately 5 percent error; in many other situations, however, these equations are almost meaningless. If, and only if, the coupled circuit is on a magnetic core, there is a reasonable chance that (9.64) will give highly significant insight into the application at hand.

In the study of iron-core transformers specifically, the ideal transformer equations (9.64) are typically developed from quite a different point of view. Here we started with equations that applied to any coupled circuit, independent of coefficient of coupling and other considerations, and then restricted those equations to lead us to (9.64).

9.10 MEASUREMENT OF L_1, L_2, M, k

The self-inductances, the mutual inductance, and the coefficient of coupling k can be calculated with reasonable accuracy in iron-core systems because, for the most part, the magnetic fluxes are constrained to the iron-core path, and this path has reasonably well-defined dimensions. In the air-core inductor, the calculation of the magnetic fluxes and the subsequent linkages is much more difficult. However, the self-inductances L_1 and L_2, the mutual inductance M, and the coefficient of coupling k can be measured rather simply where we have an *inductance bridge* or some other mechanism for measuring a single inductance.

Obviously, the self-inductance L_1 could be measured merely by examining conditions at the terminals of this winding with no current in the other winding. If a bridge is to be used, it would merely look to the impedance of the one coil alone. In the same way, the inductance of the other coil alone could be measured.

Now suppose that the two coils of a mutually coupled circuit are connected as shown in Fig. 9.12a. The total inductance L_{t_1} is some function of the two self-inductances and the mutual inductance M. Now supposing the connections to L_2 were reversed as shown in Fig. 9.12b, and the total inductance L_{t_2} is measured.

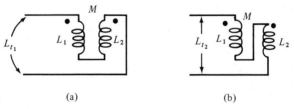

(a)   (b)

fig. 9.12   Connections to determine M and k.

These two total inductances will be different, of course. The equations for the two connections of Fig. 9.12 are

$$L_{t_1} = L_1 + L_2 - 2M \tag{9.65}$$

$$L_{t_2} = L_1 + L_2 + 2M \tag{9.66}$$

Subtracting (9.65) from (9.66), we have

$$L_{t_2} - L_{t_1} = 4M \tag{9.67}$$

and thus

$$M = \frac{L_{t_2} - L_{t_1}}{4} \tag{9.68}$$

PROBLEM
9.13   Prove that Equations (9.65) and (9.66) are true.

9.11   A GENERAL EQUIVALENT CIRCUIT

Equations (9.42) and (9.43) completely describe the magnetic coupled circuit and, of course, are highly useful in making various analyses of the application of the coupled circuit to a wide range of situations. However, you have already experienced many situations where an equivalent circuit gives certain perspective and insight that is difficult to see in mathematical forms such as these two equations.

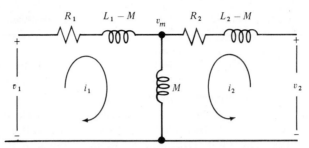

fig. 9.13   Equivalent circuit or model of a coupled circuit.

At first glance, it may not occur to you that Fig. 9.13 is an equivalent circuit or a model for a coupled circuit described by (9.42) and (9.43). In solving Prob. 9.14 you will demonstrate that this circuit is truly equivalent to the coupled circuit.

PROBLEMS

9.14 Write the two loop equations for the circuit of Fig. 9.13 and show that these equations are identical to (9.42) and (9.43), thus proving that Fig. 9.13 is truly the equivalent of the magnetically coupled circuits.

9.15 A coupled circuit is described by the following: $L_1 = 2 \times 10^{-6}$ H, $L_2 = 6 \times 10^{-5}$ H, $k = 0.7$. Calculate the following parameters of the equivalent circuit: M, $(L_1 - M)$, and $(L_2 - M)$.

Your solution to Prob. 9.15 should show that $(L_1 - M)$ is negative. This could be disturbing, but one should remember that a valid model or equivalent is not necessarily physically realizable.

PROBLEMS

9.16 Starting with the equivalent circuit of Fig. 9.13, prove (9.41). Remember that this equation applies to the situation where $I_2 = 0$, $k = 1$, and both R_1 and R_2 are negligible. Notice that under these circumstances, you can merely apply the voltage divider equation to write the ratio v_m/v_1.

9.17 In the coupled circuit of Prob. 9.15, R_1 is 2 Ω and v_1 is a sinusoidal voltage having an rms value of 2 V and an angular velocity of 10^7 rad/s. Find the phasor value V_2 where I_2 is equal to zero.

*9.12   THE PULSE TRANSFORMER—AN APPLICATION

One of the important needs for a magnetically coupled circuit is in the situation where some signal must be transmitted from one circuit to another where there can be no electrically conducting paths between the two. Refer to Fig. 9.5 and assume that there can be no conducting path between the v_1 and the v_2 circuits. Yet we must transmit a signal from one of these circuits to the other. For an example, assume that the signal at v_1 is a rectangular pulse and that we would like to deliver a pulse of approximately the same shape to a resistor connected at the v_2 terminals. Such a coupled circuit is called a pulse transformer (see Fig. 9.14). Further assume that we would like the magnitude of the pulse at the v_2 resistor to be approximately

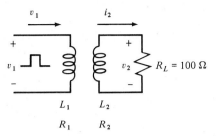

fig. 9.14   Pulse transformer and load resistance.

twice the magnitude of the input pulse. Let us analyze the magnetically coupled circuit with the view of selecting circuit parameters that will lead to satisfactory circuit performance. Our point of view here will be not one of straightforward design, where we might attempt to start with specifications and then calculate inductances, resistances, and other circuit parameters directly by means of explicit equations; rather, we will make judicious selection of pertinent parameters and then calculate circuit performance to compare with that which is desired. The formal design procedure that leads directly to circuit parameters is, of course, highly desired. Such a clean-cut approach is practical only in restricted cases. With our limited experience at this point, it seems best that we select parameters, analyze to determine performance, compare performance to that which is desired, modify parameters, and again determine the new performance. We shall proceed in that direction.

Assume that the load resistance $R_L = 100\ \Omega$ and the coupled circuit parameters are as follows:

$$L_1 = 0.005\ \text{H} \qquad L_2 = 0.05\ \text{H}$$
$$R_1 = 5\ \Omega \qquad R_2 = 20\ \Omega \qquad\qquad (9.69)$$
$$k = 0.75 \qquad M = 0.01186\ \text{H}$$

Notice that L_2 is 10 times L_1. In order that v_2 be greater than v_1, the insight from (9.63) suggests that L_2 must be greater than L_1. We will soon see that $(L_1 - M)$ in Fig. 9.13 is negative while $(L_2 - M)$ is positive, and of course R_1 is only $5\ \Omega$ compared to $R_2 + R_L = 120\ \Omega$. Because R_L is high compared to R_1, there is a strong suggestion that i_2 will have a minor affect upon v_m, and v_2 might be reasonably close to v_m. Making these two assumptions will greatly simplify the analysis and still give significant insights (not accurate results) into the performance of the real circuit. Let us pursue this simplified attack for a preliminary view. Later we will analyze the situation more precisely.

Crude Approximation

Under the assumption that v_m gives reasonable insight into the value of v_2 where i_2 is considered to be 0, the equivalent circuit of Fig. 9.13 leads us to the circuit of Fig. 9.15 if the initial $i_1(0) \neq 0$.

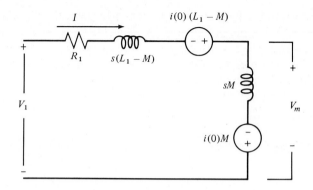

fig. 9.15   Equivalent for approximate approach.

In analyzing the situation to find v_m, we will use the s-domain approach as indicated in Fig. 9.15. If the circuit is at rest at the beginning of the input pulse, then $i_1(0)$ is zero and the circuit of Fig. 9.15 is simplified accordingly. But for a few steps in the analysis, we will assume that $i_1(0)$ might be nonzero.

By applying Kirchhoff's voltage law to the model of Fig. 9.15, we get

$$V_1 = (R_1 + sL_1)I_1 - i_1(0)L_1 \tag{9.70}$$

but we desire V_m which is

$$V_m = sM_1 V_1 - Mi_1(0) \tag{9.71}$$

When we eliminate I_1 between (9.70) and (9.71), we get

$$V_m = \frac{sM_1 V_1 - MR_1 i_1(0)}{R_1 + sL_1} \tag{9.72}$$

In our situation, $i_1(0) = 0$, and hence

$$V_m = \frac{sM V_1}{R_1 + sL_1} \tag{9.73}$$

If we are only interested in V_m during the duration of the pulse, then we may think of the pulse as being a step function. Assume that the step function has a magnitude of 1. Then it follows,

$$V_m = \frac{M}{R_1 + sL_1} = \frac{M}{L_1} \times \frac{1}{s + (R_1/L_1)} \tag{9.74}$$

and then by taking the inverse,

$$v_m = \frac{M}{L_1} e^{-R_1 t/L_1} \tag{9.75}$$

For the numerical values of (9.69), $L_1 = 0.005$ H, $L_2 = 0.05$ H, $R_1 = 5\ \Omega$, $R_2 = 20\ \Omega$, $k = 0.75$, $M = 0.01186$, it follows that

$$v_m = 2.372 e^{-1,000t} \tag{9.76}$$

The time constant for V_m is 10^{-3} s, which is two times the duration of the pulse (5×10^{-4} s). So V_m has a magnitude of 2.372 V at $t = 0$ (the beginning of the pulse) and a value of 1.439 at $t = 5 \times 10^{-4}$ s (the end of the pulse). These results give some significant insight into the performance of the coupled circuit if the assumptions previously made are reasonable (see Fig. 9.13).

Precise Solution

Consider the equivalent circuit of Fig. 9.16. Where the circuit is at rest for $t < 0$, we may write

$$V_1 = (R_1 + sL_1)I_1 - sMI_2 \qquad (i_1(0) = i_2(0) = 0) \tag{9.77}$$

$$0 = -sMI_1 + (R_2' + sL_2)I_2 \tag{9.78}$$

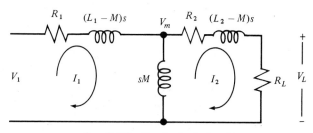

fig. 9.16   Equivalent circuit.

where $R_2' = R_2 + R_L = 20 + 100 = 120$. Solving for I_2,

$$I_2 = \frac{sMV_1}{(R_1 + sL_1)(R_2' + sL_2) - s^2M^2} \tag{9.79}$$

Expanding the denominator of (9.79) and collecting terms,

$$I_2 = \frac{sMV_1}{s^2(L_1L_2 - M^2) + s(R_1 L_2 + R_2' L_1) + R_1 R_2'} \tag{9.80}$$

where v_1 is a unit step function. For $V_1 = 1/s$, $R_L = 100$, and values of (9.69) are used,

$$I_2 = \frac{0.01186}{1.093 \times 10^{-4}s^2 + 0.85s + 600} \tag{9.81}$$

and

$$I_2 = \frac{108.5}{s^2 + 7,777s + 5.489 \times 10^6} \tag{9.82}$$

And finally,

$$v_2 = 1.748(e^{-785t} - e^{-6,992t}) \tag{9.83}$$

PROBLEM

9.18 On one set of coordinates, plot and compare v_m of (9.76) with v_2 of (9.83). Plot for $0 < t < 5 \times 10^{-4}$ s.

The performance described by (9.82) and as displayed by your plot of Prob. 9.17 might not be satisfactory, so let us increase R_L to 480 Ω and then calculate v_2.

PROBLEMS

9.19 Where $R_L = 480$ Ω and the values of (9.69), find the equation for v_2 and show that it is given by

$$v_2 = 2.331(e^{-944t} - e^{-2.422 \times 10^4 t})$$

and that a plot of this v_2 is as shown on Fig. 9.17.

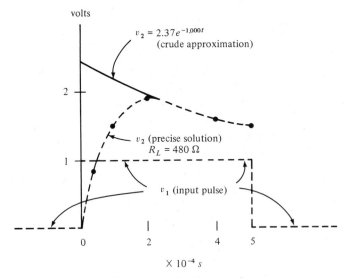

fig. 9.17 Response of pulse transformer.

9.20 By means of a computer or a calculator, compute perhaps ten different curves of v_2, each with a different set of parameters. Select some variation of parameters (R_1, L_1, R_2, L_2, k, R_L). You might choose to study the effect of varying L_1 and L_2; a classmate might wish to study the effect of varying k. Results could be compared to the advantage of all.

Obviously there are tight constraints on realizability of these parameters. For example, we must accept positive values of the resistances. Furthermore, increasing L_1 tends to increase R_1. Of course the coefficient of coupling k must be less than 1. So try to select values that seem reasonable to you. Your instructor may want to give you some constraints on selecting parameters.

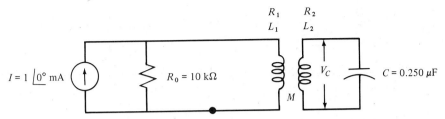

fig. 9.18   *Inductively coupled capacitor.*

9.21   As shown in Fig. 9.18, a 0.250-μF capacitor is connected to the coupled coils described by (9.69). Using the coupled circuit model of Fig. 9.16:

a.   Draw a schematic diagram of the equivalent of Fig. 9.18.

b.   Write the s-domain equation for V_C in terms of $I, R_1, R_2, (L_1 - M), M, (L_2 - M)$, R_0, and C.

c.   Transform this equation of part (b) to the ω domain.

d.   Identify the frequency at which the $C_2 - (L_2 - M)$ branch is in parallel resonance with the M branch.

e.   Calculate by means of your programmable calculator, or by means of a computer, the numerical value of $\mathbf{V}_C$ for the given numerical parameters. Calculate the $\mathbf{V}_C$ function, especially in the region of the resonance of part (d).

9.22   A proposed community power system uses a hydroelectric generating plant located at a reservoir 10 miles from the community. A two-wire transmission line is built and the total resistance of the line (both wires) is 1.0 Ω.

a.   How much power can be delivered to the community if no transformers are used in the system, the generated voltage is 125 V, the voltage delivered to the community is 115 V, and the voltage drop due to the line is 10 V? Assume unity power factor.

b.   How much power can be delivered to the community with the same generator voltage and community voltage, using the same power line, if a transformer having a 100-to-1 turns ratio is used at the generator end of the line to raise the line voltage and another identical transformer is used at the community end to lower the voltage back to 115 V?

9.23   A given sound amplifier provides power to the loudspeaker. If the amplifier has an output resistance of 100 Ω and the loudspeaker has a resistance of 8 Ω, determine:

a.   The turns ratio needed for a coupling transformer to provide maximum power to the speaker. Assume $k = 1$.

b.   The primary inductance required if the transformer is to be essentially ideal at frequencies above 100 Hz.

c.   The power delivered to the loudspeaker when the source voltage V_s of the amplifier is 100 V rms and the matching transformer is ideal.

d.   The sound power delivered to the loudspeaker if no transformer is used and $V_s = 100$ V.

Chapter 10

Polyphase Systems

Introduction   *Polyphase* means "having many phases." All of the alternating current circuits and systems that we have been discussing in this book so far have been *single-phase* systems. A single-phase source delivers one alternating current voltage appearing between two terminals or two leads. A three-phase source or generator, by far the most common polyphase source, delivers three voltages that are equal in magnitude but out of phase with respect to each other.

Nearly all electric power generation is three-phase. Nearly all ac motors rated at 10 kW or more are three-phase. Long-distance high-powered electric transmission is typically three-phase. True, we have much local distribution to residences and small commercial loads by single-phase systems, and high-voltage dc transmission is often used. But where larger blocks of ac power must be moved over considerable distances, it is nearly always done in the form of three-phase transmission. Probably, there are no single-phase ac generators having a capacity of 10,000 kW or greater in existence. On the other hand, our modern three-phase ac generators are commonly in the power range of 100,000 kW to 800,000 kW. Furthermore, large ac motors have improved operating characteristics and, at the same time, are much less expensive than a corresponding single-phase motor. Also, there are advantages of polyphase over single-phase power transmission.

10.1   A SIMPLE THREE-PHASE GENERATOR

Figure 10.1 displays a simple version of a three-phase generator. Understanding some simple principles of three-phase generation will help give meaning to the rest of this chapter. In Fig. 10.1, S_1 represents the cross-sectional view of a conductor extending *into the paper*. Likewise, F_1 is another single-conductor or *coil side*. A complete one-turn coil is formed from S_1 and F_1. In a similar fashion, S_2 and F_2 form another single-turn coil. There are three such coils on the stationary member of this simple generator. The rotating member consists of a permanent magnet

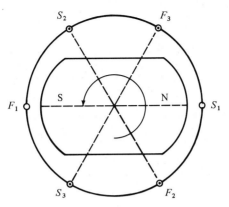

fig. 10.1   Simple three - phase generator—one turn each phase.

having a north and a south pole, as indicated in the figure. In real life, these magnetic poles are created by a direct current in a winding not shown here. The rotor is driven by a steam turbine (or some other similar motor device). The motion of the magnetic poles past the conductors induces voltages in these one-turn coils. By appropriate design, the induced voltage in a coil of many turns will be sinusoidal for most practical purposes. For the particular position of the pole structure shown in the figure, induced voltage in the coil S_1-F_1 will be a maximum. Here S_1 connotes the *start* of a winding and F_1 connotes the *finish*. If we assume positive voltage in a coil when the axis of the north pole is adjacent to the *start* side of a coil, then for the rotor position shown, the induced voltage v_1 in coil 1 is a positive maximum. After the rotor has rotated through 120°, it will be in a position such that the voltage v_2 in coil 2 is a positive maximum. Then after an additional 120° of rotation, the voltage in coil 3 will reach a positive maximum. The three sinusoids would be equal in magnitude and displaced 120° as shown in Fig. 10.2a. Notice that for this situation, voltage v_1 is leading v_2 by 120° and v_2 is leading v_3 by 120°. These three voltages represent a three-phase system. The phasors V_1, V_2, and V_3 are displayed in Fig. 10.2b. In this and similar diagrams, it is important to note that any one of the phasors can be drawn in a reference position. It was arbitrary in this case to select V_1 as the reference.

   The real three-phase generators are similar to the simple generator of Fig. 10.1 in many respects but different in other respects. In this simple generator, the coil for each phase consists of one turn only. In the practical generators of the real world, there are many turns in each phase. Figure 10.3 is a representation of a three-phase generator where there are many turns per phase. All of the turns connected in series in one phase would, of course, have two leads. It is not necessary to carry the resultant six leads (for three phases) to the device or system that consumes the power from this generator. A common lead marked 0, as shown in Fig. 10.3, reduces the number of leads to the external world to three (marked 1, 2, and 3). You may feel uncomfortable about this common connection at node 0. Do not worry about this for the moment; the point will be clarified soon.

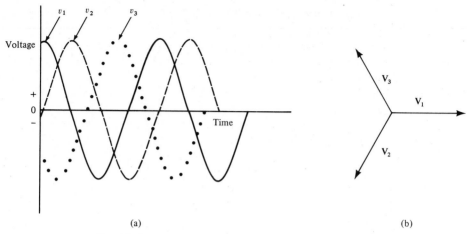

(a)   (b)

fig. 10.2   *Phasors for typical three - phase generator.*

10.2   PHASOR DIAGRAMS

Phasor diagrams are almost essential tools in analyzing many polyphase situations. We must learn how to draw these diagrams with clarity and also learn how to interpret or read others' diagrams with precision.

Refer again to Fig. 10.3. Focus on V_{10}. Notice the two subscripts 1 and 0. This has the meaning of the voltage of 1 *with respect to* 0. Double subscripts (the 1 and 0 here) are sometimes highly useful, but in other situations are merely *excess luggage*. In this situation of Fig. 10.3, the symbol V_{10} is assigned the clear meaning of the voltage at terminal 1 with respect to that at terminal 0. Or, in the time domain when v_{10} is positive, this means terminal 1 is positive with respect to terminal 0. Notice further that the + and − signs shown on Fig. 10.3 are not necessary to convey clear, precise meaning. With these things in mind, relate the phasor diagrams of Fig.

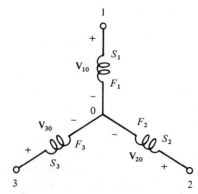

fig. 10.3   *Three - phase generator with many turns per phase.*

10.4 to the diagram of connections of Fig. 10.3. For example, in Fig. 10.4a, V_1 represents the phasor voltage of terminal 1 with respect to terminal 0 of Fig. 10.3.

There are many methods of representing three voltage phasors. Some of the possible representations are shown in Fig. 10.4. In this figure, compare (a) to (b). These two diagrams represent identically the same situation. The V_1 in (a) is identical to V_1 in (b). Likewise, V_2 in (a) is identical in both magnitude and phase compared to V_2 in (b). At first glance, the two diagrams (a) and (b) seem to have different messages, but on closer examination, they yield identically the same story. If we are concerned only with the relative magnitudes and relative phases of the three phasors, V_1, V_2, and V_3, then all five phasor diagrams of Fig. 10.4 convey nearly, but not identically, the same message. The diagrams of (a) and (b) both use V_1 as the horizontal reference phasor. On the other hand, (c), (d), and (e) employ different references. In all five cases, the three phasors (V_1, V_2, and V_3) have identically equal magnitudes, and in each representation, V_1 leads V_2 by 120°, and V_2 leads V_3 by 120°. In situations where the phase of one of these three phasors, say V_1, has meaning with respect to some other phasor not considered here, then V_1 would have to be drawn or represented in such a way as to show that phase. But on the other hand, if in Fig. 10.4 we were only concerned with representing these three phasor qualities, then any one of them can be drawn as a reference such as (a) and (b), or any one of the three could be drawn in any phase position, as long as the other two were shown in the appropriate relative position. Compare (c), (d), and (e).

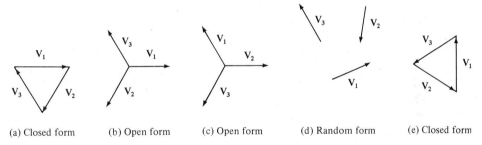

(a) Closed form   (b) Open form   (c) Open form   (d) Random form   (e) Closed form

fig. 10.4   A particular set of phasors represented in different forms.

The position of V_1 with respect to a reference is different in each of these three cases. However, the phase of V_3 is lagging V_2, which is lagging V_1 in (c) as it is in (d) and in (e). So these three diagrams, if the reference is arbitrary, convey identically the same information. It is important to note that (d) appears quite different from (c) and from (e), yet the three convey *identically* the same message. We should learn the lesson here that the position on the paper of any line that represents a phasor conveys no information. All of the information is conveyed in the relative length of the line segment and its relative angle with respect to other associated line segments.

Notice that in Fig. 10.3 we used double subscripts to indicate the sense of a voltage, such as V_{10}. This represents the voltage of 1 with respect to 0. Arrowheads are shown in Fig. 10.2 and Fig. 10.4 to indicate the sense of a voltage. Here in these figures, only one subscript is used, but this implies that the reference is clearly understood. In other words, the symbol V_1 means the voltage of 1 with respect to some reference point that is clearly understood but not specifically designated in the subscript. Figure 10.5a is a very simple representation of all of the possible phasor voltages associated with the wye connection of Fig. 10.3. Here, in Fig. 10.5a, the line segment from 2 to 1 with an implied arrowhead at 1 is interpreted to mean V_{12} or the voltage of 1 in respect to 2. This same line segment with the arrowhead at 2 has the meaning of V_{21}. See Fig. 10.5c. The representations of (b) and (c) make clear the meaning of (a). Notice the simplicity of the diagram of Fig. 10.5a that avoids the use of arrowheads, + and − signs, and double subscripts.

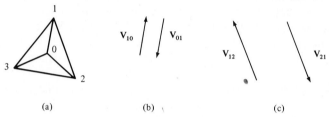

(a)   (b)   (c)

fig. 10.5   Arrows are not necessary; subscripts are not necessary.

Example 10.1
In the phasor diagram of Fig. 10.5a, the three line segments emanating from 0 are equal in magnitude and displaced from each other by 120°. This is a phasor diagram representing all of the possible voltages in the circuit of Fig. 10.3. Where $V_{10} = 10 \angle 80°$, write the time-domain equation for each of the following voltages: v_{01}, v_{10}, v_{20}, v_{30}.

Solution
Phasors are in effective values unless specified to be in maximum values. So

$$v_{10} = 10\sqrt{2} \sin (\omega t + 80°)$$

and

$$v_{01} = -v_{10} = -10\sqrt{2} \sin (\omega t + 80°) = 10\sqrt{2} \sin (\omega t - 100°)$$

$$V_{20} = 10 \angle 80° - 120°$$

so

$$v_{20} = 10\sqrt{2} \sin (\omega t - 40°)$$

and

$$v_{30} = 10\sqrt{2} \sin (\omega t - 160°)$$

• • •

The point in leaving the arrowheads off of the diagram of Fig. 10.5a lies in the fact that this diagram clearly shows all of the voltages ($\mathbf{V}_{12}$, $\mathbf{V}_{21}$, $\mathbf{V}_{23}$, $\mathbf{V}_{32}$, $\mathbf{V}_{31}$, $\mathbf{V}_{13}$, $\mathbf{V}_{10}$, $\mathbf{V}_{01}$, $\mathbf{V}_{20}$, $\mathbf{V}_{02}$, $\mathbf{V}_{30}$, and $\mathbf{V}_{03}$). Arrowheads placed on this diagram might confuse rather than clarify the picture. Too, arrowheads placed on the segment of $\mathbf{V}_{12}$ of Fig. 10.5a in order to represent both $\mathbf{V}_{12}$ and $\mathbf{V}_{21}$ would not make it clear as to which arrow was associated with which of the two voltages. So, in situations like this, it is advantageous to omit both the double subscripts and the arrowheads.

Example 10.2

In the series *RL* circuit of Fig. 10.6, we desire to represent all possible voltages on a phasor diagram without using arrowheads or double subscripts. Here the current $\mathbf{I}$

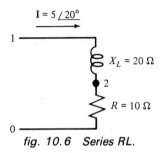

fig. 10.6   Series RL.

is given and then drawn as shown in Fig. 10.7. The arrowhead on the current seems desirable. Now consider the voltage of 2 in respect to 0. This voltage will be in phase with $\mathbf{I}$ as shown. The voltage of 1 in respect to 2 will lead the current by 90°. And, of course, $\mathbf{V}_{10}$ must be the sum of $\mathbf{V}_{20}$ and $\mathbf{V}_{12}$, as shown in Fig. 10.7.

$$\mathbf{V}_{10} = \mathbf{V}_{20} + \mathbf{V}_{12} \tag{10.1}$$

The graphical display of Fig. 10.7 also shows this same sum of complex numbers.

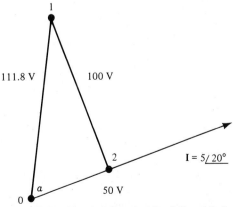

fig. 10.7   Phasors for circuit of Fig. 10.6.

•   •   •

PROBLEM

10.1 In Fig. 10.7, compute α. Now, from this diagram, write the numerical complex value of each of the following: $\mathbf{V}_{10}$, $\mathbf{V}_{12}$, $\mathbf{V}_{20}$, $\mathbf{V}_{02}$. Now validate your solution by solving for each of these voltages using the values of $\mathbf{I}$, $\mathbf{Z}$ (complex), R, and jX_L.

Example 10.3

In the two-branch network of Fig. 10.8, the left-hand branch is identical to the corresponding *RL* circuit of Fig. 10.6. Note that the current in this branch is the same for the two circuits. In the other branch, the R and the L are reversed in order. Ultimately we wish to find $\mathbf{V}_{23}$ and $\mathbf{V}_{32}$. However, as an intermediate step, draw a phasor diagram (superimposed on that of Fig. 10.7) that will show voltage 3 in respect to 0 and 1. Now from this diagram, estimate the numerical value of $\mathbf{V}_{23}$ and $\mathbf{V}_{32}$. Validate this answer by solving for these two voltages using appropriate complex algebra.

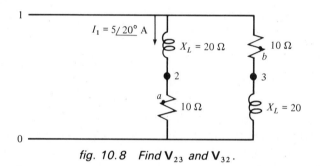

fig. 10.8   Find $\mathbf{V}_{23}$ and $\mathbf{V}_{32}$.

Solution

Figure 10.9 shows the phasor diagram for the circuit of Fig. 10.8. The triangle 0210 of Fig. 10.9 is identical to that of Fig. 10.7. Note that $\mathbf{V}_{30}$ (across X_L) leads $\mathbf{I}$ by 90° while $\mathbf{V}_{13}$ is in phase with $\mathbf{I}$. This diagram clearly shows $\mathbf{V}_{32}$ or $\mathbf{V}_{23}$. From the diagram, estimate $\mathbf{V}_{32}$ to be 115 $/\ 150°$. Calculate $\mathbf{V}_{32}$ as follows:

$$\mathbf{V}_{30} = \mathbf{V}_{20} + \mathbf{V}_{32}$$

or

$$\mathbf{V}_{32} = \mathbf{V}_{30} - \mathbf{V}_{20} = 100 \ /\ 90° + 20° + 50 \ /\ 20° - 180°$$

$$= 111.8 \ /\ 136.6°$$

● ● ●

PROBLEM

10.2 A "center tap" is available on each of the resistances of Fig. 10.8. Add the *a* and *b* nodes to the phasor diagram of Fig. 10.9. Without further complex algebra, write, by inspection, the complex value of the phasor $\mathbf{V}_{ab}$.

Now that we have practiced significantly on phasor diagrams, let us return to the three-phase problems.

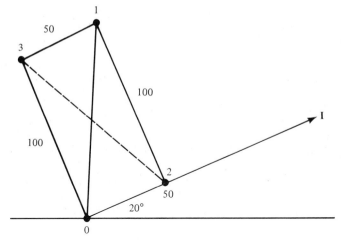

fig. 10.9 Phasor diagram for Fig. 10.8.

10.3 THE WYE CONNECTION

The three coils of a three-phase generator, motor, or transformer bank may be connected in *wye* as indicated in Figs. 10.3 and 10.10. In both of these figures, no external connection is made to the node 0 or the *neutral*. Such connections are called three-wire-wye connections. For some purposes, an external connection is made to the neutral. Then the resultant circuit is called a four-wire-wye connection. It will be shown later that the four-wire-wye connection has some limited usefulness. In the most common situation, the three separate line voltages indicated by V_L are equal in magnitude, but displaced in phase by 120°. Notice from Fig. 10.10 what is meant by the *line leads*. The three line voltages, V_L, are voltages between these line leads, as shown in Fig. 10.10. On the other hand, the voltages from the respective

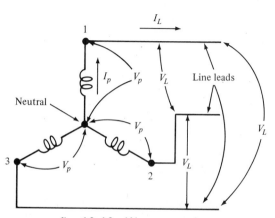

fig. 10.10 Wye connection.

line leads to neutral V_p are called the *phase voltages*, or line-to-neutral voltages. Like the three line voltages are equal in magnitude, the three phase voltages are also equal in magnitude. But, of course, the phase voltages are out of phase by 120° in respect to each other, and line voltages are also out of phase in respect to each other by 120°. See Fig. 10.5a as an example. Furthermore, none of the phase voltages are in phase with any of the line voltages.

The phase voltages (line-to-neutral) of Fig. 10.10 might be as indicated in Fig. 10.11a. Here these phase voltages are equal in magnitude and displaced 120°, as shown. The phasors of (b) differ from those of (a) only in the reference. As you know, the reference is arbitrary in many situations. In other words, (a) tells the same story as (b) unless we are specifically interested in the phase difference between one of these phasors, say V_1, in respect to some other phasor not shown here. We have chosen to use an "open diagram" in Fig. 10.11 to represent the phase voltage. See Fig. 10.4a and b for "closed" and "open" diagrams, respectively. Because the neutral, node 0, is common to the three phase voltages, the open diagram is advantageous here.

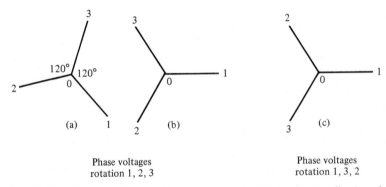

Phase voltages
rotation 1, 2, 3

Phase voltages
rotation 1, 3, 2

fig. 10.11   Phase and line voltages — wye connection (open diagrams).

Relationship Between Phase and Line Voltages

In these balanced situations where magnitudes of all line voltages are equal in respect to each other and also magnitudes of all phase voltages are equal in magnitude in respect to each other, the magnitude of line voltage, of course, is not equal to the magnitude of phase voltage. And, contrary to what one might guess, line voltage magnitude is not equal to twice the phase voltage magnitude. Let us see how this comes about. See Fig. 10.11b.

$$V_{10} = V_p \angle\, 0° \qquad V_{20} = V_p \angle -120° \qquad V_{30} = V_p \angle\, 120° \qquad (10.2)$$

On an enlarged scale, Fig. 10.12 displays V_{10}, V_{20}, and V_{12}. Line voltage V_{12} is the sum of two appropriate phase voltages as follows:

$$V_{12} = V_{10} + V_{02} \qquad (10.3)$$

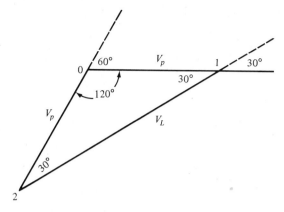

fig. 10.12 V_L and V_p.

Give special attention to the order of the subscripts. Perhaps, to make sure that you understand the order of these subscripts, you might wish to redraw Fig. 10.12 showing arrowheads and labeling the pertinent phasors with double subscripts. The order of these subscripts is based on rather simple concepts, but can be very confusing if the person does not think the situation out clearly. Now we are interested in the magnitude of $\mathbf{V}_{12}$, which is one of the line voltages. This magnitude can be taken rather easily from Fig. 10.12. Notice the resultant isosceles triangle involving the angles 120°, 30°, and 30°. It is obvious that without knowing the magnitude of $\mathbf{V}_{12}$, we do clearly know its direction. Keeping this in mind, we get its magnitude by noting that each of the phase voltages $\mathbf{V}_{10}$ and $\mathbf{V}_{02}$ have components along the resultant $\mathbf{V}_{12}$. Also, each of these phase voltages have components normal to the resultant, and of course these two separate normal components are in opposite directions and equal in magnitude, and therefore cancel each other. Then we see that each component along the resultant is $V_p \cos 30°$; therefore

$$\mathbf{V}_{12} = 2V_p \cos 30° \underline{/\ 30°} = 2 \times \frac{\sqrt{3}}{2} V_p \underline{/\ 30°} = \sqrt{3}\, V_p \underline{/\ 30°} \qquad (10.4)$$

because $\cos 30° = \sqrt{3}/2$. Notice that the magnitude of this line voltage is equal to $\sqrt{3}$ times the magnitude of a phase voltage. In this case, phase voltage $\mathbf{V}_{10}$ is 30° behind line voltage $\mathbf{V}_{12}$.

PROBLEMS

10.3 Where phase voltages in a wye system are given by the *open* diagram of Fig. 10.11b, apply the techniques used in finding $\mathbf{V}_{12}$ and proceed to find the other line voltages in this same system $\mathbf{V}_{23}$ and $\mathbf{V}_{31}$. Draw a phasor diagram showing the three phase voltages and the three line voltages. (The correct solution to Prob. 10.3 is shown in Fig. 10.13.)

10.4 Starting with the phasor diagram of Fig. 10.11c, add to this diagram the line voltage phasors $\mathbf{V}_{12}, \mathbf{V}_{23}$, and $\mathbf{V}_{31}$.

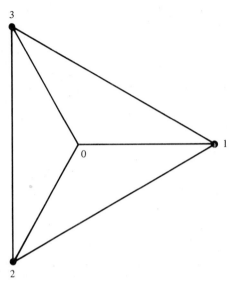

fig. 10.13   Solution to Prob. 10.3. Line and phase voltages.

Phase Rotation or Phase Sequence

The two possible phase rotations are illustrated in Fig. 10.11. The phase rotation in (a) and (b) is said to be 1-2-3, whereas the phase rotation in (c) is 1-3-2. What we have called *phase rotation* here is also called *phase sequence* by other persons in other places. Phase rotation has an important bearing on the direction of rotation of three-phase motors and also has an important bearing on certain instrument readings in situations yet to be discussed. Notice that if the three line leads were available, they could be labeled in one way to give one phase rotation, then re-labeled to give the opposite phase rotation.

PROBLEMS

10.5 Where the phase rotation is 1-2-3, and $V_{10} = 100 \angle\ 0°$ volts, write equations giving the numerical values in polar form for each of the following phasors: V_{20}, V_{30}, V_{12}, V_{23}, V_{31}, V_{13}, V_{32}, V_{02} shown in Fig. 10.14. Notice that all of these can be written directly merely by looking at the phasor diagram. Notice that for rotation 1-2-3, V_1 leads V_2, which leads V_3 each by 120°.

10.6 Given that $V_{01} = 200 \angle\ 100°$ volts and the phase rotation is 1-3-2, draw a phasor diagram showing the three symmetric phase voltages V_{10}, V_{20}, V_{30}, and the three symmetric line voltages V_{12}, V_{23}, and V_{31}. Your diagram should appear much like that of Fig. 10.14b after your diagram is rotated appropriately.

10.7 The voltage at the convenience outlets in your classroom building (perhaps) come from a four-wire three-phase wye connection of transformers. The outlets are connected from one of the line leads to neutral. If the magnitude of this phase voltage is 120 V, what must be the magnitude of the line-to-line voltage? These line voltages are available to operate three-phase motors for pumps, blowers, and the like in the building. The four

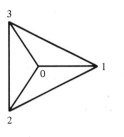

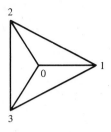

(a) Rotation: 1, 2, 3   (b) Rotation: 1, 3, 2

fig. 10.14   Phase and line voltages (closed diagrams) .

leads including the neutral also are connected (perhaps) to appropriate terminals on some of the benches in the laboratories. This makes available two levels of voltages for single-phase purposes and also the three line voltages for three-phase needs.

It is important to notice that the two phasor diagrams of Fig. 10.14 represent all possible phase and line voltages for both possible phase rotations where voltages are balanced or equal in magnitude. These and similar diagrams should be used generously in the problem situations that follow. The phasor diagram gives significant and important insight into polyphase problems, and it also helps the person avoid serious errors. Notice that balanced line voltages form an equilateral triangle, and its rotation on the paper in respect to the horizontal reference is arbitrary in most cases but may be fixed by the reference of one of these phasors to some other phasor. Then notice that the neutral or the 0 falls at the centroid of this triangle.

PROBLEM
10.8 For a balanced three-phase wye system that the phase rotation is 1-2-3, and that line voltage $V_{12} = 4,000 \underline{/\ 90°}$ volts, draw a phasor diagram showing all line voltages and all corresponding phase voltages. Give the numerical value of V_{20} in polar form. Your diagram when rotated appropriately must match that of Fig. 10.14a.

10.4 THE DELTA CONNECTION

The wye connection was formed, as shown in Fig. 10.10, by connecting one lead from each coil or each phase to a common point called the neutral. The neutral is not present in the delta connection, as shown in Fig. 10.15. In the delta connection, we have a line lead at each of the junctions between two phases or coils. The line currents in this delta are shown as I_1, I_2, and I_3. The phase currents are shown here as I_{12}, I_{23}, and I_{31}. In the most common situation (balanced), all phase currents are zero if all line currents are zero. Now let us return to the wye connection, as shown in Fig. 10.10. Notice that in a wye, the current in a phase or coil is equal to the corresponding line current. On the other hand, in a delta, a particular line current is *not* equal to any phase or coil current. Yet in a delta, a particular phase voltage is also a line voltage, while in a wye, a particular line voltage is not equal to any particular phase or coil voltage.

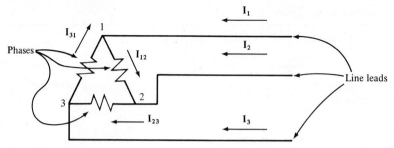

fig. 10.15   The delta connection.

Assume that three equal impedance elements are connected in delta, as shown in Fig. 10.15. The balanced applied voltages are displayed in Fig. 10.16. A particular phase current, I_{12} as an example, is given by

$$I_{12} = \frac{V_{12}}{Z_p} \tag{10.5}$$

Now if each of these impedances was a pure resistance, then the respective currents would be as displayed in Fig. 10.16a. Notice that I_{12} is in phase with V_{12} because the impedance element is a resistance. And of course, I_{23} is in phase with V_{23}, as shown. On the other hand, suppose that each of the phase elements was partly inductive such that its current lagged its voltage by 30°. Then the appropriate representation of the phase currents in respect to phase voltages is shown in (b).

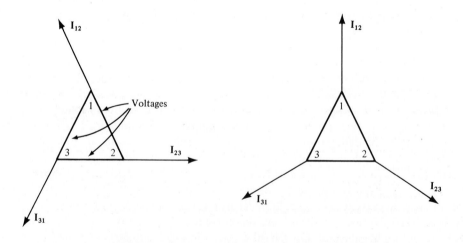

(a) Resistors in delta   (b) Inductive elements 30° lag

fig. 10.16   Delta currents and voltages.

Line Currents in a Delta

To find the line current $\mathbf{I}_1$ of Fig. 10.15, we form the sum

$$\mathbf{I}_1 = \mathbf{I}_{12} + \mathbf{I}_{13} \tag{10.6}$$

Make sure you understand the subscripts in (10.6). The three currents of this sum are displayed in Fig. 10.17.

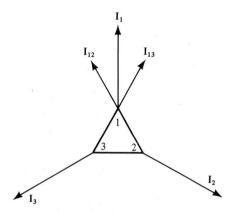

fig. 10.17   *Delta currents and voltage for resistors.*

PROBLEM

10.9   Prove that

$$I_L = \sqrt{3}I_p \tag{10.7}$$

where I_L and I_p are the respective magnitudes of line and phase currents in a balanced delta.

This equation is true regardless of the power factor or the phase angle difference between a phase voltage and the current in that phase. This is true because, for any balanced situation independent of power factor, the two phase currents contributing to a particular line current will always be equal in magnitude and displaced by 60°.

PROBLEMS

10.10   Starting with the phasor diagram of Fig. 10.17, add phasors so that all of the following phasors are shown on one diagram: $\mathbf{V}_{12}$, $\mathbf{V}_{23}$, $\mathbf{V}_{31}$, $\mathbf{V}_{13}$, $\mathbf{V}_{32}$, $\mathbf{V}_{21}$, $\mathbf{I}_1$, $\mathbf{I}_2$, $\mathbf{I}_3$, $\mathbf{I}_{12}$, $\mathbf{I}_{13}$, $\mathbf{I}_{23}$, $\mathbf{I}_{21}$, $\mathbf{I}_{32}$, $\mathbf{I}_{31}$. Organize your phasor diagram in accordance with the pattern already set in Fig. 10.17. Notice that to complete the solution to this problem, you will need only to add two more current phasors at node 2 and two more current phasors at node 3.

10.11   Each leg $\mathbf{Z}$ or phase of a balanced delta consists of a capacitor having 100 Ω in parallel with the resistor having 200 Ω as shown in Fig. 10.18. The magnitude of one of the balanced line voltages is 400 V. The phase rotation is 1-3-2. Where the phasor

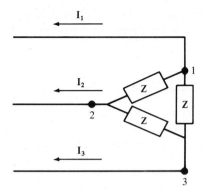

fig. 10.18   Balanced delta.

line currents are assigned outward directions as indicated in the figure, develop a phasor diagram showing all line voltages and line currents. In a polar form, give the numerical value of all three line currents and line voltages. Notice that no reference is given; therefore, the orientation of the phasor diagram on the page is arbitrary.

If correct, your phasor diagram will show that

1. The three line currents are equal in magnitude.
2. The phase rotation of line currents is 1-3-2.
3. The three phase currents are equal in magnitude and have rotation 1-3-2.
4. The phase current for any particular leg or phase leads the voltage of that leg by $\tan^{-1}(200/100) = 63.43°$.
5. The three line voltages are equal in magnitude and have rotation 1-3-2.
6. The ratio of line current magnitude to phase current magnitude is $\sqrt{3}$.

10.12 The three impedances of Prob. 10.11 are connected wye, and the system of balanced three phase voltages (400 volts line-line) of Prob. 10.11 is applied to this wye. Draw a phasor diagram showing all three phase voltages, all three line voltages, and all three line currents. Each line current in this wye should have a magnitude of one-third of the magnitude of a line current in the delta of Prob. 10.11.

10.5   SUMMARY OF WYE AND DELTA CONNECTIONS

Table 10.1 summarizes relationships between phase and line voltages and phase and line currents for both the balanced delta and balanced wye connections. It is important to notice that line and phase voltages are equal in magnitude in the balanced

Table 10.1   Wye - Delta Relationships
(Magnitudes Only)

Wye	Delta
$V_L = \sqrt{3}\, V_p$	$V_L = V_p$
$I_L = I_p$	$I_L = \sqrt{3}\, I_p$

delta, whereas line and phase currents are equal in magnitude in the balanced wye. And, of course, $\sqrt{3}$ relates phase and line voltages in the wye and also phase and line currents in the delta.

10.6 POWER FACTOR IN THREE - PHASE CIRCUITS

You will remember that in a single-phase circuit the power factor is equal to the cosine of the phase difference between current and voltage when current and voltage directions are assigned such that this angle is zero (not 180°) for resistance. Power factor in a three-phase circuit is *not* the cosine of the angle between a line current and a line voltage, but rather is the cosine of the phase angle between a current and voltage in any particular phase. For example, in Fig. 10.17 (balanced resistances), the phase angle difference between I_{13} (phase current) and V_{13} (phase voltage) is the power-factor angle. Therefore, the power factor of this circuit is (cos 0) or 1. Of course, in a balanced situation the angle between current and voltage in any particular phase is the same as the corresponding angle in any other phase. In Fig. 10.17, we see that I_{12} is in phase with V_{12}. It is important to notice that for this unity-power-factor situation, I_1 (line current) is not in phase with any particular line voltage: V_{12}, V_{23}, V_{31}. Of course, the power factor in a wye connection is also equal to the cosine of the angle between the appropriate phase current and phase voltage. Again, power factor in a wye is not the cosine of the angle between any line current and any line voltage.

PROBLEMS

10.13 A three-phase wye-connected generator delivers power to a three-phase delta-connected motor, as indicated in Fig. 10.19. The phase rotation in this system is 1-2-3. It is known that each line current has a magnitude of 100 A, the power factor in the motor is 0.85 lag (phase current lags phase voltage). Line voltage has a magnitude of 450 V. Draw a phasor diagram showing, on one diagram: line voltages, line currents, and phase currents in the motor. Clearly show the power-factor angle on this diagram. You may choose any phasor as a reference.

10.14 For the system of Fig. 10.19, draw on a single diagram: line currents, phase voltages, and line voltages for the generator. Show complex values for all phasor quantities. Clearly show the power-factor angle.

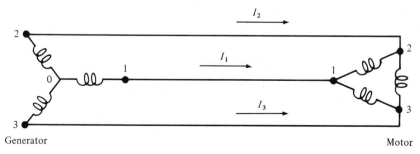

fig. 10.19   Three - phase generator delivers power to three - phase motor.

10.7   POWER IN THREE-PHASE CIRCUITS

In a balanced three-phase circuit, the power in any phase is equal to the power in any other phase of the same wye or delta. The power in one phase P_p is given by

$$P_p = I_p V_p \times \text{power factor} \tag{10.8}$$

Notice that P_p, I_p, V_p refer to magnitudes only. Or for a specific example, the power in phase 1-2 of the motor of Fig. 10.20 is given by

$$P_{12} = I_{12} V_{12} \cos\langle^{I_{12}}_{V_{12}} \tag{10.9}$$

In this equation, V_{12} refers to magnitude only, of course; I_{12} is magnitude of $\mathbf{I}_{12}$ and V_{12} is magnitude of $\mathbf{V}_{12}$. The power factor is the cosine of the angle between phasors $\mathbf{I}_{12}$ and $\mathbf{V}_{12}$.

PROBLEM

10.15   In the pattern of (10.9), write equations for P_{23} and P_{31} for the motor of Fig. 10.20.

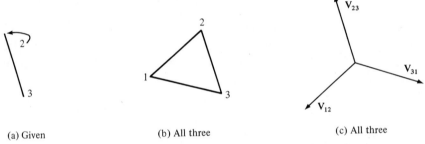

(a) Given   (b) All three   (c) All three

fig. 10.20   From one line voltage, find the other line voltages.

The total three-phase power in any balanced three-phase circuit (either wye or delta) is as follows:

$$P_3 = 3P_p = 3I_p V_p \times \text{power factor} \tag{10.10}$$

Equation (10.10) is reliable but frequently not useful because in a delta circuit very often we do not know the phase current, and in a wye circuit very often we do not know the phase voltage. So as a consequence of this, it is more useful to express three-phase power in terms of line voltage and line current, as shown in (10.11) for a delta system and in (10.12) for a wye system.

$$P_\Delta = 3I_p V_p \times \text{power factor} = 3\left(\frac{I_L}{\sqrt{3}}\right)V_L \times \text{power factor}$$

$$= \sqrt{3}I_L V_L \times \text{power factor} \tag{10.11}$$

$$P_Y = 3I_p V_p \times \text{power factor} = 3I_L \times \left(\frac{V_L}{\sqrt{3}}\right) \times \text{power factor}$$

$$= \sqrt{3}I_L V_L \times \text{power factor} \tag{10.12}$$

It is important to note that the power in a balanced three-phase system is independent of whether it is a wye or a delta. Keep in mind that these two equations (10.11 and 10.12) are in terms of effective values of current and voltage. Furthermore, you are reminded again that the power factor means the cosine of the angle between the particular phase current and its corresponding phase voltage, not the angle between the line current and line voltage.

Example 10.4

In a three-phase balanced system, we are given that the line voltage $V_{32} = 880$ $\underline{/75°}$ V, where the phase rotation is 1-3-2. Develop a phasor diagram showing all of the line voltages. Also give the numerical values of V_{12}, V_{23}, and V_{31}. Because the system is balanced, we know that the magnitude of each of the line voltages is 880 V. Then the problem is to find the appropriate phase angle of each line voltage. We also know that the three line voltages will form an equilateral triangle, where the line segment 3-2 of Fig. 10.21a is one leg of that equilateral triangle. Apex 1 might lie to the right or left of segment 3-2 depending on the phase rotation.

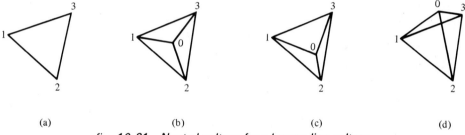

(a)   (b)   (c)   (d)

fig. 10.21   Neutral voltage from known line voltage.

Although you may be rather inexperienced with these kinds of phasor diagrams, trial and error procedure will soon lead you to the correct solution, as shown in (b). Notice that for phase rotation 1-2-3, apex 1 would have been to the right rather than to the left of segment 3-2. Having established the complete phasor diagram of line voltages, then we can readily write the numerical value of the phasor line voltages. First, V_{23} is 180° displaced from V_{32}, the given voltage; that is,

$$V_{23} = 880 \underline{/75° + 180°} = 880 \underline{/\ 105°} \text{ V} \tag{10.13}$$

Then from the known phase rotation, V_{12} must lag V_{23} by 120°; thus

$$V_{12} = 880 \underline{/105° - 120°} = 880 \underline{/-225°} = 880 \underline{/135°} \text{ V} \tag{10.14}$$

and V_{31} must lead V_{23} by 120°, as follows:

$$V_{31} = 880 \underline{/-105° + 120°} = 880 \underline{/15°} \text{ V} \tag{10.15}$$

● ● ●

PROBLEM

10.16 The phase rotation in the balanced system of Example 10.2 is changed to 1-2-3, but V_{32} is unchanged. Now develop the phasor diagram showing all line voltages, and write the numerical value of V_{12}, V_{23}, and V_{31}.

10.8 FINDING THE NEUTRAL VOLTAGE

In a balanced system where line voltages are known, as indicated in Fig. 10.21a, we sometimes need to find the neutral voltages. For the neutral voltage to exist, there must be a wye connection of impedances in order to form the neutral connection. If these three impedances are equal complex numbers, then the neutral voltages must form another balanced system, as indicated in Fig. 10.21b. From this neat symmetry, it is easy to find all three neutral voltages from a diagram such as Fig. 10.21b.

PROBLEM

10.17 For a balanced system, find the three neutral voltages V_{10}, V_{20}, and V_{30}, where $V_{21} = 44,000 \underline{/\,-40°}$ V. Phase rotation is 1-2-3.

Line voltages are typically, but not always, balanced as shown in Fig. 10.21. Furthermore, when line voltages are balanced, neutral or phase voltages are not necessarily balanced (see Fig. 10.21c and d). If balanced line voltages are applied to three unequal wye-connected impedances, then the neutral voltages will be unbalanced as illustrated by Fig. 10.21c and d. The position of the neutral 0 may lie inside of the equilateral line voltages, as shown in (c), or in some cases may lie outside, as shown in (d).

PROBLEMS

10.18 Balanced line voltages are applied to unequal impedances connected wye where $V_{21} = 44,000 \underline{/\,-40°}$ V and the phase rotation is 1-3-2. It is known that V_{20} is $37,000 \underline{/\,-90°}$ V.
   a. Find the complex value of the other two neutral voltages V_{10} and V_{30}.
   b. Support your answer to (a) by drawing an appropriate phasor diagram showing all line and neutral voltages.

10.19 Balanced voltages from a three-phase 60-Hz source are applied to a balanced-wye and a balanced-delta load, as indicated in Fig. 10.22. The three capacitors form a balanced wye and the three impedances **Z** form a balanced delta. **Z** consists of resistance of 200 Ω in series with inductance of 0.16 H. The capacitance of each capacitor is 10 μF. It is known that $I_{13} = 200 \underline{/\,-160°}$, the phase rotation of the balanced line voltages is 1-2-3. Develop a large phasor diagram showing all currents and voltages giving the numerical value of each. Apply the following consistency checks to your solution.

   1. $I_1 + I_2 + I_3 = 0$.
   2. The sum of the three capacitance currents is zero.
   3. $I_{12} + I_{23} + I_{31} = 0$.
   4. $I_1 = I_2 = I_3$ (magnitudes).
   5. $I_{12} = I_{23} = I_{31}$.
   6. Each capacitance current leads its phase voltage by 90°.
   7. $V_{10} + V_{20} + V_{30} = 0$.
   8. $V_{10} = V_{20} = V_{30}$.
   9. $V_{12} + V_{23} + V_{31} = 0$.
   10. $V_{12} = V_{23} = V_{31}$.
   11. I_1 leads I_2, which leads I_3 by 120°.
   12. I_{12} leads I_{23}, which leads I_{31} by 120°.

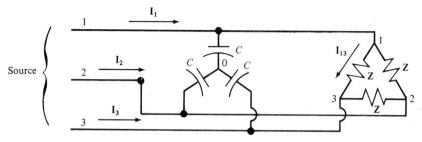

fig. 10.22   Balanced wye and delta loads from one source.

10.9   WYE-TO-DELTA AND DELTA-TO-WYE EQUIVALENTS

Every wye, be it balanced or unbalanced, has its equivalent delta—and, conversely, every delta has its equivalent wye. Since balanced circuits are typical, we will delay consideration of unbalanced wyes and unbalanced deltas.

Assume that we have a balanced delta having **Z** (complex) ohms in each leg, as shown in Fig. 10.22. In some situations the analysis can be simplified if we convert this delta to its equivalent wye. To be equivalent, the wye must produce the same complex line current, for a given set of line voltages, as its equivalent delta.

PROBLEMS

10.20   Prove that a wye consisting of three (**Z**/3) elements is equivalent to a delta consisting of three **Z** elements. See Fig. 10.23.

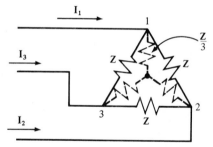

fig. 10.23   A wye that is equivalent to a delta.

10.21   Convert the wye of Fig. 10.22 to its equivalent delta. Then form one delta that is equivalent to the delta-wye combination of Fig. 10.22. Show that your new, single delta equivalent yields the same line currents (I_1, I_2, I_3) that you found in solving Prob. 10.19.

10.10   MEASUREMENT OF POWER

In the sinusoidal single-phase situation, you are familiar with the equation for average power:

$$P = IV \cos\langle^I_V = IV \times \text{power factor} \tag{10.16}$$

where P is the average power, I is the magnitude of the effective current, and V is the magnitude of the effective voltage. Let us briefly review the source of this equation. The development in Chapter 4 is briefly reviewed here. The instantaneous power (power as a function of time) is given by

$$p = iv \qquad (10.17)$$

where p is the power as a function of time, i is the current as a function of time, and v is the voltage as a function of time. Figure 10.24 is an illustration of such a situation. In this case, i and v are out of phase by 60°. Of course the average power P is the average height of the p wave. This is stated by

$$P = \frac{1}{T} \int_0^T p \; dt = \frac{1}{T} \int_0^T iv \; dt \qquad (10.18)$$

where T is period of any one of the waves. Of course, while i and v are symmetrical in respect to the horizontal axis, p is not symmetrical. The crosshatched area shown in Fig. 10.24 is a measure of the integral of (10.18). The net area between 0 and T divided by T is equal to the average power P. The average power in (a) is positive, but in (b) the average power is zero. The latter case is where the current and voltage are 90° out of phase. The circuit device that operates with zero average power as shown in (b) absorbs energy for part of the time, and then delivers the same energy back to the source circuit in the next interval of time. Either inductance or capacitance will perform this way.

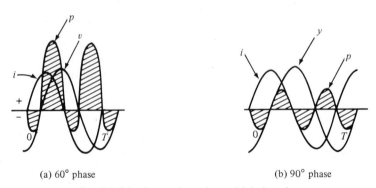

(a) 60° phase   (b) 90° phase

fig. 10.24   i, v, p in a sinusoidal situation.

PROBLEMS

10.22 Without referring back to the earlier proof in the text, prove (10.16). It is important that you note that I and V represent effective values of current and voltage, respectively, while P is the average value of power. Except in special cases, one should not attempt to relate average power to average current and/or average voltage. Also, one should not speak of effective value of power. No such term is formally defined. An attempt to use such a term will almost surely confuse the issue.

10.23   Assume that the voltage applied to a circuit element is sinusoidal of angular velocity ω and the resultant current is also sinusoidal but of triple frequency, as follows:

$$v = 10 \sin \omega t \qquad i = 5 \sin 3\omega t \qquad\qquad (10.19)$$

This would be difficult if not impossible to achieve in the practical world, but the assumed situation has important conceptual interest. Now we would like to find the average power P as a consequence of this voltage and current. We can find this average power by applying (10.18). Before doing this, let us attempt to get the answer merely by some simple sketches. On one set of coordinates, sketch both v and i. Then sketch p. Is the answer to our situation obvious? Now formally apply (10.18) by making the appropriate integration.

The important conclusion to draw from solving Prob. 10.23 is that there can be no average power in the combination of voltage of one frequency and the current of an integral frequency. Only like frequencies in current and voltage lead to average power.

PROBLEM

10.24   Assume another situation that may be very difficult to achieve in the real world, as indicated in Fig. 10.25. You will remember from your Fourier analysis that each of these waves (square and triangular) has only odd sine terms (when the coordinate system is appropriately selected). The combination of the fundamental voltage and the fundamental current will contribute to the average power. The combination of the third harmonic voltage and the third harmonic current will contribute to the average power, and so forth for each of the odd harmonics. To approximate the average power, estimate the fundamental component of the voltage and the fundamental component of the current and note that these two waves will not be in phase.

   a.   Then approximate the average power, assuming that the total power is derived from fundamental components only.
   b.   Without the use of the Fourier components of the two waves, find the precise value of the average power by applying (10.18) to separate regions where i and v are easily expressed.
   c.   Now to obtain a more accurate answer based on the use of the Fourier components, refer back to the Fourier treatment of these two waves to find the harmonic content of each wave. Calculate P so that the error by this method is not more than 2 percent.

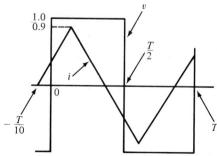

fig. 10.25   i and v leads to P(average power).

Wattmeters that Measure P

The wattmeter is an instrument that measures the average power where current and voltage are periodic, usually but not necessarily sinusoidal. These instruments perform the multiplication and the integration as indicated in (10.18). The traditional wattmeter having an indicating *needle* functions on a magnetic interaction between a fixed coil, called the current coil, and a moving coil (needle attached) called the voltage or potential coil. We will not treat the detail here, but the instantaneous force on the moving coil is sensitive to the instantaneous product of the current in the current coil and current in the potential coil. Then, in turn, the current in the potential coil is sensitive to the circuit voltage under consideration. Suffice it to say, the instrument does in fact perform the multiplication and integration indicated in (10.18). Any other more modern wattmeter (perhaps digital) would have to, of necessity, perform the corresponding multiplication and integration.

To measure the average power into an electrical load, the wattmeter having two pair of terminals is connected as indicated in Fig. 10.26. The upper two terminals on the wattmeter are called the current terminals, and these are connected such that the current to the load passes from one current terminal through the current coil in the wattmeter to the other current terminal. The terminals to the potential coil are shown as the lower two terminals in this view. This is a relatively high-resistance circuit, so the current passing through this coil ideally is negligibly small compared to the current to the load. This coil is made of relatively fine wire and a resistor is connected in series with it as indicated in Fig. 10.26. One of the current terminals is

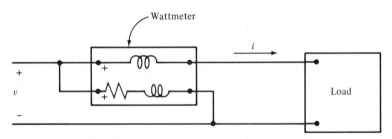

fig. 10.26   Wattmeter measures P into load.

marked with a + (or some other similar designation) and one of the potential terminals is also marked with a +. When the wattmeter is connected in respect to the source and the load as shown, the wattmeter will indicate the *positive* power flow into the load. Or, in different words, the product of these particular i and v gives a positive (upscale) deflection of the wattmeter movement if in fact the "load" absorbs power. A different view is given in Fig. 10.27a. At an instant when both i and v, assigned directions in (a), are positive, the power p at that instant will be toward the right. At that instant, the wattmeter sees positive instantaneous power p and the movement will tend to deflect upscale or positive. The inertia of the meter movement averages the instantaneous power to indicate the average power P. On the other hand as shown in (b) for this combination of current and voltage, the

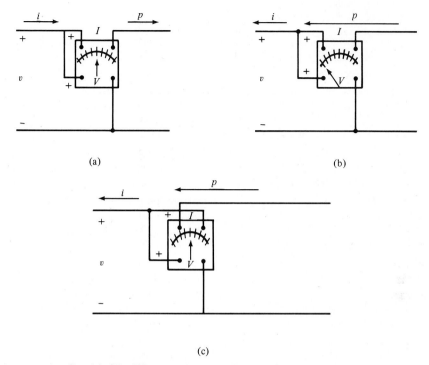

(a)

(b)

(c)

fig. 10.27   Wattmeter connections and power direction.

instantaneous power p is toward the left at the instant both this i and this v are positive. For this wattmeter connection, this i and this v—the wattmeter indication is *downscale* or against the left *endpost*. If the average power P were in fact flowing to the left as shown in (c) and the connections to the current terminals (or voltage terminals) were reversed as shown, then the wattmeter would read upscale.

Measurement of Three - Phase Power

Three leads or wires are involved in most of our three-phase circuits; occasionally a connection is made to the neutral of a wye to form a four-wire three-phase circuit. With three wires or leads connected to a three-phase load or source, as indicated in Figs. 10.28 and 10.29, the total power entering that load or source can be measured with two wattmeters connected as shown in Figs. 10.28 and 10.29. It is not obvious, but it is nevertheless true, that the total power is equal to the algebraic sum of the two wattmeter indications.

$$P_{\text{total}} = W_1 + W_3 \tag{10.20}$$

where W_1 and W_3 are the indicated power of the respective wattmeters. One must remember that either one or both of these indications might be negative. If the total power flow is to the right into a *load*, then obviously the wattmeter indication of

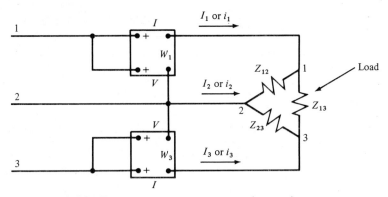

fig. 10.28 Two wattmeters to measure three - phase power.

larger magnitude must be positive. The other may be negative or positive. If one of the indications is negative for the connection shown, then to determine the magnitude of that indication, current or voltage leads must be reversed as indicated in Fig. 10.27c to observe the magnitude of the negative indication. Equation (10.20) is true whether the *box* absorbs average power or delivers average power. Or, in other words, this equation is true whether the average power flow is to the right or to the left. This equation is also true whether the situation is balanced or not. Both voltages and currents may be unbalanced. The concept is even more general than what is implied by Fig. 10.29. There are two wattmeters and three leads or wires in this situation. Consider another situation as indicated in Fig. 10.30 where there are two networks (*A* and *B*) with *n* wires, five in this case, or interconnections. Either network or both may contain one or more sources. More than one frequency might be involved. For example, network *A* might contain a dc source and a 1,000-Hz source and network *B* might contain a 5,000-Hz source. Waveforms do not need to be sinusoidal. Then under this general situation where there are *n* leads, $n - 1$ wattmeters can be used to measure the total power flow from one network into the other. The total power flow is the algebraic sum of the wattmeter indications. In Fig. 10.29, notice that wattmeter 1 senses the current in line 1; wattmeter 3 senses the

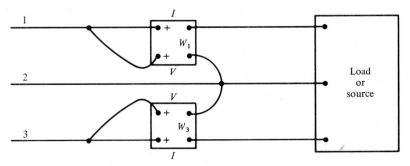

fig. 10.29 Two wattmeters — three leads.

current in line 3; line 2 is common to the two voltage circuits. So by following that pattern of connections, the three-wire situation can be extended to an *n*-wire situation. One merely needs to select one of the wires common to all of the potential circuits. One might arbitrarily select lead or wire 2 in Fig. 10.30 as the one that will be common to the four separate potential circuits.

PROBLEM

10.25   In the circuit of Fig. 10.30, draw a circuit diagram showing the connections for four wattmeters so that the total power flow from network *A* to network *B* is given by the algebraic sum of the four separate wattmeter readings.

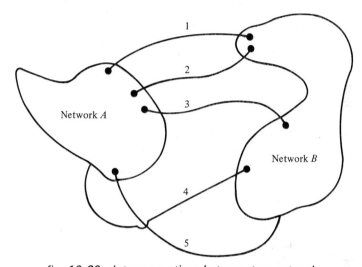

fig. 10.30   Interconnections between two networks.

It is important to note that in the situation of Fig. 10.29, or in any other more general situation, any one wattmeter indication is fictitious in that one cannot relate that indication to any particular combination of powers in the load or source. Yet as strange as it might seem, the algebraic sum of the various wattmeter indications is a measure of the total power. The proof of this concept in general is not extremely difficult, but it seems appropriate to prove a restricted case, as illustrated in Fig. 10.28. Here the load is delta, but not necessarily balanced. Furthermore, the line voltages are not necessarily balanced. The instantaneous wattmeter powers p_1 and p_3 and then the average indicated powers W_1 and W_3 for the respective wattmeters are:

$$p_1 = i_1 v_{12} \qquad\qquad p_3 = i_3 v_{32}$$

$$W_1 = \frac{1}{T} \int_0^T i_1 v_{12}\, dt \qquad W_3 = \frac{1}{T} \int_0^T i_3 v_{32}\, dt \qquad\qquad (10.21)$$

Then respective wattmeter currents i_1 and i_3 are given by

$$i_1 = i_{12} + i_{13} \qquad \text{and} \qquad i_3 = i_{32} + i_{31} \tag{10.22}$$

Remembering that $i_{31} = -i_{13}$ and $v_{32} = -v_{23}$, it follows that

$$p_1 + p_3 = i_{12}v_{12} + i_{32}v_{32} + i_{13}(v_{12} + v_{23}) \tag{10.23}$$

Then from Kirchhoff's voltage law,

$$v_{12} + v_{23} + v_{31} = 0 \qquad \text{or} \qquad v_{12} + v_{23} = v_{13} \tag{10.24}$$

Combining (10.23) and (10.24),

$$p_1 + p_3 = i_{12}v_{12} + i_{23}v_{23} + i_{31}v_{31} \tag{10.25}$$

But the instantaneous powers in the respective legs of the delta are:

$$p_{12} = i_{12}v_{12} \qquad p_{23} = i_{23}v_{23} \qquad p_{31} = i_{31}v_{31} \tag{10.26}$$

Now by comparing (10.25) with (10.26), the total power is truly identically equal to $W_1 + W_3$, the sum of the two wattmeter indications. Of course we are interested in the total average power P_t, which is given by

$$P_t = \frac{1}{T} \int_0^T (p_{12} + p_{23} + p_{31})\, dt = \frac{1}{T} \int_0^T (p_1 + p_3)\, dt \tag{10.27}$$

Then it follows that the total average power in the complete delta is

$$P_t = W_1 + W_3 \tag{10.28}$$

where W_1 and W_3 are the respective wattmeter indications.

Now if you will look back at the development of this proof, you will find that at no point did we imply that the voltages were three-phase or any other phase or magnitude relationship. Nor did we specify any sort of relationship among the currents i_{12}, i_{23}, and i_{31}.

We need to look at the wattmeter once more to be able to predict reliably its indication in terms of its respective phasor current and phasor voltage applied to its terminal (see Fig. 10.31). Where the current $\mathbf{I}$ is assigned a direction into the $+$

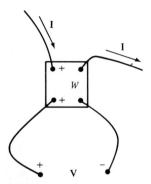

fig. 10.31   Wattmeter indication in terms of phasors.

terminal of the current circuit of the wattmeter and **V** is assigned a positive direction connected to the positive terminal of the wattmeter, the wattmeter indication is given by

$$W = IV \cos\langle_{\mathbf{V}}^{\mathbf{I}} \tag{10.29}$$

If there are any traps in calculating wattmeter indication, it has to do with this subtle point with respect to (10.29) that the current **I** is assigned a direction into the + terminal and the + side of the assigned direction of **V** is connected to the + side of the voltage circuit of the wattmeter.

Example 10.5

Two wattmeters are connected as indicated in Fig. 10.32 to measure the total power into a wye connection consisting of three equal impedances. The magnitude of each of the sinusoidal line voltages is 880 V. The power-factor angle of each impedance is 75° with the current lagging the voltage. The magnitude of the impedance of each leg is 10 Ω, the phase rotation is 1-3-2. Find the indication of each wattmeter and the total power.

$$I_L = \frac{880}{10\sqrt{3}} = 50.81 \text{ A}$$

$$P_{\text{3-phase}} = \sqrt{3} I_L V_L \cos 75° = 20.04 \text{ kW} \tag{10.30}$$

or by an alternate approach,

$$P_{\text{1-phase}} = I_p^2 R_p = \left(\frac{V_p^2}{Z_p^2}\right) \times Z_p \cos 75°$$

$$= \frac{V_p^2 \cos 75°}{Z_p} = \frac{V_L^2 \cos 75°}{3Z_p} \tag{10.31}$$

$$P_{\text{3-phase}} = \frac{V_L^2 \cos 75°}{Z_p} = \frac{880^2 \cos 75°}{10} = 20.04 \text{ kW} \tag{10.32}$$

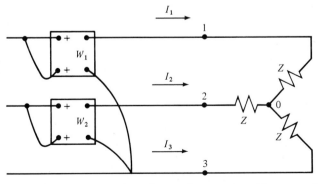

fig. 10.32   *Two wattmeters measure three - phase power. Phase rotation 1 - 3 - 2.*

• • •

Now to find the indication of wattmeter 1, review (10.29). We already have the magnitude of the wattmeter current and wattmeter voltage. These are line current and line voltage, respectively. But we do not have the phase angle difference between the appropriate wattmeter current and voltage, as indicated in (10.29). The angle we seek is the angle between I_1 and V_{13}. Make sure that you understand the reasons why this is the significant angle. Now we prepare a phasor diagram as indicated in Fig. 10.33 to establish this phase angle difference. The line voltages are shown by the equilateral triangle of Fig. 10.32. The angle of V_{10} is clearly shown to be 90°. The angle of I_1 is 75° behind voltage V_{10}. Because the power-factor angle of each impedance element is given as 75° lag, then the angle of I_1 is 15°. Now it follows that

$$W_1 = I_L V_L \cos\langle^{I_1}_{V_{13}} \qquad (10.33)$$

$$I_L = 50.81 \qquad V_L = 880 \qquad I_1 = I_L \; \underline{/\; 15°} \qquad V_{13} = V_L \; \underline{/\; 120°}$$

$$\cos\langle^{I_1}_{V_{13}} = \cos(15° - 120°) = \cos(-105°) = \cos(105°)$$

$$W_1 = 50.81 \times 880 \cos(105°) = -11.57 \text{ kW}$$

In a similar fashion, the indication of wattmeter 2 is given by

$$W_2 = I_L V_L \cos\langle^{I_2}_{V_{23}} \qquad (10.34)$$

Again from the phasor diagram, we find

$$I_2 = I_L \; \underline{/\; 135°} \qquad V_{23} = V_L \; \underline{/\; 180°}$$

$$W_2 = 50.81 \times 880 \cos(45°) = 31.62 \text{ kW}$$

$$W_1 + W_2 = -11.57 + 31.62 = 20.05 \text{ kW} \qquad \text{check}$$

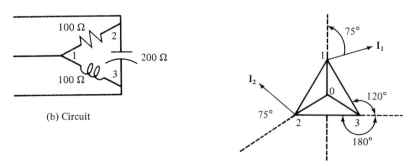

(b) Circuit

fig. 10.33 Wattmeter phasors. Angle of Z (or power factor angle) = 75° lag.

PROBLEMS

10.26 Assume that the impedance elements of Example 10.5 are connected delta. Line voltage and phase rotation in this situation are the same as in the example. Calculate the total power. The fact that line voltage is now applied to each element compared to the neutral or phase voltage in the example should lead you to an immediate value for the total power. If you are uneasy about this shortcut, calculate line current and then proceed to calculate total power. In terms of wattmeter current and wattmeter voltage, calculate each wattmeter indication. Check the sum of these two against the total power.

10.27   Repeat Prob. 10.26, but for phase rotation 1-2-3.

10.28   A three-phase induction motor delivers 590 kW through its shaft to mechanical load. Its efficiency is 0.92. Its power factor is 0.88 lag. Show a diagram of connections to measure the input power. Calculate the two separate wattmeter indications. Calculate the total power input in terms of output and efficiency. Check the sum of the two wattmeter indications against this total power. Are you concerned about the fact that the line voltage is not given in this problem?

10.11   *POWER FACTOR AND TWO WATTMETERS*

It is significant that the connection of the two wattmeters, as we have discussed here, involved only line currents and line voltages. Obviously in a balanced wye if the neutral were available, one wattmeter could be used to measure the power in one phase, and the total power would be three times the power in one phase. Or in the case of a delta, if it were practical to open one phase in order to insert the current circuit of the wattmeter, the power in that phase could be measured directly. In the typical situations it is not practical to measure power in one phase alone, but rather all measurements need to be made in terms of line current and line voltage. Sometimes we need to know the power factor in a balanced load. Assume that only line currents and line voltages are available for purposes of these measurements. It now turns out that power factor can be determined, in balanced three-phase systems, with the two wattmeter connections as shown in such diagrams as Figs. 10.29, 10.28, and 10.32. Furthermore, we need not ask ourselves whether the load is wye or delta, or some combination of these two. In Fig. 10.34, the phasor diagram (c) applies equally well to the wye shown in (a) or the delta shown in (b). The particular phase

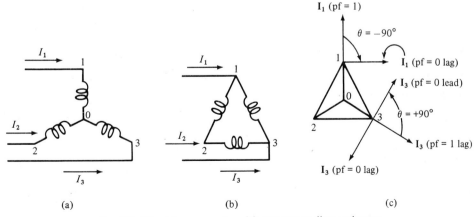

fig. *10.34*   *Line currents with respect to line voltages.*

rotation is 1-3-2 and the reference phasor happens to be V_{32}. This phasor diagram applies directly to the wye where the neutral 0 is available. Notice that V_{10} is at an angle 90° and for unity power factor, I_1 would have the same angle, as indicated in (c). Now if the power factor were 0 lag, then I_1 would be 90° behind V_{10} or at angle 0 for this reference system. Then if we consider I_3, of course it will be in phase with V_{30} at unity power factor. For zero power factor lead and lag, respectively, I_3 takes the indicated positions. Thus it becomes very easy to represent the phase position of line currents with respect to line voltages in a balanced wye. After a little thought, it will become obvious that the diagram of (c) applies equally well to the delta of (b). Let us check this assertion out for unity power factor. For unity power factor, I_{13} is in phase with V_{13}. The phase angle here is 120°. Likewise, I_{12} is in phase with V_{12}. Now, of course, $I_1 = I_{12} + I_{13}$, and this sum has an angle of 90°. We could also argue that every delta has its equivalent wye and that the phase angle in each impedance element of equivalent wye is identical to the phase angle of that impedance in the delta. In transforming from delta to wye, the magnitude of impedances has to be adjusted, but the phase of impedances will not change. So we can safely conclude that we need not know whether the load or source is wye or delta. The one diagram as shown in (c) will represent either situation.

Assume that the wattmeter connection of Fig. 10.29 is made in the line leads of a balanced load, either wye or delta. The phase angle seen by wattmeter 1 in this case is the angle between I_1 and V_{12}. Let θ represent the power-factor angle such that θ is negative for current lagging voltage; see Fig. 10.34c. Now for the wattmeter connection of Fig. 10.28 and the phase rotation of Fig. 10.34, it follows that wattmeter 1 indication is given by

$$(I_1, V_{12})W_1 = I_L V_L \cos(\theta + 30°) \tag{10.35}$$

Notice that this wattmeter sees I_1 and V_{12}, and the phase diagram tells us that the phase angle difference between this current and voltage is $\theta + 30°$. On the other hand, wattmeter 3 sees I_3 and V_{32}, and the phase angle between these two is $\theta - 30°$:

$$(I_3, V_{32})W_3 = I_L V_L \cos(\theta - 30°) \tag{10.36}$$

PROBLEM

10.29 Starting with (10.35) and (10.36), prove that the sum of the two wattmeter readings is equal to the total power in the load.

In solving Prob. 10.29, you demonstrated again that

$$W_1 + W_3 = \sqrt{3}\, I_L V_L \cos\theta \tag{10.37}$$

PROBLEM

10.30 Similar to that of $(W_1 + W_3)$, find an expression for $(W_1 - W_3)$ and then demonstrate that the following relation is true:

$$\theta = \tan^{-1}\frac{(W_1 - W_3)\sqrt{3}}{W_1 + W_3} \tag{10.38}$$

Note that θ of (10.38) is the power-factor angle and it is given in terms of the two wattmeter readings W_1 and W_3. With these measurements, it is not necessary to know line current and/or line voltage in order to establish the power factor, the cosine of θ.

10.12   POWER IN UNBALANCED SITUATIONS

The measurement of power in unbalanced wyes and deltas introduces some awkwardness but no fundamental stumbling block. Consider the unbalanced situation of Fig. 10.35. The line voltages are shown in (a) and have the values indicated. Of course the impedances are unbalanced.

(a) Line voltages
$\mathbf{V}_{31} = 219 \,\underline{/17°}$
$\mathbf{V}_{32} = 70 \,\underline{/65°}$

(b) Circuit

fig. 10.35   Unbalanced delta.

PROBLEMS

10.31 The unbalanced voltages of Fig. 10.35a are applied to the balanced delta of Fig. 10.35b. Connect two wattmeters like that shown in Fig. 10.29 to measure the total power in this unbalanced delta. In this unbalanced delta, all of the average power, of course, is absorbed in the 100-Ω resistor. From the data given, find $\mathbf{V}_{12}$ and then calculate the total power in this delta.

10.32 For the indicated wattmeter connection in the unbalanced delta of Fig. 10.35, determine the two wattmeter indications. The sum of these two indications should be the total power, as you determined from Prob. 10.31. Add phasors to the diagram of (a) to show the two wattmeter currents $\mathbf{I}_1$ and $\mathbf{I}_3$.

10.33 In the situation of Prob. 10.32 interchange the capacitor and the inductor and again find W_1 and W_3.

10.34 Given the impedance elements of Fig. 10.35b, rearrange these elements in a wye such that the resistor is connected from line 1 to neutral, the inductor from line 3 to neutral, and the capacitor from line 1 to neutral. The line voltages for this new system are the same as those specified in Fig. 10.35. The wattmeters are still connected the same, as shown in Fig. 10.30. Now under these circumstances, find the total power, W_1, and W_3.

We immediately recognize that this is a more difficult problem compared to the unbalanced delta situation. Though we know the line voltages, we do not immediately know the neutral voltage. However, once the neutral voltage is determined, the problem could proceed without *snags*. Then in your first step, determine the neutral voltage. Here it is convenient to think of one of the line leads as the reference, even though you might be tempted to think of the neutral as a reference. Remember, it is the neutral voltage that we wish in respect to one of the line leads. So, for example, select line lead 1 as the reference, and then proceed to find the voltage $\mathbf{V}_{01}$. To eliminate the problems connected with simultaneous equations, apply the nodal-superposition approach learned earlier in the text.

Having found voltage $\mathbf{V}_{01}$, represent this voltage on the phasor diagram of (a) so that all line-to-neutral voltages as well as line-to-line voltages are displayed on the one diagram. Then proceed to determine total power and then W_1 and W_3.

Mathematics of Complex Numbers

A.1   WHY COMPLEX NUMBERS IN ENGINEERING?

In the minds of many students, there seems to be confusion about the use of "imaginary" numbers in describing "real" physical systems. Although a typical student has encountered the quantity $\sqrt{-1}$ in high-school mathematics courses, there is often a vagueness about what this "imaginary" number really is and how it can be used in engineering design and analysis. Consequently, we shall first consider an example of a simple physical system, a mass on a spring, and try to show how imaginary numbers come to be used in describing such systems. This appendix is intended to be a review of the mathematics of complex numbers, not a rigorous and complete development of the mathematics.

An Example

Figure A.1 shows a diagram of a mass on a spring, with the definition that x is the displacement of the mass from its equilibrium position, that is, the position when the mass is hanging on the spring with no motion. Assuming that frictional forces can be neglected, the basic differential equation obtained from Newton's law and Hooke's law is

$$m \frac{d^2x}{dt^2} + kx = 0 \tag{A.1}$$

where k is the spring constant. Let's find an expression for x when the mass is held at some initial displacement and released at $t = 0$, and when the initial conditions are

$$x = x_0 \qquad \text{at } t = 0 \tag{A.2}$$

$$\frac{dx}{dt} = 0 \qquad \text{at } t = 0 \tag{A.3}$$

The second equation is a statement that the velocity is zero at $t = 0$.

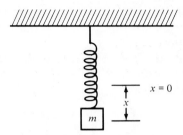

fig. A.1   A mass suspended on a spring. The displacement from the rest position is x.

Following the standard methods of solving linear differential equations (see Chapter 3), we use the trial solution

$$x = Ae^{st} \tag{A.4}$$

Upon substitution of (A.4) into (A.1), we find that s must satisfy

$$s^2 = -k/m$$

From this equation, we find the two values of s:

$$s_1 = j\sqrt{k/m}$$
$$s_2 = -j\sqrt{k/m}$$

where j is the familiar complex quantity defined by $j = \sqrt{-1}$. Consequently, the solution is

$$x = Ae^{j\sqrt{k/m}\,t} + Be^{-j\sqrt{k/m}\,t} \tag{A.5}$$

The initial conditions must be used to find the arbitrary constants A and B. Differentiating (A.5) to find the velocity gives us

$$\frac{dx}{dt} = j\sqrt{k/m}\,(Ae^{j\sqrt{k/m}\,t} - Be^{-j\sqrt{k/m}\,t}) \tag{A.6}$$

For (A.5) and (A.6) to satisfy (A.2) and (A.3), respectively, it must be true that

$$x_0 = A + B$$
$$0 = A - B$$

Simultaneous solution of these two equations yields

$$B = A$$
$$A = x_0/2$$

The final expression for the displacement is

$$x = x_0\left(\frac{e^{j\sqrt{k/m}\,t} + e^{-j\sqrt{k/m}\,t}}{2}\right) \tag{A.7}$$

From (A.7), it is clear that the expression for the displacement includes the imaginary operator j, yet we know that the displacement is a measurable quantity, and hence the values of x must be real.

We can gain insight into this puzzle by solving (A.1) another way, thus arriving at an equivalent but different expression for x. Since we know from experience that the motion of the mass is an oscillation, let's use the trial solution

$$x = C \cos \alpha t + D \sin \alpha t \tag{A.8}$$

from which we get

$$\frac{d^2 x}{dt^2} = -\alpha^2 (C \cos \alpha t + D \sin \alpha t)$$

Substitution into (A.1) gives us

$$(-\alpha^2 + k/m)(C \cos \alpha t + D \sin \alpha t) = 0$$

Equation (A.8) will therefore satisfy (A.1) if

$$\alpha^2 = k/m$$

Then solving for the two values of α, we have

$$\alpha_1 = \sqrt{k/m}$$
$$\alpha_2 = -\sqrt{k/m}$$

and C and D are arbitrary constants to be found from the initial conditions. Using either α_1 or α_2 in (A.8) and requiring (A.8) to satisfy the initial conditions gives the same result:

$$x = x_0 \cos \left(\sqrt{k/m} \, t \right) \tag{A.9}$$

With the solution in this form, we see that the displacement is a sinusoidal function of time, with all values of x being real.

The Euler Formulas

If we have done our work correctly, we would expect that since (A.7) and (A.9) are equivalent expressions, it must follow that

$$\cos \sqrt{k/m} \, t = \frac{e^{j\sqrt{k/m}\,t} + e^{-j\sqrt{k/m}\,t}}{2}$$

Although we have not given a formal proof, the relation nevertheless is true. A similar relation can be obtained by taking the derivatives of (A.7) and (A.9); we then get

$$\sin \sqrt{k/m} \, t = \frac{e^{j\sqrt{k/m}\,t} - e^{-j\sqrt{k/m}\,t}}{2j}$$

The generalizations of these results are

$$\cos \theta = \frac{e^{j\theta} + e^{-j\theta}}{2} \tag{A.10}$$

$$\sin \theta = \frac{e^{j\theta} - e^{-j\theta}}{2j} \tag{A.11}$$

which are called *Euler's formulas.* By adding and subtracting (A.10) and (A.11), we can get two other very useful results:

$$e^{j\theta} = \cos \theta + j \sin \theta \tag{A.12}$$

$$e^{-j\theta} = \cos \theta - j \sin \theta \tag{A.13}$$

These last two relations are also sometimes called Euler's formulas, or Euler's identities.

Perhaps now you can see from our simple example and Euler's formulas how imaginary numbers enter into the analysis of physical systems. The expression (A.7) for x contains j's, but the sum of the two terms containing j's is real. Euler's formulas show us how complex numbers (numbers that can be written as the sum of a real number and an imaginary number) can be used in combination to describe physical systems. It always happens when complex numbers appear in expressions describing physical quantities that they appear in combinations like the one in (A.7) which gives real numbers.

Advantages in Using Complex Numbers

Why ever use complex numbers? Why not use the form in (A.9), which contains no complex numbers, in preference to the form in (A.7), which does? The answer is that the use of complex numbers can result in significant advantage in analysis and design, as pointed out in Chapter 4 in the development of phasor transforms. The advantage stems from relations like

$$\frac{d}{dt} e^{j\omega t} = j\omega e^{j\omega t} \tag{A.14}$$

where the derivative of the exponential is equal to the exponential itself times a constant, in contrast to

$$\frac{d}{dt} \cos (\omega t) = -\omega \sin \omega t$$

where the derivative is a different function. As you will note from the derivations in Chapter 4, it is the characteristic form of the derivative in (A.14) that makes the definition of impedance possible. Thus if we did not use complex numbers, we could not define impedance—a very powerful concept.

A.2 REPRESENTATION OF COMPLEX NUMBERS

There are two principal ways to represent complex numbers, the rectangular form and the polar form. Each has its advantages and disadvantages, as pointed out below.

Rectangular Form

A rectangular complex number is a number which can be written as the sum of a real and an imaginary part, such as

$$\mathbf{A}_1 = 3 + j4$$

$$\mathbf{A}_2 = -2 + j3$$

$$\mathbf{A}_3 = 1 - j3$$

Notice that complex numbers such as $\mathbf{A}_1$, $\mathbf{A}_2$, and $\mathbf{A}_3$ are clearly shown in boldface type. In the expression for $\mathbf{A}_1$, 3 is said to be the *real component* of $\mathbf{A}_1$ and $j4$ is said to be the *imaginary component* of $\mathbf{A}_1$. Any complex number can always be written in the form

$$\mathbf{A} = a + jb \qquad (A.15)$$

where a and b are real numbers; a is the real component of A and jb is the imaginary component of $\mathbf{A}$.

A convenient graphical representation of complex numbers in rectangular form is shown in Fig. A.2. The axes represent the *complex plane*, with the horizontal axis representing real components and the vertical axis representing imaginary com-

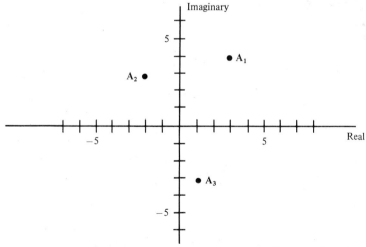

fig. A.2 *Graphical representation of complex numbers in the complex plane. Shown are $\mathbf{A}_1 = 3 + j4$, $\mathbf{A}_2 = -2 + j3$, and $\mathbf{A}_3 = 1 - j3$.*

ponents. Any point in the complex plane can be located by specifying the proper real component and the proper imaginary component, and any complex number can be represented by a point in the complex plane.

Polar Form

The second way to represent complex numbers is called the *polar form*. Any point in the complex plane can also be located by specifying the magnitude and angle of a directed line segment from the origin to the point (Fig. A.3). The angle is specified with respect to the positive real axis. The polar form of $\mathbf{A}_1$ shown in Fig. A.3, which is based on the location of the point by a vector, is

$$\mathbf{A}_1 = 5e^{j53.1^\circ}$$

where 5 is the magnitude (length) and 53.1° is the angle between the directed line and the real axis.

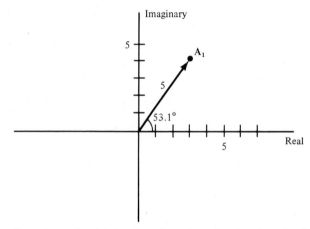

fig. A.3 *Location of a point in the complex plane by the magnitude and angle of a vector from the origin to the point.*

Any complex number can be represented in polar form like

$$\mathbf{A} = Ae^{j\theta} \tag{A.16}$$

where A means the magnitude of $\mathbf{A}$ (regular type for real numbers; boldface for complex), and θ is the angle of the complex quantity (Fig. A.4). A shorthand notation often used in engineering for (A.16) is

$$\mathbf{A} = Ae^{j\theta} = A \underline{/\theta}$$

where the magnitude and the angle are written as shown, with no exponential. The two terms are defined to be equivalent, but sometimes we prefer the exponential notation for some mathematical manipulations.

From a mathematician's point of view, the angle in the exponent should be

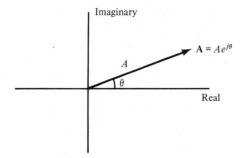

fig. A.4   Representation of any complex number in polar form.

written in radians, like $e^{j\pi/4}$ or $e^{j0.6}$, but it has become the practice in engineering to write the angle either in radians or in degrees, like $e^{j47°}$. The convention that we shall adopt is that if the degree sign is not present, the angle is in radians; if the degree sign is present, the angle is obviously expressed in degrees.

We should also point out that

$$e^{j(\theta + 2\pi)} = e^{j\theta}e^{j2\pi} = e^{j\theta}$$

since $e^{j2\pi} = 1$. Thus $(\theta + 2\pi)$ is said to be *equivalent* to θ. In terms of the complex plane, this simply means that there is no way to tell how many times you have gone completely around the origin in measuring an angle from the real axis. We usually specify the angles as being between $-360°$ and $360°$ or between -2π and 2π radians. Also note that $240°$ and $-120°$ are equivalent angles.

Conversion From One Form to the Other

It is often desirable to convert a complex number from the rectangular form to the polar form and vice versa. The conversion relations are easily obtained from the graphical representations, as shown in Fig. A.5. We can get the polar form of $\mathbf{A}_1$

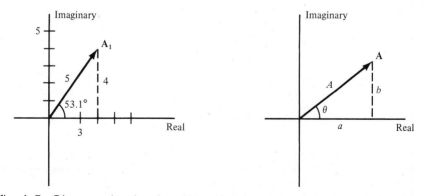

fig. A.5   Diagrams showing the relations between the rectangular and polar forms.

from the rectangular form by noting that A_1 is given by

$$A_1 = \sqrt{3^2 + 4^2} = 5$$

and θ_1 is given by

$$\theta_1 = \tan^{-1} \tfrac{4}{3} = 53.1°$$

The rectangular form can be obtained from the polar form by noting that

$$\text{Re}(\mathbf{A}_1) = 5 \cos 53.1° = 3$$

and

$$\text{Im}(\mathbf{A}_1) = j5 \sin 53.1° = j4$$

Where $\text{Re}(\mathbf{A}_1)$ means the real component of $\mathbf{A}_1$ and $\text{Im}(\mathbf{A}_1)$ means the imaginary component of $\mathbf{A}_1$.

In general, the corresponding conversion relations for any complex number $\mathbf{A}$ represented by (A.15) or (A.16) are

$$A = \sqrt{a^2 + b^2} \qquad (A.17)$$

$$\theta = \tan^{-1}(b/a) \qquad (A.18)$$

$$a = A \cos \theta \qquad (A.19)$$

$$b = A \sin \theta \qquad (A.20)$$

which relations can easily be obtained from the right triangle shown in Fig. A.5.

Note that (A.19) and (A.20) can also be obtained by using Euler's formula (A.12) and (A.13):

$$A = A(\cos \theta + j \sin \theta) = A \cos \theta + jA \sin \theta$$

Hence,

$$a = A \cos \theta \qquad b = A \sin \theta$$

Hand calculators having moderate capability will convert polar complex numbers to rectangular complex numbers (or the reverse) in one operation. You should become familiar with the capability of your own calculator.

In applying (A.18), you must be careful to get θ in the correct quadrant. For example, let's write $\mathbf{A}_2 = -2 + j3$ in polar form. Then

$$\theta_2 = \tan^{-1}\left(\frac{3}{-2}\right) \qquad (A.21)$$

Now, if you make the indicated division and find the arctangent as suggested by (A.21), you might get $\theta_2 = -56.3°$, depending on your calculator. But if you look at Fig. A.6, you can see that θ_2 is certainly not $-56.3°$. The problem is that the minus sign goes with the denominator (the 2) in (A.21), but this procedure did not distinguish between $3/(-2)$ and $-(3/2)$. On the other hand, if your calculator is programmed to make this conversion from rectangular to polar coordinates, probably it will give the correct angle, $123.7°$ in this case.

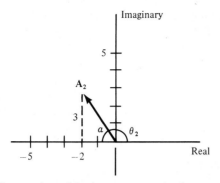

fig. A.6   *Conversion of **A**₂ from rectangular form to polar form.*

You always should sketch the rectangular complex number and make a mental approximation of the equivalent polar number as a check against your calculator solution.

Conjugate Numbers

The conjugate of a complex number $\mathbf{A} = (a + jb)$ is designated as A^* and is given by $A^* = (a - jb)$. The conjugate of a complex number is obtained by changing the sign of the imaginary component of the number; or stated another way, by replacing j with $-j$. Thus the conjugate of $2 + j3$ is $2 - j3$. The conjugate of $-4 - j3$ is $-4 + j3$.

The conjugate of a complex number in polar form can also be obtained by replacing j with $-j$. Thus $(3e^{j\pi/4})^* = 3e^{-j\pi/4}$. In general, if $\mathbf{A} = Ae^{j\theta}$, then $\mathbf{A}^* = Ae^{-j\theta}$. You can see how changing the sign of the imaginary component in rectangular form corresponds to changing the sign of the angle in polar form by referring to Fig. A.7.

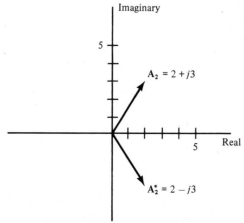

fig. A.7   **A**₂ *and* **A**₂* *in the complex plane.*

The conjugate of a complex number in any form can always be obtained by replacing each j with $-j$. For example, if

$$C = \frac{3 + j5}{(2 - j6)e^{j\pi/6}}$$

then

$$C^* = \frac{3 - j5}{(2 + j6)e^{-j\pi/6}}$$

Although the expression for C could be reduced to the form

$$C = -0.17 + j0.91$$

and the conjugate found by changing the sign of the imaginary component, there is no need to reduce the expression just to find the conjugate. (We will describe ways of reducing expressions in the next section after we discuss multiplication and division of complex numbers.)

Also note that

$$(A^*)^* = A$$

That is, the conjugate of the conjugate of the number is equal to the number itself. Furthermore, taking the conjugate of a number never changes its magnitude.

Equal Complex Numbers

Two complex numbers are said to be *equal* if and only if both their real and imaginary components are respectively equal. That is,

$$A = a + jb \qquad\qquad (A.22)$$

and

$$B = c + jd \qquad\qquad (A.23)$$

are equal if and only if $a = c$ and $b = d$.

In polar form, this means that if

$$A = Ae^{j\theta} \qquad\qquad (A.24)$$

$$B = Be^{j\psi} \qquad\qquad (A.25)$$

then $A = B$ if and only if $A = B$ and $\theta = \psi$. For example, $3e^{j\pi/3}$ and $3e^{j7\pi/3}$ are equal complex numbers because $7\pi/3 = \pi/3 + 2\pi$, and the two angles are equivalent.

PROBLEMS

A.1 Locate the following numbers in the complex plane:

a. $-3 + j2$
b. $-4 - j4$
c. $2 - j3$
d. $6 + j1$
e. $4e^{j10°}$

f. $4e^{j120°}$
g. $6e^{j200°}$
h. $5e^{-j30°}$
i. $2e^{j\pi/6}$
j. $7e^{-j3\pi/4}$

A.2 Write the following complex numbers in polar form:
   a. $1.5 + j2.4$ d. $-3 - j2$
   b. $3 - j2$ e. $-3 + j4$
   c. $-4 + j5$ f. $x + jy$

A.3 Write the following complex numbers in rectangular form:
   a. $3e^{j\pi/4}$ e. $Re^{j\phi}$
   b. $2e^{j60°}$ f. $e^{-j\pi/3}$
   c. $6e^{j115°}$ g. $5 \angle 420°$
   d. $4 \angle 120°$ h. $2 \angle 30°$

A.4 Write the expression for each of the following complex numbers without using the conjugate sign (*).

   a. $(6 + j5)^*$ d. $\left[\dfrac{(-5 - j2)}{(3 - j4)(6 + j5)} e^{-j30°} \right]^*$

   b. $(3e^{j30°})^*$

   c. $[(3 + j1)^* + 2e^{j\pi/4}]^*$ e. $\left[\dfrac{e^{j30°} + e^{-j30°}}{2} \right]^*$

A.5 In each case, show whether the given angles are equivalent:
   a. $75°$ *and* $795°$ d. $80°$ *and* $790°$
   b. $\pi/3$ *and* $4\pi/3$ e. $-35°$ *and* $325°$
   c. $-\pi$ *and* $+\pi$ f. $-280°$ *and* $80°$

A.3 THE ALGEBRA OF COMPLEX NUMBERS

Addition and Subtraction

The addition of rectangular complex numbers is easily performed. The rule is easy. Two complex numbers are added by adding the real parts and adding the imaginary parts. Thus

$$(2 + j3) + (5 - j4) = 7 - j1$$

In the general case, with A and B defined as in (A.22) and (A.23),

$$\mathbf{A} + \mathbf{B} = (a + c) + j(b + d) \tag{A.26}$$

The graphical method corresponding to (A.26) is to add the directed lines that represent **A** and **B** graphically. The two methods commonly used to add directed lines graphically are to add them heel-to-toe, or to complete the parallelogram. Both methods are illustrated in Fig. A.8.

Subtraction of two complex numbers is an easy extension of addition. To find $\mathbf{A} - \mathbf{B}$, simply change the sign on both the real and imaginary components of **B** and add. Thus,

$$(2 + j3) - (5 - j4) = (2 + j3) + (-5 + j4) = -3 + j7$$

In the general case

$$\mathbf{A} - \mathbf{B} = (a - c) + j(b - d)$$

where **A** and **B** are defined in (A.22) and (A.23).

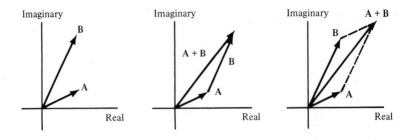

(a) Complex numbers to be added   (b) Heel-to-toe method   (c) Parallelogram method

*fig. A.8 Graphical addition of **A** and **B**. In (b), **A** and **B** are placed heel - to - toe. In (c), the parallelogram is drawn. Either method can be used for graphical addition.*

Graphical subtraction is shown in Fig. A.9. Changing the sign of **B** corresponds to reversing the direction of **B** without changing its magnitude. **A** and negative **B** are added graphically to get **A** − **B**.

To add two numbers expressed in polar form, first convert them to rectangular form and then add them. You will have occasion to add polar numbers many many times. Perhaps your calculator has a routine that (1) converts each polar number to rectangular form, (2) adds the two in rectangular form, and (3) then converts the rectangular sum to polar form. If so, become familiar with this capability. If not and if your calculator is programmable, you will find it advantageous to construct an appropriate program for future use.

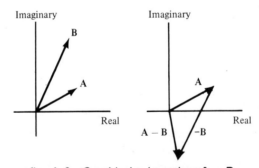

*fig. A.9 Graphical subtraction: **A** − **B**.*

Multiplication and Division

On the other hand, it is easier to multiply and divide complex numbers if they are expressed in polar form. The product of $3e^{j30°}$ and $4e^{j20°}$ is

$$(3e^{j30°})(4e^{j20}) = (3)(4)e^{j30°}e^{j20°} = 12e^{j50°}$$

The exponents are added in multiplication just as they are added in $(x^2)(x^3) = x^5$.

Similarly, the division of $3e^{j30°}$ by $4e^{j20°}$ is

$$\frac{3e^{j30°}}{4e^{j20°}} = \frac{3}{4} e^{j10°}$$

The exponents are subtracted in division just as they are subtracted in $(x^2)/(x^3) = x^{-1}$.

In the general case,

$$\mathbf{A} \times \mathbf{B} = AB e^{j(\theta + \psi)} \qquad (A.27)$$

and

$$\frac{\mathbf{A}}{\mathbf{B}} = \frac{A}{B} e^{j(\theta - \psi)} \qquad (A.28)$$

Although the procedure is not as convenient, complex numbers can be multiplied and divided when they are in rectangular form. For example,

$$(2 + j3)(5 - j4) = (2)(5) - j(2)(4) + j(3)(5) - j^2(3)(4)$$
$$= 10 + j7 + 12$$
$$= 22 + j7$$

where j is treated like any other number in the multiplication. Note that since $j = \sqrt{-1}$,

$$j^2 = -1$$
$$j^3 = -j$$
$$j^4 = 1$$

etc.

In the general case,

$$\mathbf{A} \times \mathbf{B} = (a + jb)(c + jd)$$
$$\mathbf{A} \times \mathbf{B} = (ac - bd) + j(bc + ad) \qquad (A.29)$$

Division of complex numbers in rectangular form is a little more involved. For example, to evaluate

$$\frac{2 + j3}{5 - j4}$$

we multiply by $(5 + j4)/(5 + j4)$, which does not change the value of the fraction,

$$\frac{2 + j3}{5 - j4} = \frac{2 + j3}{5 - j4} \cdot \frac{5 + j4}{5 + j4} = \frac{-2 + j23}{5^2 + j(20 - 20) + 4^2} = -\frac{2}{41} + j\frac{23}{41}$$

The reason for multiplying by $(5 + j4)/5 + j4)$ is that the resulting product in the denominator is real. That is, $(5 - j4)(5 + j4) = 5^2 + 4^2$ because the j terms add to zero. The process of multiplying by the proper quantity to get a real denominator is called *rationalizing* the denominator. The general relation

$$\mathbf{A} \times \mathbf{A}^* = a^2 + b^2$$

is important to remember for rationalizing and for other purposes.

In the general case,

$$\frac{\mathbf{A}}{\mathbf{B}} = \frac{(a + jb)}{(c + jd)} = \frac{ac + bd}{c^2 + d^2} + j\frac{bc - ad}{c^2 + d^2} \tag{A.30}$$

Equations (A.27) and (A.28) (polar form) are usually more convenient than (A.29) and (A.30) (rectangular form) for finding products and quotients. Graphical methods for finding products and quotients are not very convenient.

Powers of Complex Numbers

In polar form, a complex number can easily be raised to a power. For example,

$$2(e^{j20°})^3 = 2^3 e^{j60°} = 8e^{j60°}$$

In general,

$$(Ae^{j\theta})^n = A^n e^{jn\theta} \tag{A.31}$$

according to the laws of exponents.

To find the power of a rectangular number, merely convert the rectangular number to its equivalent polar form, then apply (A.31).

A.4 THE CALCULUS OF COMPLEX NUMBERS

There is an extensive amount of mathematical theory related to functions of complex variables, including many theorems on differentiation and integration of functions of complex variables. Fortunately, for what we need in this text, we can differentiate and integrate by treating j as a constant. Primarily we will use relations like

$$\frac{d}{dt}(e^{j\omega t}) = j\omega e^{j\omega t}$$

and

$$\int_a^b e^{j\omega t}\,dt = \frac{1}{j\omega}e^{j\omega t}\Bigg]_a^b$$

As pointed out in Chapter 4 and in Section A.1, the first relation is the basis for the usefulness of the phasor transform and is the underlying reason for the use of complex numbers.

PROBLEMS

A.6 Find the sum of $3e^{j30°}$ and $2e^{j60°}$.

A.7 Find the indicated sums and differences:
- a. $(6.8 + j5.2) + (3.2 + j9.5)$
- b. $(20 + j30) - (-40 - j60)$
- c. $(\alpha + j\beta) - (3 + j2)$

A.8 Find the indicated products:
- a. $(6.8 + j5.2)(3.2 + j9.5)$
- b. $(70e^{j\pi/4})(10e^{j\pi/3})$
- c. $(10e^{j30°})(15e^{j10°})(18e^{-j80°})$
- d. $(2 + j3)(4 + j5)(-6 + j3)$

A.9 Find the following quotients:
- a. $\dfrac{(2 + j1)}{(5 + j4)}$

- b. $\dfrac{(6e^{j30°})(8e^{j20°})}{(4e^{j70°})(3e^{-j10°})}$

- c. $\dfrac{(5 + j1)(2 + j3)}{(6 + j2)}$

A.10 Evaluate the following:
- a. $\mathrm{Re}[5e^{j30°}]$
- b. $\mathrm{Im}[5e^{j30°}]$
- c. $\mathbf{B} + \mathbf{B}^*$ where $\mathbf{B} = \alpha + j\beta$
- d. $\mathbf{B} - \mathbf{B}^*$ where $\mathbf{B} = \alpha + j\beta$
- e. $(3 + j4)(3 - j4)$
- f. $(3 + j4)(3 + j4)$

A.11 Write **A** in the form of $a + jb$ and in the form of $Ae^{j\theta}$ if

$$\mathbf{A} = \frac{3 + e^{j\pi/4}}{(4 + j2)(6e^{j30°} + 5 + j3)}$$

A.5 SUMMARY OF SOME USEFUL RELATIONS

Table A.1 summarizes some useful relations for easy reference.

TABLE A.1   Some Useful Relations

Definitions	$\mathbf{A} = a + jb$ $\qquad \mathbf{A} = Ae^{j\theta}$ $\qquad \mathbf{A} = A \angle \theta$ $\mathbf{B} = c + jd$ $\qquad \mathbf{B} = Be^{j\psi}$ $\qquad \mathbf{B} = B \angle \psi$
Addition, subtraction	$\mathbf{A} + \mathbf{B} = (a + c) + j(b + d)$ $\mathbf{A} - \mathbf{B} = (a - c) + j(b - d)$
Multiplication	$\mathbf{A} \times \mathbf{B} = A \times Be^{j(\theta + \psi)}$ $\mathbf{A} \times \mathbf{B} = (ac - bd) + j(bc + ad)$
Division	$\dfrac{\mathbf{A}}{\mathbf{B}} = \dfrac{A}{B}\, e^{j(\theta - \psi)}$ $\qquad \dfrac{\mathbf{A}}{\mathbf{B}} = \dfrac{ac + bd}{c^2 + d^2} + j\,\dfrac{bc - ad}{c^2 + d^2}$
Rationalization	$\dfrac{1}{c + jd} = \dfrac{c - jd}{(c + jd)(c - jd)} = \dfrac{c - jd}{c^2 + d^2}$
Sum and difference of a number and its conjugate	$\mathbf{A} + \mathbf{A}^* = 2a$ $\qquad \mathbf{A} - \mathbf{A}^* = j2b$
Product of a number and its conjugate	$\mathbf{A} \times \mathbf{A}^* = a^2 + b^2$
Magnitude of a number	$A = \sqrt{\mathbf{A}\mathbf{A}^*} = \sqrt{a^2 + b^2}$
Real and imaginary parts	$\mathrm{Re}[\mathbf{A}] = a$ $\qquad \mathrm{Im}[\mathbf{A}] = jb$
Euler's formulas	$e^{j\theta} = \cos\theta + j\sin\theta$ $\qquad e^{-j\theta} = \cos\theta - j\sin\theta$ $\cos\theta = \dfrac{e^{j\theta} + e^{-j\theta}}{2}$ $\qquad\qquad \sin\theta = \dfrac{e^{j\theta} - e^{-j\theta}}{2j}$
Conversion relations	$A = \sqrt{a^2 + b^2}$ $\qquad a = A\cos\theta$ $\theta = \tan^{-1}(b/a)$ $\qquad b = A\sin\theta$
Identities	$e^{j\pi/2} = j$ $\qquad e^{j2\pi} = 1$ $\qquad j^3 = -j$ $e^{j\pi} = -1$ $\qquad j^2 = -1$ $\qquad j^4 = 1$
Complex number raised to the nth power	$\mathbf{A}^n = (Ae^{j\theta})^n = A^n e^{jn\theta}$

Techniques for More Complex Networks

CIRCUIT REDUCTIONS BY NODE ELIMINATIONS OF THE CONVERSION OF STAR NETWORKS TO MESH NETWORKS

Especially in Chapters 2 and 6, and elsewhere to a lesser degree, circuit reductions involving impedance or admittance elements have been used extensively to achieve desirable simplifications. We have combined series and parallel combinations of these impedance-admittance elements to reduce network complexity. However, there are many situations such as in Fig. B.1 where there are no admittance (or impedance) elements in series or in parallel with other passive elements (admittance or impedance). However, by means of a new circuit reduction technique, we will soon have the ability to reduce a network such as that of Fig. B.1 to the simpler network of Fig. B.2. This reduction technique will lead us to the values I_1 and V_5, where these two are identical, respectively, in the two networks. Of course, I_{50} of Fig. B.2

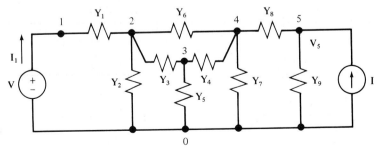

fig. B.1   *No passive elements in series or parallel.*

461

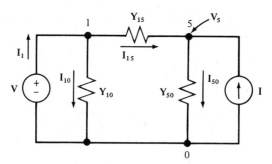

fig. B.2   Equivalent to that of Fig. B.1 values.

(current in Y_{50}) is not the current in Y_9 of Fig. B.1. Furthermore, I_{10} and I_{15} of Fig. B.2 have no simple interpretation in the network of Fig. B.1. The reduction technique that leads to the equivalent of a simpler circuit such as that of Fig. B.2 is straightforward and does not employ simultaneous equations.

The circuit reduction technique that we are about to develop converts what is called a *star* network to an equivalent *mesh* network. Figure B.3a displays a five-element star network. In Fig. B.1, Y_3, Y_4, and Y_5 form a three-element star, while Y_1, Y_2, Y_3, and Y_6 form a four-element star. Figure B.3b displays a five-node mesh that is equivalent to the five-element star network of Fig. B.3a. At first glance, it appears that replacing a star with its equivalent mesh complicates, rather than simplifies, the situation. However, we will soon see that this conversion really can reduce and simplify networks.

Let's search for relationships that will allow us to calculate the mesh admittance elements Y_{12}, Y_{13}, $Y_{14}, ..., Y_{23}, Y_{24}, ..., Y_{34}, Y_{45}$, etc. in terms of the star admittance elements Y_1, Y_2, $Y_3, ..., Y_n$ that make the mesh network equivalent to the star network. Now, each of these networks is described by its own admittance matrix.

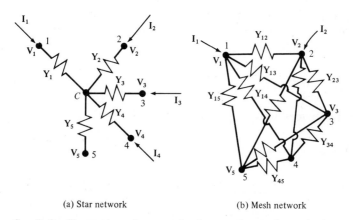

(a) Star network   (b) Mesh network

fig. B.3   Star and mesh networks that can be made equivalent.

The elements in the admittance matrix that relate node voltages and impressed currents are:

$$\begin{bmatrix} \mathbf{Y}_{11} & \mathbf{Y}_{12} & \mathbf{Y}_{13} & \cdots & \mathbf{Y}_{1n} \\ \mathbf{Y}_{21} & \mathbf{Y}_{22} & \mathbf{Y}_{23} & \cdots & \mathbf{Y}_{2n} \\ \mathbf{Y}_{31} & \mathbf{Y}_{32} & \mathbf{Y}_{33} & \cdots & \mathbf{Y}_{3n} \\ \mathbf{Y}_{41} & \mathbf{Y}_{42} & \mathbf{Y}_{43} & \cdots & \mathbf{Y}_{4n} \\ \vdots & \vdots & \vdots & & \vdots \\ \mathbf{Y}_{n1} & \mathbf{Y}_{n2} & \mathbf{Y}_{n3} & \cdots & \mathbf{Y}_{nn} \end{bmatrix} \tag{B.1}$$

where $\mathbf{Y}_{11}$ is the self-admittance for node 1, $\mathbf{Y}_{22}$ is the self-admittance for node 2, $\mathbf{Y}_{12}$ is the mutual admittance between nodes 1 and 2, and so on. If you are a little hazy on the significance of this matrix and the meaning of each element, you may wish to refer back to page 242. For the star and mesh networks to be equivalent, the corresponding elements in the matrices for the two networks must be identical. In other words, where the two networks are equivalent,

$$\mathbf{Y}_{11s} = \mathbf{Y}_{11m}$$

$$\mathbf{Y}_{12s} = \mathbf{Y}_{12m} \tag{B.2}$$

$$\mathbf{Y}_{13s} = \mathbf{Y}_{13m}$$

etc.   for all elements

where $\mathbf{Y}_{11s}$, $\mathbf{Y}_{12s}$, etc., are elements in the star matrix and $\mathbf{Y}_{11m}$, $\mathbf{Y}_{12m}$, etc., are corresponding elements in the mesh matrix. Notice that $\mathbf{Y}_{12m} = -\mathbf{Y}_{12}$, where $\mathbf{Y}_{12}$ is the value of the physical admittance element between nodes 1 and 2 of Fig. B.3b and $\mathbf{Y}_{12m}$ the corresponding element in the **Y** matrix of that network. The value of the element $\mathbf{Y}_{12s}$ is not so obvious. We can find $\mathbf{Y}_{12s}$ by "killing" all nodal voltages in the star except $\mathbf{V}_1$. Then we proceed to find the current that enters the star at node 2 due to $\mathbf{V}_1$ acting alone. Let this current be $\mathbf{I}_{21}$. As a first step, we find $\mathbf{I}_{11}$, the current entering node 1 due to $\mathbf{V}_1$ alone. This current encounters admittance $[\mathbf{Y}_1\,\textcircled{S}\,(\mathbf{Y}_s - \mathbf{Y}_1)]$ where $\mathbf{Y}_s$ is the sum of all of the star admittances in parallel, or $\mathbf{Y}_s = \mathbf{Y}_1 + \mathbf{Y}_2 + \mathbf{Y}_3 + \cdots + \mathbf{Y}_n$, where there are n elements in the star. (Recall that $\textcircled{S}$ means "in series with.") So,

$$\mathbf{I}_{11} = [\mathbf{Y}_1\,\textcircled{S}\,(\mathbf{Y}_s - \mathbf{Y}_1)]V_1 = \frac{\mathbf{Y}_1(\mathbf{Y}_s - \mathbf{Y}_1)V_1}{\mathbf{Y}_1 + \mathbf{Y}_s - \mathbf{Y}_1} \tag{B.3}$$

The current $\mathbf{I}_{11}$ divides at node C so we can use the current divider relation to get

$$\mathbf{I}_{21} = -\frac{\mathbf{Y}_2}{(\mathbf{Y}_s - \mathbf{Y}_1)} \times \mathbf{I}_{11} = -\frac{\mathbf{Y}_1\mathbf{Y}_2(\mathbf{Y}_s - \mathbf{Y}_1)V_1}{\mathbf{Y}_s(\mathbf{Y}_s - \mathbf{Y}_1)} = \frac{-\mathbf{Y}_1\mathbf{Y}_2V_1}{\mathbf{Y}_s} \tag{B.4}$$

or

$$Y_{12s} = \frac{\mathbf{I}_{21}}{\mathbf{V}_1} = -\frac{\mathbf{Y}_1\mathbf{Y}_2}{\mathbf{Y}_3} \tag{B.5}$$

Remembering that $Y_{12m} = -Y_{12}$ and combining (B.2) and (B.5), we have

$$-Y_{12} = Y_{12m} = Y_{12s} = \frac{-Y_1 Y_2}{Y_s}$$

or

$$Y_{12} = \frac{Y_1 Y_2}{Y_s} = \frac{Y_1 Y_2}{Y_1 + Y_2 + Y_3 \cdots Y_n} \tag{B.6}$$

Equation (B.6) gives the value of the Y_{12} admittance of the mesh in terms of the elements $Y_1, Y_2, Y_3, \ldots, Y_n$ in the equivalent star.

By the same procedure, we can easily show that

$$Y_{13} = \frac{Y_1 Y_3}{Y_s}$$

Or for any admittance in the mesh

$$Y_{km} = \frac{Y_k Y_m}{Y_s} \qquad k \neq m \tag{B.7}$$

where k and m refer to any two different nodes in the star network and its equivalent mesh network.

It is important to notice the contiguous relationship among the admittance elements Y_{km}, Y_k, and Y_m as illustrated in Fig. B.4.

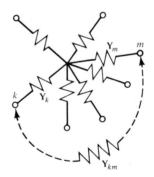

fig. B.4   Contiguous relationship among Y_k, Y_m, and Y_{km}.

After seeing this contiguous relationship, one will find it much easier to recall the necessary numbers to calculate a given element in the mesh by referring to the circuit diagram rather than by referring to (B.7).

Example B.1
a.   Find the admittances Y_{12}, Y_{23}, Y_{31} of Fig. B.5b that make this delta equivalent to the wye of Fig. B.5a.

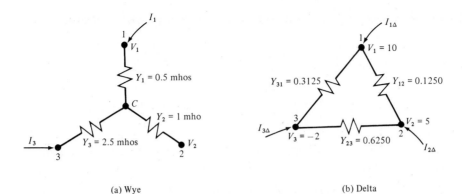

(a) Wye   (b) Delta

fig. B.5   A wye (three - element star) and its equivalent delta (three - node mesh).

Solution

$$Y_{12} = \frac{0.5 \times 1}{0.5 + 1 + 2.5} = \frac{0.5}{4} = 0.1250 \text{ mhos}$$

$$Y_{23} = \frac{1 \times 2.5}{4} = 0.6250 \text{ mhos}$$

$$Y_{31} = \frac{2.5 \times 0.5}{4} = 0.3125 \text{ mhos}$$

b.   Assume the following voltages are applied to both the wye and the delta. Solve for the three wye currents and the three delta currents.

$$V_1 = 10 \text{ V} \qquad V_2 = 5 \text{ V} \qquad V_3 = -2 \text{ V}$$

For the delta currents,

$$I_{1\Delta} = (V_1 - V_3)Y_{31} + (V_1 - V_2)Y_{12} = (10 + 2)\,0.3125 + (10 - 5)\,0.125 = 4.375$$

$$I_{2\Delta} = -5 \times 0.125 + 7 \times 0.625 = 3.750$$

$$I_{3\Delta} = -7 \times 0.625 - 12 \times 0.3125 = -8.125$$

(a) Two-element star   (b) Two-node mesh

fig. B.6   Single admittance in a two - node mesh equivalent to a two - element star.

According to Kirchhoff's current law, these three currents must add to zero as

$$I_{1\Delta} + I_{2\Delta} + I_{3\Delta} = 4.3750 + 3.750 - 8.125 = 0$$

c.   Find the three currents to the nodes 1, 2, 3 of the equivalent star. First find the nodal voltage V_C.

$$V_C = \frac{V_1 Y_1 + V_2 Y_2 + V_3 Y_3}{Y_1 + Y_2 + Y_3} = \frac{10 \times 0.5 + 5 \times 1 + (-2) \times 2.5}{4}$$

$$= \frac{5}{4} = 1.250$$

$$I_{1Y} = (V_1 - V_C)Y_1 = (10 - 1.25)\,0.5 = 4.375$$

$$I_{2Y} = (5 - 1.25) \times 1 = 3.750,$$

$$I_{3Y} = (-2 - 1.25) \times 2.5 = -8.125$$

We see that the respective currents at the nodes of the star are equal to the corresponding currents at the mesh nodes, so the two networks are equivalent.

• • •

Example B.2
Two series elements can be considered a two-element star as shown in Fig. B.6a. By applying the general concept of star-to-mesh conversion, we find

$$Y_{12} = \frac{Y_1 Y_2}{Y_s} = \frac{Y_1 Y_2}{Y_1 + Y_2}$$

the old familiar expression for the single admittance of two admittances in series.

• • •

Node Elimination

By referring back to Fig. B.2, we see that by replacing a star with its equivalent mesh, one eliminates a node, node C in this figure. We will soon see how successive elimination of nodes can greatly simplify a rather complex network. Refer again to Fig. B.1. By eliminating node 3 (convert the Y_3, Y_4, Y_5 star to an equivalent mesh), the resultant circuit is simplified as shown in Fig. B.7.

Nodes 2 and 4 can also be eliminated for further reduction or simplification. We will see how this works out in the following example.

Example B.3
By node eliminations, reduce the network of Figs. B.1 and B.8 by eliminating node 3. In Fig. B.8, each element is resistive and the applied voltage V and current I are steady. Of course the principles discussed here apply to admittance in both the $j\omega$ and s domains. Since the arithmetic is much simpler for pure conductances, we have used that case in this example.

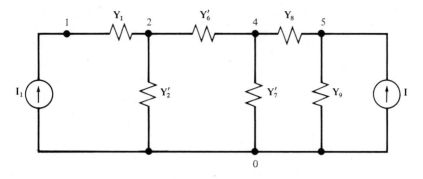

where $\mathbf{Y}_2' = \mathbf{Y}_2 + \mathbf{Y}_{35}$; $\mathbf{Y}_6' = \mathbf{Y}_6 + \mathbf{Y}_{34}$; $\mathbf{Y}_7' = \mathbf{Y}_7 + \mathbf{Y}_{45}$

fig. B.7   Equivalent to Fig. B.1 after node 3 has been eliminated.

Now proceed to find the delta that is equivalent to the Y_3, Y_4, and Y_5 wye in order to eliminate node 3. For this node, $Y_s = 0.2 + 0.3 + 0.1 = 0.6$. Then the new equivalent delta will have elements

$$Y_{34} = \frac{0.2 \times 0.3}{0.6} = 0.1$$

$$Y_{45} = \frac{0.1 \times 0.3}{0.6} = 0.05$$

$$Y_{53} = \frac{0.1 \times 0.2}{0.6} = 0.0333$$

Notice that the subscripts 3 and 4 in $\mathbf{Y}_{34}$ refer to admittance elements Y_3 and Y_4, rather than nodes 3 and 4. Of course $\mathbf{Y}_{34}$ is in parallel with Y_6 so that when these two are combined, the new value of $Y_6' = 0.5 + \mathbf{Y}_{34} = 0.5 + 0.1 = 0.6$. In a similar way, Y_7 is modified by the parallel $\mathbf{Y}_{45}$, making the new value of $Y_7' =$

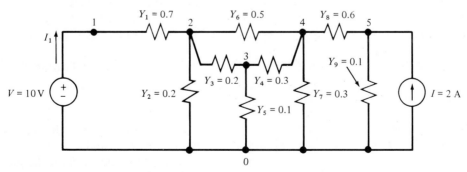

fig. B.8   Admittances in mhos for the original network of Fig. B.1. All elements resistive; V and I are steady.

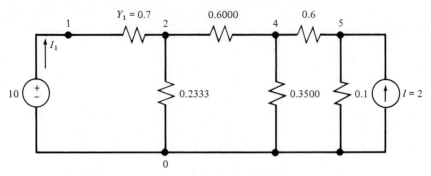

fig. B.9   Equivalent to Fig. B.8 after eliminating node 3.

$0.3 + Y_{45} = 0.35$. So by eliminating node 3 of the circuit of Fig. B.8, we arrive at the simpler, equivalent network of Fig. B.9.

Now to further the simplification, eliminate node 2 from Fig. B.9.

• • •

PROBLEMS

B.1   Show that the circuit of Fig. B.10 is equivalent to that of Fig. B.8 by eliminating node 2 in Fig. B.9.

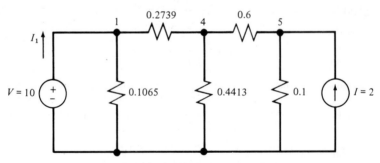

fig. B.10   Equivalent to Fig. B.9.

B.2   By eliminating node 4 of Fig. B.10, show that the circuit of Fig. B.11 is equivalent to that of Fig. B.10.

The circuit of Fig. B.11 is a remarkable reduction from the original circuit of Fig. B.8. Now we can readily find V_5 and I_1. By nodal superposition,

$$V_5 = \frac{2 + 10 \times 0.1250}{0.1250 + 0.3013} = 7.624 \text{ V}$$

Then,

$$I_1 = 10 \times 0.1984 + (10 - 7.624)\, 0.1250 = 2.281$$

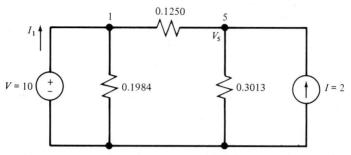

fig. B.11   Equivalent to Figs. B.8, B.9, and B.10.

Note that we have found these values without the use of simultaneous equations. If we had approached the problem of finding I_1 and V_5 by the use of nodal methods alone, we would have been required to solve a set of five simultaneous equations. If we tried to find them by loop currents, we would have had six simultaneous equations to solve. With some effort, still without simultaneous equations, we can use the known values of V_5 and I_1 to find all nodal voltages and branch currents in the original circuit of Fig. B.8. Let's proceed in that direction.

On Fig. B.12 (the original circuit of Fig. B.8), we will record all nodal voltages and branch currents as we find their values in succession. At the moment we know $V_1 = 10$ (given), $I_1 = 2.281$, $V_5 = 7.624$, and $I = 2$ (given). Now, in succession we find

$$I_{50} = V_5 \times 0.1 = 0.7624$$

$$I_{54} = 2 - I_{50} = 1.238$$

$$V_4 = V_5 - \frac{I_{54}}{0.6} = 7.624 - \frac{1.238}{0.6} = 5.561$$

$$I_{40} = V_4 \times 0.3 = 1.668$$

$$V_2 = V_1 - \frac{I_1}{0.7} = 10 - \frac{2.281}{0.7} = 6.741$$

$$I_{24} = (6.741 - 5.561)0.5 = 0.5900$$

$$I_{20} = V_2 \times 0.2 = 1.348$$

$$I_{23} = I_1 - I_{24} - I_{20} = 2.281 - 0.5900 - 1.348 = 0.3430$$

$$I_{43} = I_{24} + I_{54} - I_{40} = 0.5900 + 1.238 - 1.668 = 0.1600$$

$$I_{30} = I_{43} + I_{23} = 0.1610 + 0.3430 = 0.5040$$

$$V_3 = \frac{I_{30}}{0.1} = 5.040$$

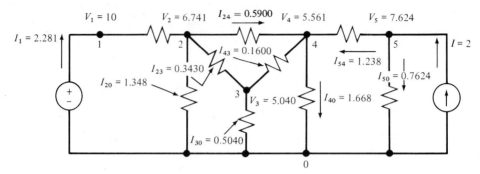

fig. B.12   Showing branch currents and nodal voltages.

Make a consistency check by summing the voltages around the 0-4-5-0 loop to see if the sum is zero:

$$\frac{0.7624}{0.1} - \frac{1.668}{0.3} - \frac{1.238}{0.6} = 0.0007$$

This is a rather good check considering the amount and nature of the calculations.

PROBLEMS

B.3   Because in coming problems you must make a significant number of node eliminations, prepare and test a program for your calculator, or for your interactive computer facility, that will eliminate a node having 2, 3, or 4 branches. *Suggestion:*   Plan your algorithm for the four-element star-to-mesh conversion, which will almost surely serve for three- and two-element stars. For the four-element star having nodes 1, 2, 3, and 4 in addition to the center node that is to be eliminated, there will be six mesh elements, $\mathbf{Y}_{12}$, $\mathbf{Y}_{13}$, $\mathbf{Y}_{14}$, $\mathbf{Y}_{23}$, $\mathbf{Y}_{24}$, and $\mathbf{Y}_{34}$. You may find it convenient to calculate the mesh elements in this order.

B.4   Referring back to the network of Fig. B.8, we eliminated node 3, node 2, and then node 4. Other orders could have been used.

For the network of (B.8), eliminate node 2 followed by node 3. After eliminating these two nodes, your equivalent (allowing for small arithmetic differences) should be identical to that of Fig. B.10.

B.5   In the network of Fig. B.13, there are three voltage generators as shown.

a.   Find the value of V_3 that makes $I_3 = 0$ where V_1, V_2, and all admittances are given, as shown in Fig. B.13. This is the kind of problem where node elimination is advantageous. By eliminating nodes 4, 5, and 6, you will be able to find, rather easily, the necessary value of V_3.

b.   Using your value of V_3, find a sufficient number of branch currents to apply some consistency check on your value of V_3.

B.6   Find the value of V_3 in Fig. B.13 such that I_3 is equal to 1.0.

Appropriate Simultaneous Equations

Earlier we learned that one should be wise in choosing equations for simultaneous solution for: nodal voltages, loop currents, or branch currents. If the analyst chooses

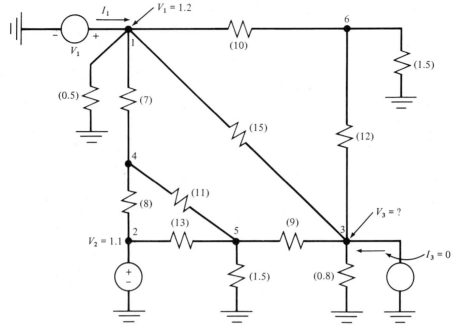

fig. B.13   Find V_3 for $I_3 = 0$. Admittances are in parentheses (mhos). Voltages are in volts.

to solve for *n* unknowns by simultaneous equations, he or she must, of course, find *n* independent equations. To quite a degree, the number of unknowns for simultaneous equation solution is a matter of choice depending on what reductions or simplifications the analyst wishes to apply before facing the simultaneous equation part of the total solution. For example, in the circuit of Fig. B.14 and without circuit reductions, there are two unknown nodal voltages, V_1 and V_2, three unknown loop currents, I_1, I_2, I_3, and five unknown branch currents. However, one might simplify first. One might, as an example, eliminate node 1 by a star-to-mesh conversion as

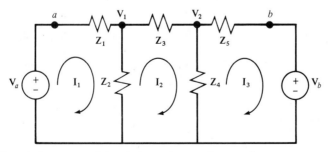

fig. B.14   Two unknown nodal voltages; three unknown loop currents; five unknown branch currents.

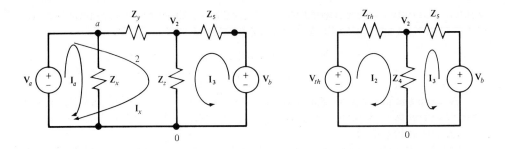

(a) Reduced by star-to-mesh   (b) Reduced by Thévenin

fig. B.15   Reduction of the circuit of Fig. B.14.

previously discussed. This leads to the reduced circuit of Fig. B.15a, while a Thévenin reduction that replaces the V_a, Z_1, Z_2, and Z_3 network with the V_{th}, Z_{th} network leads to the network of Fig. B.15b. There are two unknown nodal voltages in the original network of Fig. B.14 and one in each of the networks of Fig. B.15. In general, the designer has the option of reducing the number of unknowns (nodal voltages, loop currents, or branch currents) by means of circuit reduction.

Consider the loop current approach in respect to Figs. B.14, B.15a, and B.15b. There are three unknown loop currents in the circuit of Fig. B.14. These three could be found by the solution of three simultaneous equations. In Fig. B.15a, there are two unknown loop currents, I_x, I_3, and one *known* loop current, $I_a = V_a/Z_x$. And, of course, there are two unknown loop currents I_2 and I_3 in the circuit of Fig. B.15b.

One should be careful not to assume an excess number of unknown nodal voltages on one hand or unknown loop currents on the other hand. Figures B.15a and B.16 illustrate this point. By the way that loop currents are assigned in Fig. B.15a, we see that we need not include I_a in the familiar set of simultaneous equations because it is immediately known. This leaves only two simultaneous equations. If loop currents were assigned differently as in Fig. B.16, then it might appear that there are three unknown currents and that three simultaneous equations must be

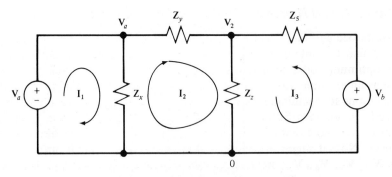

fig. B.16   Do not write three loop equations.

written. But only two are needed to find I_2 and I_3 by simultaneous solution because the known V_a is applied to the I_2 loop, so that we can write

$$V_a = (Z_y + Z_z)I_2 + Z_z I_3$$

$$V_b = + Z_z I_2 + (Z_z + Z_5)I_3$$

To keep the number of loop equations to a minimum, one could identify known branch currents and then assign a loop current to each of these known branch currents, as in Fig. B.15a.

Much the same point of view in respect to known currents is applied where there are current sources as well as voltage sources. See Fig. B.17. Here the source

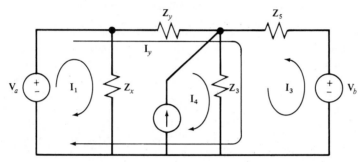

fig. B.17   Known currents I_1 and I_Y.

current I_4 is known and is assigned a loop. Again we do not write an equation for the loops traced by the known currents I_4 and I_1. The pertinent simultaneous equations are

$$(V_a - I_4 Z_3) = (Z_y + Z_3)I_y + Z_3 I_3$$

$$(V_b - I_4 Z_3) = Z_3 I_y + (Z_3 + Z_5)I_3$$

Loop Currents and Nodes in More Complex Networks

As networks become more complex, the method of assigning the optimum loop currents to find the unknown currents or, on the other hand, the method of assigning the optimum nodes to find nodal voltages becomes more difficult. Consider Fig. B.18 as an example. Consider the problem of selecting the minimum number of nodal equations for simultaneous solution. First, the choice of the optimum reference node is *not* obvious. Suppose, however, we refer V_1, V_3, V_4, and V_5 to node 2 as reference. Then note that the known voltages V_a and V_b are *not* referred to node 2 as reference. Now it might appear that there are four remaining unknown nodal voltages V_1, V_3, V_4, V_5; however, V_5 is not independent of V_1. We see that $V_5 = V_1 - V_a$ and $V_4 = V_3 - V_b$. So we need to solve only two simultaneous equa-

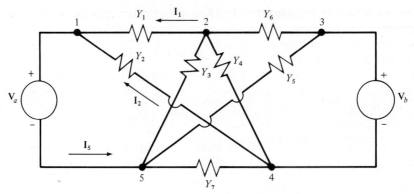

fig. B.18   A more complex network.

tions. Then the simultaneous solution might be in terms of unknown voltages V_1 and V_3 (two equations). Or, alternately, we will select V_4 and V_5 as the unknown voltages where V_2 is the reference.

PROBLEMS

B.6   For Fig. B.18, write two simultaneous, nodal equations for V_4 and V_5 with node 2 as the reference for these two voltages. Remember that the voltage at node 3 is $(V_4 + V_b)$; the voltage at node 1 is $(V_5 + V_a)$. You might find it difficult to apply the "nodal-superposition" approach. It might be safer but more tedious to find the current in each branch at each unknown node such as

$$I_5 = I_1 + I_2 = -(V_5 + V_a)Y_1 + (V_4 - V_5 - V_a)Y_2$$

Complete the effort to arrange the two simultaneous equations in the standard format.

Let us view the circuit of Fig. B.18 in light of simultaneous loop equations. Perhaps we can view the circuit more easily if it is redrawn as shown in Fig. B.19.

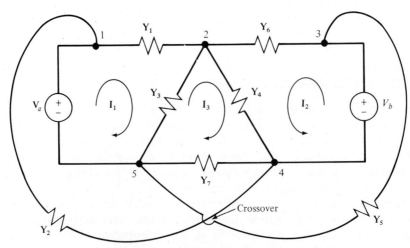

fig. B.19   A different version of Fig. B.18.

This circuit is said to be nonplanar because the circuit cannot be drawn without a "crossover."

We begin to assign loop currents such as I_1, I_2, and I_3. After I_1 is assigned as shown, then I_2 is appropriate because it passes through at least one element (source or admittance) not previously traversed by a current. After I_1 and I_3 have been assigned as shown, the shown assignment of I_3 is appropriate because neither I_1 nor I_2 passes through Y_7, which is traversed by I_3. Before we finish the task of loop current assignments, consider the following two guidelines.

For optimum assignment of loop currents:

1. Each new loop current must traverse at least one element not traversed by previously assigned loop currents.
2. After all loop currents have been assigned, any particular loop current cannot be the only current in any two elements or branches. As an example of this principle, one cannot complete the loop current assignments in Fig. B.19 by assigning a loop current to the path 145321 because this current is the only current in both Y_2 and Y_5. This assignment creates an ambiguity: the branch current in Y_2 is required to be the branch current in Y_5.

There is a formal mathematical technique that can be applied advantageously to complex networks in order to optimize the procedure for finding the needed simultaneous equations. This technique, called *circuit topology*, is not particularly useful in the analysis of networks having only simple to moderate complexity. Circuit topology is treated in advanced textbooks on networks.

PROBLEMS

B.7 Complete the task of assigning the optimum loop currents to the circuit of Fig. B.19. Check your final assignment against the principles stated above. (*Hint*: There are five loop currents.)

B.8 In standard format, write the five simultaneous, loop equations for Fig. B.19. For convenience, use impedances rather than admittances ($Z_n = 1/Y_n$).

Index